EFCE Event No. 361

Innovation in Process Energy Utilisation

A three-day symposium organised by The Institution of Chemical Engineers (South Western Branch) and The Institute of Energy (South Wales and West of England Section) and held at the University of Bath, 16 – 18 September 1987.

Organising Committee

D
621. 4
1 NN

A. Rogers (Chairman)	British Steel
C. Bower	British Coal
J. R. Cattle	ISC Chemicals
R. G. Herapath	University College of Swansea
D. P. Jenkins	British Steel
M. A. Patrick	University of Exeter

THE INSTITUTION OF CHEMICAL ENGINEERS
SYMPOSIUM SERIES No. 105

ISBN 0 85295 210 4

PUBLISHED BY THE INSTITUTION OF CHEMICAL ENGINEERS

Copyright © 1988 The Institution of Chemical Engineers

First edition 1988 – ISBN 0 85295 210 4

MEMBERS OF THE INSTITUTION OF CHEMICAL ENGINEERS (Worldwide)
SHOULD ORDER DIRECT FROM THE INSTITUTION

Geo. E Davis Building, 165–171 Railway Terrace, Rugby, Warks CV21 3HQ.

Australian orders to:
R M Wood, School of Chemical Engineering and Industrial Chemistry,
University of New South Wales, PO Box 1, Kensington, NSW, Australia 2033.

NON MEMBERS ORDERS SHOULD BE DIRECTED AS FOLLOWS:

UK, Eire and Australia	The Institution of Chemical Engineers, Geo. E. Davis Building, 165–171 Railway Terrace, Rugby, Warks CV21 3HQ.
	or
	Hemisphere Publishing Corporation, 79 Madison Avenue, New York, NY 10016–7892, U.S.A.
Rest of the World	Hemisphere Publishing Corporation, 79 Madison Avenue, New York, NY 10016–7892, U.S.A.

FOREWORD

In every industry great efforts have been made to improve process energy utilization. This was initially stimulated by the massive increases in oil prices and the consequential effects on the cost of other energy sources. To-day, innovative techniques are becoming standard production methods as all industries minimise energy utilization to compete in highly competitive market places.

Across the various process industries, these innovations in utilization of energy use are dedicated to the specific process. However the basic tenets are very similar and can find application elsewhere.

The objective of this conference is to highlight these techniques with the express purpose of cross-fertilising ideas between industries. The emphasis on unit operations and cost reduction is expected to have wide application and relevance to all groups working in these areas.

A. Rogers

CONTENTS

* This paper is printed out of page sequence

† Figures for these papers were not received with the preprints but are printed at the end
of all the papers

A STEAM HEAT PUMP FOR INCREASING THE ENERGY EFFICIENCY IN DISTILLATION

E Chalmers*, D L Hodgett** and A A Mitchell***

The first heat pump for distillation in the UK was installed in a malt whisky distillery in 1986. Novel aspects of the system are the modifications to the process to enable continuous and efficient operation of the heat pump across a number of batch stills, the use of steam as the working fluid in a closed cycle and compression of the steam by a twin-screw compressor. The heat pump is incorporated into the distillery steam distribution system and operates in parallel with the steam boiler, allowing complete flexibility of operation. The system, after 2 months, achieved over 90% of its design performance and was in continuous operation. The installation demonstrates the large savings in prime energy which can be achieved by the optimised integration of heat pumps within unit operations. The principle, and much of the system design, is capable of being applied to continuous distillation columns and to other unit operations such as drying, steam stripping and solvent recovery.

1. INTRODUCTION

It is estimated that distillation accounts for 67 million GJ of energy use per annum in the UK (1), much of which is provided by steam and most of which is rejected to the environment by cooling towers. Only in very large installations e.g. petroleum refineries and other petrochemical plants, is heat recovery practised by cascading of columns, by heat integration with other unit operations or by compression of the top product, so that there is still considerable scope for energy conservation.

Although there has been a trend in recent years away from steam heating to direct firing (because of the energy savings resulting from the elimination of the boiler and steam distribution system), even greater savings are possible if steam heating is retained. This is particularly true for columns distilling aqueous mixtures since these operate around 100°C and low pressure steam is the ideal heating medium for them. Since they reject heat at or just below 100°C this heat can be recovered to produce sub-atmospheric steam which can then be compressed to the pressure necessary to heat the column. A closed cycle heat pump with water/steam as the working fluid is thereby formed. This, using heat exchangers designed to give low approach temperatures, can have a high enough Coefficient of Performance (COP) - the ratio of heat flow in the steam delivered at the reboiler to the electrical power drawn by the compressor motor - to ensure a large prime energy saving and a good economic return to the user.

*   Robert Morton (DG) Ltd, Burton-on-Trent
** Electricity Council Research Centre, Capenhurst
*** International Distillers and Vintners Ltd, Keith

1

This paper describes a steam heat pump which has been installed in the Auchroisk malt whisky distillery of International Distillers and Vintners Ltd (IDV). It is a result of a four year R&D programme at the Electricity Council Research Centre (ECRC) and is licenced to Robert Morton (DG) Ltd (RMDG).

2. SYSTEM DESIGN

2.1 Concept

The production of malt whisky involves the fermentation of a wort obtained by the mashing of malted barley to produce a liquor of about 6-8% by mass of ethanol. This is usually double batch distilled in copper pot stills, the first of which increases the ethanol mass fraction to 20% and the second increases it to about 60%. It is then matured in oak casks for at least 3 years (2). The first distillation (the wash still) uses about 11 MJ/litre pure alcohol and the second (the spirit still) about 7.5 MJ/litre compared to 4.6 MJ/litre for mashing and 5.1 MJ/litre for charge preheating before distillation. See Figure 1 for a Sankey diagram of the process.

The first stage of distillation is more attractive for heat pumping because the lower alcohol content means it has a higher heat demand and a higher bubble point than the second distillation. The bubble point is the minimum temperature at which condensation takes place in the condenser so that the evaporating temperature of the heat pump will be higher than in the second distillation.

Most malt whisky distilleries heat the stills with steam at about 3 bar and cool the condenser with river or cooling tower water. Since the still contents boil initially at 94°C and condense between 94°C and 82°C the inherent temperature difference across which an ideal heat pump would work is 12K. By using 3 bar steam and river water this is increased to beyond 150K. Several distilleries have used the condensers to provide water at up to 90°C for mashing and preheating but it can be seen from Figure 1 that the supply of hot water from both the first and second distillations would exceed the demand by a large margin.

In the installation at Auchroisk the heat pump operates across the four first stage stills while the second stage still condensers are used to provide hot mashing water, to preheat the charges to the first stage stills (Figure 2) and, in winter, to heat the buildings. Since the stills and condensers are remote from the boilerhouse, where the compressor is located, a pumped recirculating water system is used to cool the condensers rather than direct reboiling clean steam in the condensers. This has the added advantage of keeping the cold side above atmospheric pressure and therefore any leakage will be of water into the product side rather than vice-versa.

Water at 80°C is pumped to the condensers (Figure 3) and returns at about 85°C to a flash vessel located above the compressor, where its pressure is reduced to 0.5 bar producing steam and water at 80°C. The water is pumped back to the condensers while the steam is compressed to 1.7-2.0 bar by an oil-free twin-screw compressor. The steam is then fed to the heating surface inside the still to boil the ethanol-water mixture which then condenses in the condenser giving up its latent heat to the recirculating water. The steam condensate is also returned to the flash vessel. The distillate is cooled to 25°C in separate subcoolers using by river water.

The only energy inputs to the system are the electrical power to the compressor motor and the recirculating water pumps, the compressor cooling oil pump and the compressor water injection pump. The boiler provides steam to the stills when the compressor is off and it also provides steam for the final preheating of the still charge until there is sufficient heat being collected from its condenser for the heat pump to operate. Heat rejection takes place in the subcooler, the compressor oil cooler and in a trim cooler which balances the system. The latter is also used to reject the heat collected in the condenser when the compressor is off and the stills are heated by boiler steam.

A comparison of Figure 2 with Figure 1 shows that the theoretical specific fuel consumption is reduced from 0.98 to 0.35 litres oil/litre pure alcohol with an additional electricity consumption of 0.5 kWh per litre. The heat pump is responsible for a reduction of 0.38 and the other heat recovery for 0.25 litres/litre. The 0.5 kWh is equivalent to 0.15 litres of oil/litre of alcohol if an electrical generation and distribution efficiency of 30% is assumed.

2.2 Process Modifications

Four major changes were necessary to the distillery to make the system both technically and economically viable. These were:

1. a reduction in steam pressure in the still heaters from 3 bar to less than 2 bar
2. replacing the condenser by one capable of being cooled by 80°C water
3. addition of a subcooler to ensure 25°C distillate (a process requirement) even when full condensation does not take place in the condenser
4. the sequential operation of the 4 stills to give a reasonably constant heat supply from the condensers.

Reduction in the steam pressure was achieved by replacing the existing stainless steel pan heaters by copper coils. Retaining pan heaters would have entailed more than doubling the area, resulting in a loss of useful volume and difficulties in cleaning.

The new condenser design was the result of development at ECRC. Existing condensers in most distilleries are vertical all copper shell and tube heat exchangers with the cooling water, in a single upwards pass, inside the tubes and with the condensation taking place downwards in the shell. Typically about 2/3 of the condenser area is used for condensation and 1/3 for subcooling. No design code exists for the condensers and most appear to be dimensioned empirically. A process requirement was that any change to the condenser design to achieve the new working condition should be minimal, as the condenser is considered to be a major influence on the product quality. A computer model of the condenser was therefore developed. This model took account of film condensation of a mixture on the vertical tube and either laminar or turbulent flow of water with large temperature gradients across the section inside the tube. The model was verified by operating a 1/40 scale (i.e. 3 tubes instead of 120) condenser in the laboratory (3).

The subcooler has to cool the distillate from approximately 81°C to 25°C using river water which can vary in temperature from 2°C to 20°C and must be able to condense a proportion of the vapour when the condenser fails to condense all of it. This can occur when the still is either driven too fast (by the still operator) or if there is not enough cooling water flow, e.g. when four stills are in simultaneous operation instead of three. Other requirements were that the subcooler should drain freely ir._o the receiver and that there should be negligible pressure drop on the distillate side to avoid pressurising the still and condenser. For these :asons a horizontal subcooler with the distillate in the shell and cooling water in the tubes was chosen, even though this had an inferior overall heat transfer coefficient compared with the more normal cooling water in shell design. The distillate inlet and outlet and baffles were arranged so that the subcooler shell was about 2/3 full of liquid and 1/3 of the heat transfer area was available to condenser vapour. The subcooler was therefore able to cope with up to 5% vapour under steady conditions and 10% during transients.

The final change to the distillery was to the sequence of operations. Prior to the beginning of the project the stills had been operated fairly randomly with a consequent large variation in steam demand. Since the second distillation was the production rate determining operation and these had a cycle time of 8 hours it was decided to fix the cycle time for the first distillation at the same time. If each still was operated so that the distillation time was 5.25 hours and the still starting times were staggered by 1.75 hours then at any one time 3 out of the four stills would be distilling. During the remaining 1.75 hours each still would be discharged (0.25 hour), recharged (0.25 hour), and brought up to its bubble point (0.5 hour).

The theoretical model of the still and condenser and measurements on site gave an average steam demand for the three stills of 1575 kW and a heat rejection from the three condensers of 1432 kW. The compressor suction volumetric flowrate is 7800 m^3/h and discharge mass flow 2.5 tonne/h. The compressor absorbed power is 290 kW and the electric motor is a 355 kW, 415 V induction motor. This is started by an autotransformer starter which limits the current to 1400 amps. The system COP (related to the motor power) is 5.0 at 2.0 bar discharge pressure and 0.47 bar suction pressure.

2.3 System Operation and Control

The system is controlled as follows. When only one still is in operation the compressor is off and the still is heated by the boiler and the condenser heat is rejected to the river via the trim cooler. When two or more stills are on, the compressor operates. When two stills are in operation then approximately 1/3 of the compressor discharge flow is returned to the flash vessel via an automatic bypass valve and about 500 kW is rejected by the trim cooler. When three stills are in operation the system is in approximate balance and only about 100 kW is rejected by the trim cooler. When four stills are in operation the boiler supplies the extra steam needed and the trim cooler rejects the excess heat recovered by the system. The trim cooler is controlled by the temperature of the recirculating water i.e. when it rises above 80°C the trim cooler valve opens. The compressor bypass valve opens on falling suction pressure.

3. INSTALLATION AND COMMISSIONING

The system was ordered in January 1986 and installed during the s_x week summer shutdown in July/August 1986. During the last week of August the system was started and operated with only water in the stills and from the beginning of September with the normal product. At first the distillery was operated at reduced capacity but after approximately 6 weeks the distillery and heat pump were operating continuously. By the beginning of November the heat pump was operating at approximately 90-95% of capacity.

The most important commissioning problems were:

1. Air. The presence of air in the steam would result in higher power consumption, reduced heat transfer in the stills (and hence operation of the compressor at reduced capacity), and reduced water flow to the condensers. Initially the system was designed with steam injection to the flash vessel and compressor inlet to eliminate air from the system before starting. However it was found that the steam heating heated the compressor unevenly. Consequently the system is started with air in the flash vessel and compressor and with the discharge open to atmosphere. This has resulted in greater problems in air elimination as the air was entrained in the recirculated water to the condensers. Subsequently air eliminators have been fitted and the air content in the system has now been reduced to the point where there is no noticeable increase in compressor power, but after a cold start-up e.g. after the Christmas holiday, the process of air elimination takes some days.

2. Poor condensate return. Prior to the introduction of the heat pump system the condensate had been returned to the boiler via a Crane condensate pump. With the heat pump system it is returned to the flash vessel and at first problems were met with flashing in the condensate line causing backup in the still heaters and reduced heat transfer rates. The fitting of a valve in the line at the entrance to the flash vessel to maintain the line pressure above the saturation pressure of the condensate solved this problem.

3. Fluctuations in flash vessel level. The level in the flash vessel is controlled by a differential pressure controller which opens a valve to admit boiler feed water on low level and opens another to allow the excess to drain to the feed tank. At first the level fluctuated wildly causing the compressor to trip on high level (risk of liquid in compressor) or on low level (risk of vapour in the recirculating water pumps). Addition of a sight glass, recalibration of the controller and resetting of the automatic valves solved the problem.

4. Saturated discharge steam. The quantity of water injected at suction to control the compressor discharge superheat to 4K is small (approximately 200 litres/hour). It has proved very difficult to control this to the required accuracy (3%) with a response time sufficiently fast to match the compressor. At first the system delivered saturated steam but the superheat is now controlled normally between 2 K and 10 K, with occasional excursions outside this band. This has been achieved by reducing the water pressure upstream of the water injection valve, by having a small bypass around the valve and by using more, but smaller, nozzles.

4. PERFORMANCE

The system performance has been monitored continually since September 1986 using a datalogger which collects data at 1/2 hour intervals and transmits all the data collected each day to a microcomputer at ECRC, Capenhurst which then analyses the results and draws graphs of the most important channels and derived results. The datalogger also continuously displays the four most important channels and can print out all the channels every minute on demand. There are 60 data channels which monitor the steam flow from the compressor, the total steam flow, the water flows through one condenser and subcooler, the trim cooler, the oil cooler, the water injection and the water overflow from and makeup to the flash vessel and temperatures and pressures around the system. The state of the control valves, the compressor and auxilliary powers are also logged.

Figures 4 to 9 show the results obtained on the 13 November 1986, which are typical of those obtained during November and December. Figure 4 shows the operation of the stills (up represents the still steam control valve is open, down is closed). The still run-time is about 5 to 6 hours, the occasional short duration "off" conditions occur during the "bringing-in" of the stills, i.e. when the steam flow has to be regulated to avoid foaming. Figure 5 shows the cumulative total of the number of stills in operation, which is normally three but rises to four when a still is being brought up to the boil and falls to two when one still is finished before another is ready to be started. Once, for a short interval only, no stills are in operation (i.e. at 16.5 hours). The compressor did not trip out because this period was shorter than the preset delay time in the system control.

Figure 6 shows the compressor discharge and suction pressures throughout the day. These remain fairly constant around 1.7 to 2 bar and 0.46 to 0.52 bar) even though the number of stills in operation fluctuates. Figure 6 also shows the steam pressure in still 4, which during each 7 hour cycle rises from 1 bar (still off) to 1.5 to 1.8 bar when the still is on. Figure 7 shows the heat flow in the total steam delivered to the stills and that delivered by the compressor. This demonstrates that the compressor supplies all the steam needed excluding the peaks in steam demand which occur when a still is being heated up (when no heat is available in its condenser). Figure 8 gives the power absorbed by the compressor drive motor. This varied between 325 and 295 kW during the day and is strongly dependent on the discharge pressure (see Figure 6).

Figure 9 shows the variation of the COP throughout the day. It lies between 4 and 4.8 compared to the design value of 5. This COP does not include the auxilliary power needed for pumps etc, (approximately 35 kW) as this is approximately equal to the power saved on the boiler fans, the Crane pump and pumps used to pump the river water up to the header tank.

5. ECONOMICS

The total installed capital cost of the system was £275,119. The energy prices in January 1986, when the machine was ordered were 12.5 p/l for heavy fuel oil and 3.3 p/kWh for electricity. With an annual operating time of 6500 hours the (46 weeks at 142h/week) the payback period was 3.5 years. At current energy prices (10p/l and 3.2p/kWH respectively the payback period is 5½ years. There was a 15% grant from the Energy Efficiency Demonstration Scheme (EEDS).

6. DISCUSSION AND CONCLUSIONS

The installation has demonstrated that a heat pump can produce large energy savings in distillation. The heat pump has achieved over 90% of its design performance and operated continuously, with only occasional short stops

6

caused in general by operator inexperience, after only 2 months of commissioning. It has been shown that heat pumps using water as the working fluid can be successfully applied in unit operations which result in heat rejection above 80°C. The system described could equally be applied to continuous distillation columns - in fact such an installation would be simpler than the one described, since the wide variations in load found in batch distillation would not have to be accommodated. The use of steam as the working fluid has significant advantages over other fluids - primarily the ease of integration with the existing steam services. The system also demonstrates the larger energy savings that can be achieved if steam heating of unit operations is retained.

7. REFERENCES

1. Drying, Evaporation and Distillation, Energy Technology Support Unit Market Study No. 3, Energy Publications, 1985, ISBN 0 905332 43 1
2. Moss, M S & Hume, J R, 1981, The Making of Scotch Whisky, James & James
3. Hodgett, D L, 1985, A Heat Pump for Batch Distillation, Proc. Commission E2, Trondheim (Norway), International Institute of Refrigeration, Paris

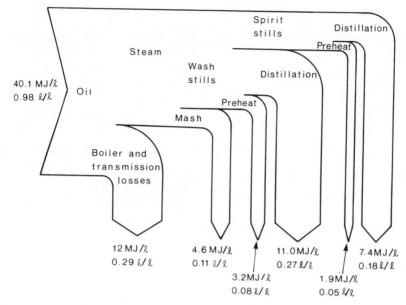

Figure 1. MALT WHISKY DISTILLERY WITH NO HEAT RECOVERY

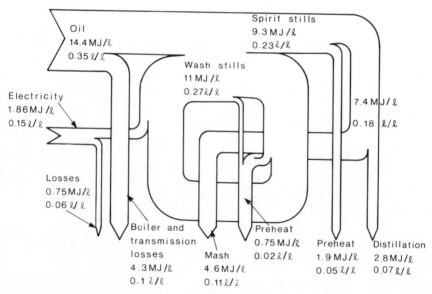

Figure 2. MALT WHISKY DISTILLERY WITH HEAT PUMP AND HEAT RECOVERY

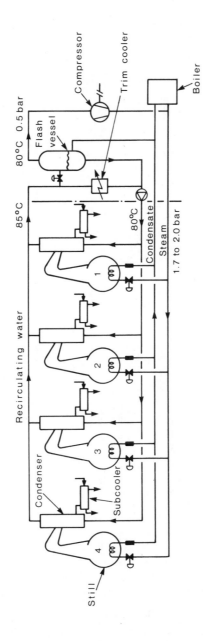

Figure 3. SCHEMATIC DIAGRAM OF INSTALLATION

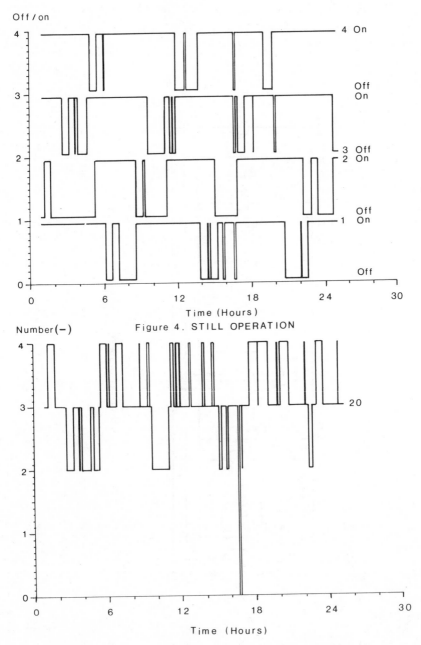

Figure 4. STILL OPERATION

Figure 5. NUMBER OF STILLS DRAWING STEAM

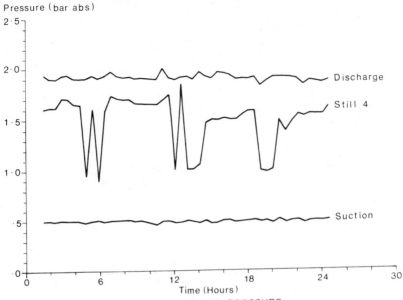

Figure 6   STEAM PRESSURE

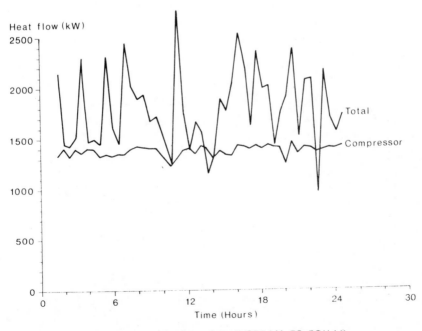

Figure 7   HEAT FLOW IN STEAM TO STILLS

11

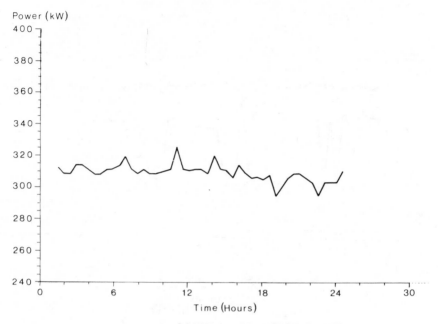

Figure 8   COMPRESSOR MOTOR POWER

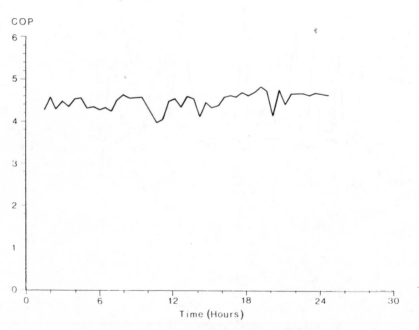

Figure 9 COEFFICIENT OF PERFORMANCE

ADSORPTION HEAT RECOVERY IN DRYING

E Masanja* and C L Pritchard**

Drying is an energy-intensive process, but is usually inefficient. Most of the energy lost leaves in exhaust streams; most of this (typically 80 percent) in the form of latent heat. Effective heat recovery from these streams requires that both the latent and sensible heats be recovered and upgraded to useful temperatures, since direct heat recovery is only possible at very low temperatures. This study considers adsorption heat recovery using magnesium chloride as an adsorbent. The study shows that, using a feed stream of 30-60 deg.C and 50 percent RH, an outlet stream of 80-100 deg.C and 10 percent RH can be achieved, suitable for recycling to a drying process.

INTRODUCTION

Drying or drying-related unit operations account for a high proportion of the process industries' energy consumption: eg 80-90 percent of the energy used in Building brick industry; 92 percent of that used in the Cement industry (1). It is estimated that about 5 percent of the total energy consumption in the UK is used in drying (2) and perhaps 10 percent in the USSR (3).

Despite being energy-intensive, most drying processes are inefficient, (i.e., the proportion of the heat supplied which is actually used to reduce the moisture content of the solids is small). This value varies from about 65 percent in the building brick industry to as little as 7 percent in some pharmaceutical processes. Most of the lost energy leaves in exhaust streams. The form of this lost energy depends on the stream temperature and humidity. Although there are diverse drying process in industry matched with a diversity of drying equipment, the energy lost is mostly (typically 80 percent) in the form of latent heat. There are inherent problems in heat recovery from moist gas streams. Firstly, only the sensible heat is in general accessible for recovery for feed preheating. Indeed heat recovery equipment is often designed to avoid condensation and its attendant problems of corrosion and moisture removal. Commonly used heat recovery equipment exemplifies this approach:

Heat Wheels (Rotary regenerators). In these the hot humid exhaust gases give up their heat to a metal or ceramic matrix contained in a disc which is rotated through the exhaust duct. The heated portion of the disc rotates into the adjacent duct through which the cold entering gas passes, and gives up its heat. Although heat wheels have a high heat exchange efficiency (typically 90 percent), their application is normally restricted to low temperature ranges (5). Other disadvantages include high pressure drops, fouling, and contamination by condensed moisture. In principle, a hygroscopic matrix can be used to allow recovery of latent heat and provide a means of humidity control (6).

* Department of Chemical and Process Engineering, University of Dar es Salaam.
** Department of Chemical Engineering, University of Edinburgh.

Runaround Coils. In these, a separate heat-transfer fluid is circulated through a pair of heat exchangers, one mounted in the inlet and the other in the exhaust duct. This introduces two gas-phase heat transfer resistances, with associated capital cost and temperature difference penalties. Fouling and corrosion with acidic gases (eg those containing SOx, NOx and fluorides) can be problems.

Heat-pipe Recuperators transfer heat between exit and inlet streams by a series of heat pipes. These have the advantage over runaround coils of a very high heat-transfer coefficient on the working fluid side, and no moving parts. But their performance is still dominated by the gas phase heat transfer resistances, and they cannot recover latent heat.

Fluid-bed Recuperators exploit the high heat-transfer coefficients and capacity of flidised beds to collect and retain heat from the waste stream. Hot particles from the bed are transferred continuously to a second bed through which the incoming gases are passed; and the cooled particles are recirculated to the first bed. This system is complicated to control and can contaminate the gas stream by carry-over of fines from the beds. It is generally only applicable to very large heat recovery systems.

All of these heat recovery methods are restricted to recovering sensible heat in the leaving gas stream; and this is generally of low grade (available at low temperatures) compared to what is actually needed in drying processes. Effective recovery therefore requires that both the latent heat (the largest proportion) and sensible heats be recovered and that these be upgraded to useful temperatures.

Closed cycle heat pumps employing refrigerative dehumification have been demonstrated successfully and are especially useful where humidity control is also desired (4). However, these units generally have high capital cost, and available refrigerants limit their useful operating range to 40-80 deg C, thus restricting their widespread dissemination.

An alternative method of recovery is to exploit the reversible exothermic conversion of a salt to its hydrate, in order to return the latent heat directly to the circulating air stream. During the conversion, the heat generated in the packed bed of salt is carried away by the air flowing through the bed, and part of the heat is stored in the bed, raising its temperature significantly.

Although this process depends for its success on the source of waste heat for regeneration of the salt, it can eliminate both heat losses and the need for expensive heat transfer surfaces. This paper reports a laboratory investigation of adsorption heat recovery from a paper mill exhaust stream (typically 50 deg.C and 50 percent RH) using magnesium chloride as an adsorbent.

THEORY

Adsorption is the accumulation of substance, usually gas, on a solid surface. It can be physical or chemical in nature. Physical adsorption is rapid, reversible and significant only at relatively low temperatures, giving single or multiple layers of weakly-bonded adsorbed material on the surface of the solid. Chemical adsorption is slow, irreversible, and may be accompanied by dissociation and consequent strong bond formation.

Adsorption is invariably exothermic, and the heat evolved in the adsorption of water may be up to three times its latent heat of vapourisation. When adsorption is carried out in a packed bed in plug flow, the process can be considered to be taking place on infinitely many, infinitesimal adsorbing zones along the bed length. The first zone in contact with the inflowing adsorbate, say zone 1 (see figure 1) starts to adsorb. When it approaches saturation the following zone, say zone 2, begins to be loaded with adsorbate, and so on. Thus a concentration band or wave appears to be moving along the bed length (Fig 1).

Since heat is given off during adsorption, the bed and fluid temperatures at the adsorption zone rise and, like the concentration wave, the temperature wave also moves along the bed length as adsorption proceeds. Depending on the magnitude of the specific heat of the fluid and that of the absorbent, the temperature wave may lead or lag behind the concentration wave. The temperature wave will lead the concentration wave if the fluid specific heat is higher than that of the adsorbent (7). By monitoring the fluid (air) temperature and humidity (both at inlet and outlet); bed weight; and bed temperature at several locations along the bed (ten in this investigation), the bed temperature variations (waves) and physical property data can be used to analyse the adsorption process and establish the key adsorption process parameters.

A Dispersion–Concentric model (D–C model) was used to analyse the heat generation and the unsteady heat transfer from the adsorbent particles to the flowing fluid (air) and to the surroundings. The following assumptions were made:
- The heat transfer fluid (air) is in dispersed plug flow.
- Adsorbent particles are spherical, and intraparticle fluid concentration and temperature are radially symmetric.
- Intraparticle diffusion rate may be expressed in terms of an effective diffusivity, De, which is constant.
- Heat loss may be expressed in terms of the temperature difference between the bulk fluid and the ambient temperature.

A mass balance over an element of bed, length dx, with fluid velocity Ux gives

$$\left(\frac{\partial c}{\partial t}\right)_x = \mathcal{D}_{ax}\frac{\partial^2 c}{\partial x^2} - U_x\frac{\partial c}{\partial x} - \left(\frac{\partial c_i}{\partial r}\right)_R \cdot \frac{a\,D_e}{\epsilon_b} \qquad (1)$$

where c_i is the adsorbate concentration in the fluid within a particle.
A mass balance at radius r within a spherical particle gives:

$$\epsilon_p\left(\frac{\partial c_i}{\partial t}\right)_r = D_e\left\{\frac{1}{r^2}\frac{\partial}{\partial r}\left(r^2\frac{\partial c_i}{\partial r}\right)\right\} - \rho_s\frac{\partial c_{ad}}{\partial t} \qquad (2)$$

The adsorption equilibrium relation may be written:

$$\left(c_{ad}\right)_r = F\left((c_i)_r\right) \qquad (3)$$

and the mass transfer rate at the particle–fluid interface is given by:

$$D_e\left(\frac{\partial c_i}{\partial r}\right)_R = K_F\left(c - (c_i)_r\right) \qquad (4)$$

Similarly, for fluid temperature T_f and particle temperature T_s, a heat balance gives:

$$\frac{\partial T_f}{\partial t} = \alpha_{ax}\frac{\partial^2 T_f}{\partial x^2} - U_x\frac{\partial T_f}{\partial x} - \frac{a}{\epsilon_b}\cdot\frac{K_s}{C_f\rho_f}\left(\frac{\partial T_s}{\partial r}\right)_R - \frac{2h_o(T_f - T_o)}{R_\ell\,\epsilon_b\,C_f\,\rho_f} \qquad (5)$$

- where the last term gives heat losses from the bed to the ambient surroundings at To. If the heat of adsorption is $-\Delta Ha$, and heat transferred by the **fluid** within the particles is negligible, then heat transfer within the particles is given by:

$$\rho_s C_s\frac{\partial T_s}{\partial t} = K_s\left\{\frac{1}{r^2}\frac{\partial}{\partial r}\left(r^2\frac{\partial T_s}{\partial r}\right)\right\} + (-\Delta Ha)\frac{\partial c_{ad}}{\partial t} \qquad (6)$$

With no accumulation of heat at the fluid–particle interface:

$$K_s\left(\frac{\partial T_s}{\partial r}\right) = h_p\left(T_f - T_s\right)$$

For axial thermal dispersion where heat transfer by radiation is negligible:

$$\alpha_{ax} = \frac{Re}{\epsilon_b} \cdot \rho_f C_f + 0.5 D_p U$$

where

$$Re/_{R_f} = 0.707 \left(k_s/_{k_f} \right)^{1.11}$$

and for mass transfer: $\quad \mathcal{D}_{ax} \, \epsilon_b/_{.D_v} = 20 + 0.5 Sc.Re$

These parameters are then used in the solution of equations (1) – (7) to predict new temperature waves for different values of De and Ka. These values of De and Ka are varied so as to minimise the mean root square error between the experimental and computed temperature profiles. When the error is minimised, the De and Ka values can be taken as good estimates of the adsorption parameters, and used to calculate the mean heat of adsorption -ΔHa. Details of this procedure are published elsewhere (8, 9).

EXPERIMENTS

Air was contacted with hot circulating water in a humidifying column to produce a steady stream of saturated (100 percent RH) air at controlled temperature. A Commodore "PET 4032" was used for data acquisition using the IEEE bus for temperature and flow measurements, and the user port for weight measurements. Air flow was measured by both a Rotameter tube and an orifice plate. A differential pressure transducer was used to log the pressure drop across the orifice plate into the "PET". Temperatures were measured using type K, grade 2 thermocouples. These have a standard accuracy of +-2.2 deg.C in the 0-270 deg.C range. During regeneration experiments, a precision digital temperature controller was used to control the regeneration air temperature. A top pan electronic balance was used to monitor the bed weight to +- 0.1g and the reading was logged into the computer. A humidity probe was used to measure humidity at the bed outlet. The Magnesium Chloride adsorbent was crushed and sieved, and samples in the size range 1 < d < 3.1mm were used. The initial moisture content was determined using a vacuum testing machine. For flowsheet see figure 2.

RESULTS AND DISCUSSION

In experiments, a feed stream of saturated air at 30-50 deg.C (Humidity 0.025– 0.087 kg/kg) and air inlet velocities of 0.326-0.456 m/s were used. The maximum outlet temperature attained was in the range of 80-100 deg.C. This represents an air temperature lift of 30-60 deg C. The outlet humidities were correspondingly reduced; thus, where this hot and effectively dry air stream can be recycled to a drying process, the method can be used for energy recovery.

Some results are given in figures 3 and 4. In figure 3, curves of outlet humidity, w_2 and outlet temperature, T_2 are plotted against time for several saturated air feed temperatures. Similar plots, but for different air velocities, are given in figure 4.

Fig. 3 shows that higher feed temperatures result in shorter breakthrough times (time for outlet humidity to attain a maximum value). The abnormal rise in outlet humidity in curve A is probably the result of a partial breakthrough caused by uneven packing of the bed. Fig. 4 shows that higher velocities are associated with higher outlet temperatures, owing to higher particle-to-fluid heat transfer rates.

The parameter estimation technique gave values of De = 2 x 10(-8) m2/s; Ka = 0.5 m/s and -ΔHa in the range 1100 – 5400 kJ/kg. The uncertainty in the estimate of heat of adsorption by this method is a consequence of the disparity between the thermal and concentration waves observed in the bed.

Magnesium chloride has several disadvantages as an adsorbent. It is very hygroscopic and is corrosive when wet. Its solubility, which increases with temperature, causes loss of porosity and increased pressure drop when a critical loading is exceeded. In experiments, this appeared to correspond to the formation of the hydrate $MgCl_2 \cdot 4H_2O$.

Although in the experiments the outlet temperature started to fall after about 55 minutes, in the actual process adsorption would be continued until either the outlet temperature fell below acceptable levels, or the absorbent reached its critical loading, whichever came first. This time evidently depends not only on the air inlet humidity and temperature, but also on the air flow rate (Fig. 4).

Stable, non-soluble, high temperature adsorbents such as molecular sieves can be equally useful in this field of heat recovery; as has been demonstrated in solvent recovery. Trial runs using molecular sieves, M.S 5A, (16-40 mesh particle size) resulted in outlet temperatures as high as 110 deg.C and practically zero outlet humidity, from a feed stream of 50 deg.C, 100 percent RH.

A PRACTICAL HEAT RECOVERY SYSTEM

An effective adsorption heat recovery unit for drying processes would operate as a heat pump. As with any heat pump, a fraction of high grade energy, (usually small) has to be employed to pump the heat to higher levels :a consequence of 2nd law of thermodynamics. In an adsorption heat recovery system, this high grade energy is the high-temperature heat required to regenerate the adsorbent by driving off the adsorbed moisture. This heat is applied indirectly through a jacket or embedded coil. The vapour driven off from the bed is condensed in a heat exchanger where it heats a portion of the recycle air; thus its latent heat is recovered. A practical arrangement of a 2-bed adsorption drying system is proposed in Fig 5. The basis for calculation of the illustra tive material and heat flows in Fig 5 is described in Appendix 1. In this example, heat supplied to the regeneration beds at 820 kW (63 percent of the requirement of conventional drying), permits the recovery of 50 percent of the latent heat and 100 percent of the sensible heat of the drying air stream.

CONCLUSION

Adsorption heat recovery has the potential to recover both the latent and sensible heats from moist gas streams. The capital cost of adsorption systems is in general lower than that of heat pumps using refrigerative dehumidification. Adsorption systems must be carefully controlled to avoid loss of capacity by agglomeration of the adsorbent particles.

REFERENCES

1. Department of Energy, "UK Energy Use and Energy Efficiency in UK. Manufacturing Industry to the year 2000", Vol. 2, H.M. Stationary Office, London.

2. Sharman, F W, and Johnson, R H, "The Electricity Council Research Centre job No. 4143", 1986, p.1.

3. Keey, R B, "Drying Principles and Practice", Pergamon Press, 1972, p 1.

4. Lawton, J, "Closed and Open Cycle Heat Pumps for Drying", in I Chem E North-Western Branch, "Heat Pumps : Energy savers for the Process Industries", Symposium papers, 1981 No. 3.

5. Payne, A, "Managing Energy in Commerce and Industry", Butterworths, 1982, p 159.

6. Turner, W C, "Energy Management Handbook", John Wiley and Sons, 1982, pp 259-260.

7. Masanja, E, "Adsorption Heat Recovery", M.Phil Thesis, Edinburgh University, 1986.

8. Kaguei, S, Yu, Q, and Wakao, N, "Thermal waves in an adsorption column: parameter estimation." Chem Eng Sci 7, 1985, pp 1069 – 1076.

9. Masanja, E and Pritchard, C L, "Determination of Heat of Adsorption by a parameter estimation technique." In preparation.

10. Wakao, N, and Kaguei, S, "Heat and Mass Transfer in Packed Beds", Gordon and Breach, New York, 1983, pp139-293.

APPENDIX : DESIGN CALCULATIONS

BASIS: Paper dryer producing 1.5 t/h paper at 6 percent moisture from a feedstock containing 50 percent water by weight. Outlet airstream at 50 deg.C and 50 percent R.H.

1.5 t/h paper @ 6 percent moisture $\equiv$ 1.41 t/hr dry weight.
Original moisture content 50 percent implies 1.41 t/hr water fed, & 0.09 t/hr removed on paper.
∴Moisture removed in airstream = 1.32 t/hr
Outlet humidity = 0.04 kg/kg dry air
∴ Airflow rate = 1.32/0.04 t/hr bone dry air = 33 t/hr.
This represents

$$\frac{25 \times 1.32 \times 1000 \times 22.4 \times 323}{3600 \times 28.8 \times 273}$$

= 8.43 m3/s of air at outlet conditions.
At a superficial velocity of 1 m/s this requires a bed diam. of 3.3m.
For a cycle time of 1 hour and a bed loading of 0.2 , the adsorption of 1.32 t moisture requires 6.6 t. of dry adsorbent. At bulk density of 1500 kg/m3 this represents a layer 0.26 m. thick.
To prevent agglomeration and promote an even distribution of air throughout the bed, the adsorbent would be arranged in shallow trays, with heat transfer coils (for regeneration) embedded in each.

Part of the exhaust air is passed through the adsorber beds; part is recycled to the dryer through the recovery heat exchanger, where its humidity is unaffected. Recycling 50 percent of the exhaust produces an inlet humidity to the dryer of 0.02 . To achieve the required drying duty, its temperature would have to be 96 deg.C , corresponding to a R.H. of 4 percent.
Experimental work on adsorption indicates a 50 deg.C temperature rise of air in the adsorber. The recycled air would thus be mixed with adsorber air at 100 deg.C and zero R.H.
An enthalpy balance on the mixed stream indicates that the recycle heat exchanger must raise the temperature of the recycle air to 93 deg.C. The enthalpy of the recycle air is thus 200 kJ/kg.
The inlet enthalpy is 155kJ/kg, so heat duty of recycle exchanger =

$$\frac{33 \times 1000 \times (200 - 155)}{2260 \times 1000}$$

$\equiv$ 0.657 t/h of condensing steam at around 100 deg.C.
This represents half of the latent heat which should be available from the

condensation of water driven off from the regenerating bed. The regeneration heat requirement will be that required to drive off adsorbed water, or approximately

1.32 x 2260 x 1000 kJ/hr
= 2.94 GJ/hr or 816 kW.

We compare this with the heat required for drying, based on a "once-through" system fed with fresh air at 20 deg.C. and 50 percent R.H. (0.007 kg/kg dry air):
Air requirement = 1.32/(0.04 - 0.007) = 40 t/h
Heat requirement = 40 x 1000 x (155 - 38) = 4.68 GJ/h
On this basis, saving = 1.74 GJ/h or 37 percent of the heat requirement.
If simple heat recovery were added to the conventional system, and the exit air cooled to its dewpoint, the gross heat requirement would be reduced by

40 x 1000 x (155 - 140) = 0.6 GJ/h.

Thus adsorption heat recovery could save around three times as much heat as conventional systems. In the process described, to supply this quantity of heat (1.74 GJ/h) using steam at £18 per tonne. would cost £14 per hour or approx. £100,000 per year.

Apart from the adsorbers, the major item of equipment is the recycle heat exchanger. A finned-tube exchanger, with steam condensing in the tubes, might achieve a heat-transfer coefficient of 750 W per m2.K based on the bare tube area.
With steam condensing at 100 deg.C., this gives a heat exchanger area of 25 m2.

Based on these equipment sizes, a payback period of 1 - 2 years appears realistic for an adsorption heat recovery system.

SYMBOLS USED

a	– Interfacial area per unit volume (m2/m3)
C	– Concentration (kmol/m3)
Cad	– Adsorbed adsorbate concentration (kmol/m3)
Ci	– Intraparticle adsorbate concentration (kmol/m3)
Cf	– Fluid specific heat (J/kg.K)
Cs	– Solid specific heat (J/kg.K)
Dax	– Axial fluid dispersion coefficient (m2/s)
De	– Effective diffusivity (m2/s)
hp	– Particle-to-fluid heat transfer coefficient (W/m2.K)
ho	– Overall heat transfer coefficient between fluid and external (ambient) surroundings (W/m2.K)
ks	– Solid thermal conductivity (W/m.K)
kf	– Fluid thermal conductivity (W/m.K)
Kf	– Fluid mass transfer coefficient (m/s)
R	– Particle radius (m)
Rt	– Tube (bed) radius (m)
r	– Radial distance variable in particle (m)
Tf	– Temperature in bulk fluid phase (oC)
To	– Temperature of surroundings (ambient) (oC)
Ts	– Temperature in adsorbent particle (oC)
t	– Time (s)
U	– Fluid velocity based on cross-sectional area of bed (m/s)
Ux	– Interstitial mean axial velocity of fluid (m/s)
w	– Humidity of airstream (kg/kg)

x	– Axial distance (m)
α_{ax}	– Axial fluid thermal dispersion coefficient (m2/s)
ρ_f	– Fluid density (kg/m3)
ρ_s	– Solid density (kg/m3)
ε_b	– Bed voidage (–)
ε_p	– Intraparticle voidage (–)
$-\Delta Ha$	– Heat of adsorption (J/kmol)

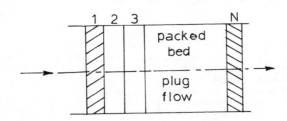

Fig 1: Adsorption Zones

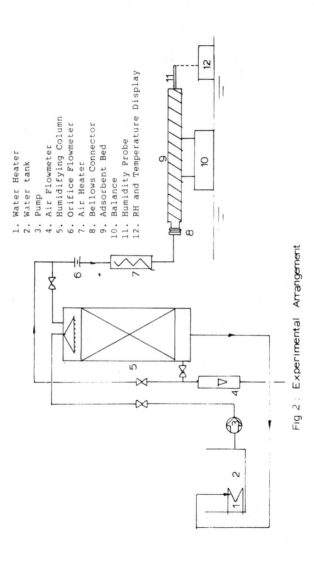

1. Water Heater
2. Water tank
3. Pump
4. Air Flowmeter
5. Humidifying Column
6. Orifice Flowmeter
7. Air Heater
8. Bellows Connector
9. Adsorbent Bed
10. Balance
11. Humidity Probe
12. RH and Temperature Display

Fig 2 : Experimental Arrangement

Fig. 3 : Experiment Results: Adsorption.

Effect of Inlet Temperature. (U=0.393m/s)

A: T_1 = 54.4 Deg. C

B: T_1 = 48   Deg. C

C: T_1 = 44.4 Deg. C

D: T_1 = 34.2 Deg. C

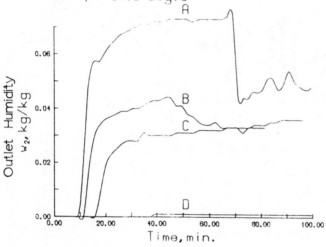

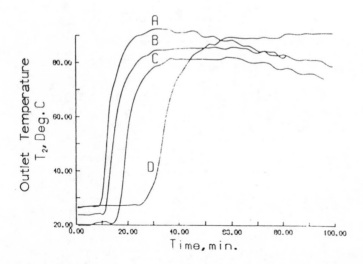

Fig. 4 :Experiment results:Adsorption
Effect of Velocity
A: U = 0.456 m/s; T_1 = 47.8 Deg.C
B: U = 0.393 m/s; T_1 = 52.3 Deg.C
C: U = 0.326 m/s; T_1 = 52 Deg.C
Date:A:-26/03/86;B:-26/02/86;C:-28/11/85

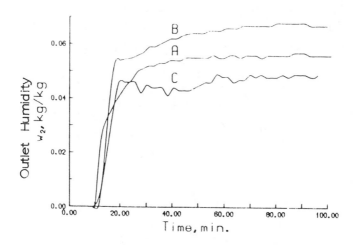

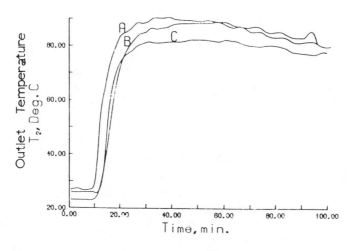

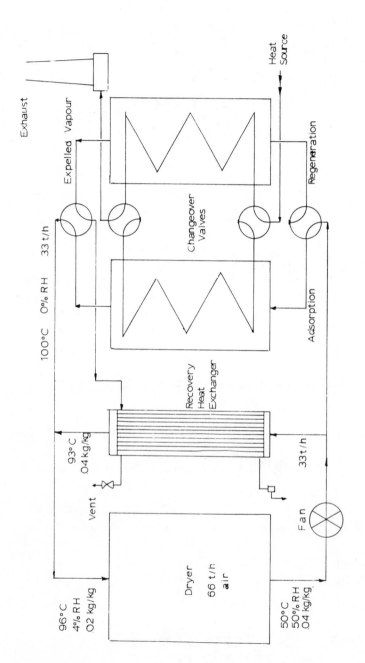

Fig 5 : Adsorption Heat Recovery Scheme

HEAT RECOVERY FROM MEAT RENDERING CONTACT DRYER
BY MEANS OF STEAM COMPRESSION HEAT PUMP

R.H.Green*   and I. J. Sharpe**

Steam compression is a technique by which waste
steam can be increased in pressure and re-used as
a process heat source. Contact dryers have been
identified as an attractive area of use for this
technique and the design for a particular
application within the meat rendering industry is
discussed.

The system offers significant energy cost
savings, with the potential for a 70% reduction
in the energy consumption of an existing dryer.

INTRODUCTION

It is estimated that British industry consumes 103.8 million GJ
of energy per annum in drying processes[1]. Much of this
energy goes to waste in the form of exhausted water vapour and
there is considerable scope for heat recovery.

Heat pump techniques offer the opportunity to recover the latent
heat contained within vapour streams and are now used
commercially for batch drying operations at temperatures up to
70°C. However, this represents a relatively small proportion
of the energy consumed within drying and there was clearly a
need to develop heat pumps suitable for continuous dryers
operating at elevated temperatures.

There has been an on going program of experimental work at the
Electricity Council Research Centre (ECRC), on the use of heat
pumps with contact dryers since 1981. A small pilot plant
dryer, (20-50 kg/h evaporation rate), has been operated in
conjunction with a closed cycle refrigerant heat pump and both
open and semi-closed cycle steam compression systems. The
former system is not regarded as showing such economic promise
as the latter two and is not considered here. All aspects of
control and operation of the systems have been fully proven.

*   Electricity Council Research Centre,
    Capenhurst, Chester
** East Midlands Electricity, Coppice Road,
    Arnold, Nottingham.

This paper discusses the transfer of this technology to a
full-scale demonstration plant and describes a particular project
within the meat rendering industry which has been successful in
attracting financial support under the EEC Energy Demonstration
scheme.

STEAM COMPRESSION

Heat Pumps

A heat pump extracts heat energy from a low temperature source
and upgrades it to a higher level by means of a vapour
compression cycle. A domestic refrigerator can be considered as
a heat pump. The heat, which is extracted from the interior and
the food, is upgraded by the energy consumed by the compressor
and is transferred to the room via the heat exchanger at the back
of the refrigerator. The work done by the compressor being
proportional to the temperature lift achieved.

In this case the desired effect is to extract heat in order to
maintain a low temperature. However, with the escalation in fuel
costs in recent years, there is an increasing incentive to use
this principle to provide heat at a higher temperature from low
temperature waste heat.

In the past, the temperature achievable with heat pumps has been
limited to relatively low levels, However,with recent
developments in compressor technology, it is now possible to
produce high temperature steam. This is particularly relevant to
processes where low pressure steam is being driven off from the
product and steam is also used for process heating. Where this
occurs, it is possible to compress the waste steam, which is
normally vented to atmosphere and use it to provide steam at a
higher pressure for re-use within the process.

This concept of steam compression, or mechanical vapour
compression (MVC), as it is sometimes known, has been
successfully applied to evaporators and, more recently, to a
distillation column $^{(2)}$. However, as yet the principle has not
been demonstrated on an industrial scale on a dryer.

Two types of steam compression system are considered here.

Open Cycle

In an open cycle system, the vapour from the process is injected
directly into the compressor where it is compressed to the
process steam pressure. This then gives up its heat on
condensing within the process heat exchanger (Figure 1).

Closed Cycle and Semi-Closed Cycle

In a closed cycle system (Figure 2), vapour from the process is
condensed in a heat exchanger where it evaporates the working
fluid within the heat pump circuit. The working fluid is then
compressed and gives up its heat at a higher temperature within
the process. However, where the working fluid is steam, it is
possible to mix the compressed steam with supplementary steam
from boiler plant. This semi-closed cycle is described in more
detail later.   26

CONTACT DRYERS

Thermal dryers for particulate materials fall broadly into two
categories, convection dryers and conduction or contact dryers
(Figure 3). Convection dryers are by far the most common being
relatively cheap and, if used with high temperatures, can achieve
reasonable thermal efficiencies. Contact dryers are less common
because of their higher capital cost despite, in general, giving
higher thermal efficiencies. In principle they require little or
no through flow of air or inert gas and can use any heating
temperature provided it is above the evaporating temperature of
the moisture. This makes them inherently more suitable for use
with heat pumps[3] than the convection dryers, since it is a
relatively easy matter to match drying conditions to the
requirements for an efficient heat pump i.e. a low temperature
lift. Convection dryers frequently use very high inlet gas
temperatures which would be both impossible and undesirable to
match with a heat pump. Reducing the temperature to improve the
situation would lead to very high
air flows and consequently large air handling costs.

Some of the areas in which contact dryers find favour at present
are as follows:-

1.   The drying of heat sensitive materials. Any danger of
        exposure to very high temperature must be avoided. In
        such situations the dryer is often used under vacuum.

2.   Where the air handling costs are to be kept to a minimum
        e.g. solvent recovery or odour control.

3.   The use of a cheap low temperature energy source such as
        waste steam from another process.

4.   Sterilisation. All the solids must reach a certain
        temperature and be sustained at that temperature for some
        time.

The desire to use a heat pump for energy saving reasons
introduces similar constraints to situations 1, 2 and 3 above and
again the choice points to a contact dryer with the extra capital
cost being offset by the energy cost savings. In the long term
it is the authors' view that such a combination of contact dryer
and heat pump will find favour in a large number of cases where
convection dryers are at present used.

However in the short term there is a need to demonstrate the
system on an industrial scale and clearly the most sensible
approach is to modify an existing process. One of the main
industries to use contact dryers is the meat rendering industry
(for reasons of sterlisation and odour control), and consultation
with a customer within the East Midlands Electricity Board has
shown that the operating conditions are such that a retrofit of
the heat pump is a possibility. It is for this reason that our
work has concentrated on this area.

MEAT RENDERING

The meat rendering industry treats waste material from abbatoirs and meat processors. The raw materials, either offal or bone, are milled before being dried and sterilised to produce tallow and bonemeal. The drying operation is highly energy intensive, with the moisture content of the product being reduced from typically 60% down to 5%. Traditionally, batch drying systems have been employed, but in recent years there has been an increasing use of continuous contact dryers.

Typical of these dryers is the rotating disc contact dryer. In this design of dryer, the product is heated by direct contact with the surface of the metal discs which are steam heated. Dryer sizes vary between sites, but evaporation rates of the order of 4 tonnes per hour are typical. The water vapour driven off from the product is drawn through a spray condenser in order to remove contaminants. All of the heat associated with the vapour goes to waste.

Process steam pressures are normally of the order of 6 bar and typically 1.3. to 1.4 kg. of process steam are consumed for each 1 kg of moisture evaporated from the product. Process steam is provided from fuel fired boiler plant.

The nature of the industry requires that the process runs at a high load factor and plants normally operate in excess of 6000 hours per annum.

SYSTEM DESIGN

Type of system

The open cycle heat pump has both a lower capital cost and a lower running cost than the semi-closed cycle system (no additional heat exchanger is required and consequently the overall temperature lift is lower). It is thus the obvious first choice and indeed an open cycle system has been proposed for a similar application elsewhere [4].
However, the semi-closed system offers a number of advantages which in the opinion of the authors, outweigh these considerations.

(i)   The vapour leaving the dryer is highly corrosive and will attack any mild steel sections of the dryer, steam piping and compressor.

(ii)   The vapour is also contaminated with fats and small solid particles which are likely to foul the internal heat transfer surfaces of the dryer and block the narrow condensate return passages.

(iii)   The dryer exhaust must be at a slightly subatmospheric pressure (a few inches water gauge), to prevent the leakage out of unpleasant odours. This leads to a danger of air ingress into the system. In the open cycle system the air is compressed along with the vapour and will tend to accumulate within the heating discs leading to a rapid deterioration in dryer performance.

The semi-closed cycle is, on the other hand, much less
sensitive to air ingress and will always supply steam to
the dryer at the desired pressure. The energy efficiency
of the process does however deteriorate with increased
air ingress.

(iv)   The semi-closed cycle system is easier to control than
        the open cycle system as the heat exchanger acts as a
        buffer against small fluctuations in dryer performance.

        For these reasons the semi-closed cycle system has been
        chosen for this application. A full schematic diagram of
        the proposed installation is shown in Figure 4.

Size of System

The dryer is nominally rated for the following duty:-

Drying rate   = 4000 kg/h water vapour   removed

Condensing temperature   = 160°C

Steam usage   = 1.35 kg steam/kg evaporated.

To allow for some variation in dryer load without altering the
heat pump conditions and to ensure that the compressor never
produces an excess of steam, it was decided on consultation with
the customer, to design the system for a base load drying rate of
3500 kg/h and hence a steam demand of 4725 kg/h. This permits
any variations in feed rate and moisture contents, resulting in
variations in evaporation rate, to be allowed for by varying the
amount of steam supplied by the boiler. The boiler also provides
the steam under start up conditions. It is important that the
heat pump is not oversized since this both raises the cost of
each unit of steam delivered, making the system less economically
viable, and complicates the operating procedure. The oversizing
of heat pump systems has been a major reason for the poor
performance of a number of projects in the past (Ref.5).

Component Design

Heat Exchanger

The function of the heat exchanger is to condense the
contaminated vapour from the dryer and raise clean steam. The
design requirements are that it:-

(i)   Can achieve a small overall temperature difference, so as
        to keep the running costs down, whilst not having a
        prohibitively high capital cost

(ii)   Can be readily cleaned since the vapour from the dryer
         contains some solids and fats.

(iii)   Will readily vent any incondensibles present.

It is desirable to have the contaminated vapour condensing within
the tubes of a shell and tube heat exchanger since the tubes are
relatively easy to clean. One promising design for this which
has been the subject of experimental work at ECRC is the kettle
reboiler, consisting of a flooded horizontal tube bundle placed
in an enlarged shell.

Fouling of the tubes is an area of concern and on-site tests with
a single tube condenser rig have been performed to quantify this
effect. As a result of these tests, short-term build up of
fouling is not regarded as a problem, but some provision for
cleaning the tubes should be made. It is envisaged that this
will be carried out during weekend shutdowns.

The heat exchanger design is perhaps the major area of technical
risk involved with the system. Experimental work both at ECRC
and on the user's site have reduced some of the uncertainties and
a design procedure for the heat exchanger has been determined.
Further work is needed on corrosion testing of materials to
establish the cheapest material for construction. It is assumed
here that stainless steel will be required.

Compressor

The compressor will be of the twin-screw type for a duty of this
size. To prevent contamination of the steam with the lubricant
the compressor is of the oil free type. The pressure lift is
rather high and if attempted with a single stage machine would
lead to a low efficiency. It is proposed to use a two-stage
machine which will achieve a higher overall efficiency than the
single stage option and the consequent lower running costs will
quickly offset the higher capital cost.

The effects of air ingress

The presence of incondensibles in the contaminated vapour stream
from the dryer has a detrimental effect on the system energy
performance and should be minimised. The effects of air ingress
are as follows:-

(i)   The temperature at which heat can be recovered from the
         vapour steam reduces as the air content increases.
         Figure 5 shows the inlet and outlet temperatures required
         of a heat exchanger for the present duty. The upper curve
         shows the inlet dew point which decreases rapidly as the
         ratio of steam-air falls below 10:1. The lower curve
         shows the corresponding heat pump evaporating temperature
         assuming a minimum temperature difference of $5^{\circ}C$.

(ii)   The heat transfer coefficient of the condensing vapour
         decreases as the amount of air increases.

(iii)   The reduced evaporating temperature increases the overall
         temperature lift of the heat pump and consequently the
         required compressor power increases.

(iv)   The required compressor volumetric throughput increases
         with increasing air ingress as the evaporating
         temperature and hence density decreases.

From Figure 5, it is desirable to achieve steam to air ratios
above 10:1 to avoid the design becoming highly sensitive to
further small changes in air content. In situations where a high
steam to air ratio cannot be achieved then consideration would
have to be given to reducing the heat pump duty to maintain a
high evaporating temperature.

The two main problem areas from the point of view of minimising
air ingress are the solid feed and discharge. Air ingress is
kept to a minimum at these points by a combination of good
component selection and control of the pressure within the
dryer. On the pilot plant at ECRC a double disc valve
arrangement was found necessary for the discharge since other
systems e.g. a rotary valve, had too high a leakage rate as a
proportion of the drying rate. With this arrangement negligible
air ingress was achieved. On the actual meat rendering plant
discharge is via a screw conveyor and steam to air ratios in
excess of the minimum requirement have been achieved.

ECONOMIC ANALYSIS

The system requires no specialist operators and the economic
analysis therefore rests on a comparison of the annual energy
cost savings achieved by the heat pump system with the additional
capital and maintenance costs incurred. A full analysis should
be based on a discounted cash flow over the life of the project.
However, this will be influenced by such factors as the company's
cost of capital, taxation, etc. and a simple payback has
therefore been used in this analysis.

Energy Costs

For the purpose of this analysis, the following energy costs are
assumed:-

         Steam   £10/1000 kg
         Electricity 3.4p/kWh

         A steam cost of £10/1000kg is based on a gas price of
         30p/therm and a steam system efficiency of 70%.

Energy Cost of Existing Dryer

In operating at a drying rate of 4000 kg/h the dryer will consume 5400 kg of process steam per hour. The energy cost is therefore £54/h.

Energy Cost of Dryer with Steam Compression Retrofit

The steam compression system is based on a 760 kW (electrical) compressor handling 4725 kg of steam per hour. The balance of 675 kg/h being provided by the boiler plant. It is assumed that the thermal efficiency of the boiler plant will fall when operating at this reduced demand such that the steam cost rises to £11.70/1000 kg.

The electrical consumption of the compressor therefore represents a cost of £25.84/h, with supplementary steam costing £7.90/h. The total energy cost is therefore £33.74/h.

Energy Cost Saving

The energy cost saving resulting from the installation of the steam compression system is therefore £20.26/h.

Simple Payback

The dryer operates 6,600 hours per annum, giving an annual energy cost saving of £133,700. Based on projected installed capital and maintenance costs, the system offers a simple payback in approximately two years.

Fuel Price Sensitivity

During 1986, fuel prices fell sharply and, consequently, the energy cost savings outlined above may not currently be realisable. However, it should be remembered that this analysis is based on a retrofit installation. Where a new or replacement dryer is to be installed, the system optimisation should include the effect of condensing temperatures on both running costs and capital costs. Decreasing the condensing temperature will improve the heat pump performance, but will lead to an increase in dryer size. In this situation, a more favourable economic case can be made, even with lower fuel prices.

CONCLUSIONS

The operation of a contact dryer with both open and semi-closed cycle steam compression systems has been demonstrated on a pilot plant scale at ECRC. Meat rendering has been selected as one possible area for a full scale demonstration plant - allowing the system to be retrofitted to an existing process. The economics of the proposed semi-closed cycle system have been shown to be very favourable giving a payback period of approximately two years against a fossil fuel price of 30 pence/therm.

Although the case presented is that for a particular sector of industry, the process will also be applicable to drying operations within the brewing, distilling, food, chemical and other industries.

REFERENCES

1.   Energy Technology Support Unit, Market Study No.3.
        "Drying Evaporation and Distillation".

2.   D.L. Hodgett, 1987 Conference Proceedings, "Innovation in
        Process Energy Utilisation", I.Chem.E.Conference.

3.   R.H.Green, 1982, "Energy Performance of Heat Pumps and
        Continuous Dryers", Conference Proceedings Third
        International Drying Symposium, England.

4.   A.E.Ruggles, 1985, "Contaminated Vapor Rotary Screw
        Compressor", Thermo Electron Corp., USA, Report No.
        TE4326-198-85.

5.   R.Gluckman, 1987, "Practical Performance of Industrial
        Heat Pumps", Proceedings of Institute of Refrigeration
        1987/88

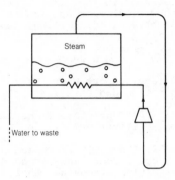

Figure 1. Open Cycle Heat Pump

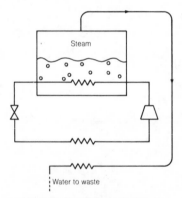

Figure 2. Closed Cycle Heat Pump

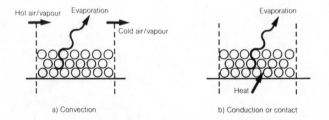

a) Convection

b) Conduction or contact

Figure 3. Dryer Types

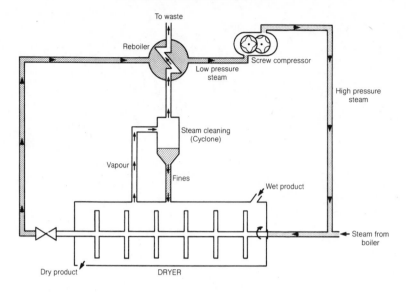

Figure 4. Schematic Diagram of System

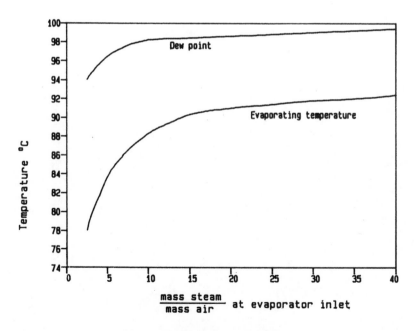

**Figure 5. The effect of air concentration on
heat exchanger design temperatures**

35

THE ECONOMICS OF MEMBRANE SEPARATIONS IN THE PROCESSING

OF BIOLOGICAL MATERIALS

Dr. W. R. Bowen*

Membrane separation processes offer effective and efficient
means of processing biological materials.   The nature and
scope of these processes are first outlined.   The general
costs of installing and running membrane plant are then
presented.   The economics and energy efficiency of membrane
processes are compared with those of the competing techniques
or evaporation, centrifugation and freeze-drying.   Finally,
the developments necessary for better understanding and
improvement of the efficiency of the membrane processes are
outlined.

1.   INTRODUCTION

Effective and efficient product separation is the key to economic operation in
many of the process industries.   However, certain types of materials are
inherently difficult and expensive to separate.   Prominent examples include:

1.   The solid-liquid separation of finely dispersed solids, especially those
which are compressible, have a density close to that of the solvent, have high
viscosity, or are gelatinous.

2.   Separation of low molecular weight non-volatile organics or pharmaceuticals
and dissolved salts.

3.   Separation of biological materials, which are very sensitive to their
environmental conditions.

The processing of all of these categories of material has become of increased
importance in recent years with the growth of the newer biotechnological
industries and with the increasingly sophisticated nature of processing in the
food industries.

Since the 1960's the new technology of membrane separation processes has
developed rapidly.   Such processes have been widely applied to a wide range of
conventionally difficult separations.   They potentially offer the advantages of
ambient temperature operation, relatively low capital and running costs, and
modular construction.   The present paper begins by outlining the nature and
scope of membrane separation processes.   The paper then presents comparisons
of the economics of membrane processes and competitive techniques in the
separation of biological materials.   Here it will be found that the membrane

*Colloid and Interface Group, Department of Chemical Engineering,
 University of Wales, Swansea, SA2 8PP, U.K.

processes can offer significant economic advantages, including benefits in process energy utilisation. Finally, some developments which should lead to the enhancement of the efficiency of membrane separation processes will be considered.

2. THE NATURE OF MEMBRANE SEPARATION PROCESSES

When difficulties arise in the processing of utilisation of biological materials, there is one question which is of paramount importance - how does nature solve the problem? The solution which nature has developed is likely to be both highly effective and energy efficient. Nature separates biologically active materials by means of membranes. We are all surrounded by a membrane, our skin, and membranes control the transport of materials at all levels of life, down to the outer layers of bacteria and sub-cellular components. A membrane may be defined as an interphase separating two phases and selectively controlling the transport of materials between those phases. It is an interphase rather than an interface because it occupies a finite element of space.

2.1 Classification of membrane processes

Industrial membrane processes may be classified according to the size-ranges of materials they are able to separate and the driving force used in separation. There is always a degree of arbitrariness about such classifications, but the distinctions which are typically drawn are shown in Table 1. This paper is primarily concerned with the pressure driven processes, microfiltration (MF), ultrafiltration (UF) and reverse-osmosis (RO). These processes are already used on a large scale in the processing of non-biological materials. For example, reverse-osmosis is used world-wide for the desalination of brackish water with about 1,000 plants in operation[1].

Table 1   Classification of membrane separation processes for liquid systems.

Name of process	Driving force	Separation size range	Examples of materials separated
Microfiltration	Pressure gradient	10 μm - 0.1 μm	Small particles, large colloids, microbial cells.
Ultrafiltration	Pressure gradient	0.1 μm - 5 nm	Macromolecules, proteins, emulsions
Reverse osmosis (hyper-filtration)	Pressure gradient	< 5 nm	Small organics, desalination
Dialysis	Concentration gradient	< 5 nm	Treatment of renal failure
Electrodialysis	Electric field gradient	< 5 nm	Desalination, small organics
Electrofiltration	Combined pressure and electric field gradient	10 μm - 5 nm	As for microfiltration and ultrafiltration

Plants with capacities of up to 10^5 m^3/day of potable water are planned. Again, it is now, for example, standard practice to include an ultrafiltration unit in paint plants in the car industry. Recovery of paint from wash waters can result in savings of 10-30% in paint usage and a pay-back time of 1-2 years. In both of these examples the economics of the processes are reasonably well established. However, the economics of these processes in the treatment of biological materials is less well-documented. Dialysis has one major use, hemodialysis of patients with renal failure where it is most appropriate to use such a gentle process. However, it is a relatively slow process, not really suited to large scale industrial operation. Electrodialysis is used world-wide for the desalination or concentration of brackish water. In 1981 there were about 300 plants in operation. Economics presently favour reverse-osmosis rather than electrodialysis for these processes. It may have a future role in the desalination of protein solutions. This paper considers the recovery of materials from solution, and membrane processes for gas separation are therefore excluded from this Table 1.

2.2 The nature of synthetic membranes

The membranes[2] used for microfiltration, ultrafiltration or reverse-osmosis are most commonly composed of polymeric materials, cellulosic, polysulphone or polyamide. They may be homogeneous or heterogeneous in structure and are typically about 100 μm thick. Microfiltration membranes are generally symmetric in structure, but ultrafiltration and reverse-osmosis membranes have an asymmetric structure with a finely porous layer of 1-2 μm thickness backed by a more open pore structure giving less resistance to fluid flow. They are generally produced by a phase-inversion technique, it being in principle possible to produce membranes with a mean pore size anywhere in the ranges indicated in the Table 1. More recently, ceramic membranes have become widely available for microfiltration and ultrafiltration. These are generally produced by sintering fine particles of alumina or zirconia at high temperatures. They can have better chemical resistance and a narrower pore size distribution than polymeric membranes. They are also fully steam sterilisable, an important advantage in the processing of biological materials. Stainless steel membranes are now available for microfiltration.

2.3 The principle of cross-flow filtration

When a process stream is treated by a membrane, there is an increase in concentration of the rejected solute at the retentate side of the membrane. This is an example of the phenomenon known as 'concentration polarization'. It is this concentration-polarized layer which forms the major resistance to permeation in most pressure-driven membrane separation processes. In order to reduce the effect of concentration polarization, these processes are normally used in the cross-flow mode. That is, the feed stream is pumped at a relatively high velocity, typically 2-8 m s^{-1}, tangentially across the membrane face. (Flow may be laminar or turbulent, depending on the design and application). The higher the cross-flow velocity then the thinner the boundary layer thickness and hence the smaller the effect of concentration polarization. Cross-flow filtration maintains a high overall liquid removal rate as no filter cake is formed, gives high solute retention as small pore size membranes can be used, and maintains the feed in a mobile form, being especially useful for viscous but shear thinning dispersions. The flow channel typically has a thickness of 1-25 mm and may be up to several metres in length. The flow channel may be defined by two parallel flat membrane sheets or else may be the bore of a tube or fibre. Membrane sheets may be packed together to form modules as in plate-and-frame filters. Membrane tubes may be grouped into modules similar in pattern to tubular heat exchangers. In all cases, apart from hollow fibres, the membranes will be held in place by rigid macroporous supports. To give some idea of the size of modules, those produced by major membrane equipment manufacturers have membrane areas of $0.5-20m^2$. These may then be grouped together to produce a plant of any size. The applied pressure in the case of microfiltration is normally about 2 atmospheres, for ultrafiltration it is 1-10 atmospheres, and for reverse-osmosis 10-80 atmospheres.

3.   THE ECONOMICS OF MEMBRANE SEPARATION PROCESSES

If an industrialist is considering the installation of a membrane plant he will wish to evaluate in detail the economics of the membrane option for a particular process, comparing costs with those of competitive processes. There have undoubtedly been very many studies of this kind, though in most cases the results of the analysis have not been published for reasons of commercial confidentiality.   They would in any case not always be of general interest due to the different economic circumstances encouraging or discouraging investment in different locations, differing accounting procedures, and often highly specialized markets for products.   Nevertheless, a number of interesting studies have been published and will be considered later in this section. However, this section begins by adopting a more general approach.

3.1 The general cost of installing and running membrane plant

A more generally useful approach is to arrive at average costs for installing and running membrane plant[3].   The costs can be expressed in a way which is independent of membrane permeation rate, the only operational parameter which differs significantly from one application to another.   By then introducing membrane permeation rates together with product yields and product values it is possible to estimate the economic viability of new separations.

The present capital cost of installing membrane plant is about £1,000 per m^2 of membrane area.   The cost depends on location, types of materials used and the extent of instrumentation and automation.   As membrane plants are normally modular in construction, this capital cost will be relatively unaffected by the size of the plant.   The annual operating costs, based on 300 days operation, for a membrane plant, are then as shown in Table 2.

Table 2.   General costs of installing and running membrane plant

Nature of cost	Cost (£) per m^2 membrane	Percentage of overall cost	
		MF/UF	RO
1. Capital charge	200	42	28
2. Membrane replacement	100	21	14
3. Power	60 (MF,UF) 300 (RO)	12	41
4. Chemical cleaning	45	9	6
5. Labour	60	12	8
6. Maintenance	20	4	3
Total	£485 (MF,UF) £725 (RO)	100	100

Thus, on this basis, the annual running cost per m^2 of membrane area is ∿ £485 for microfiltration and ultrafiltration, and $\sim$ £725 for reverse-osmosis. The bases of the costs in Table 2 need some explanation. The capital charge allows depreciation over 5 years. It is assumed that the membranes are polymeric and need replacing every year. Ceramic membranes are about four times as expensive as polymeric membranes but they also last longer, in some cases for more than four years continuous operation without replacement. The energy requirement consists of two main items, one due to the feed pump which pressurises the system and the other due to the recirculation pump which maintains the cross-flow velocity. The first item is proportional to the feed volume and applied pressure, the second a function of cross-flow velocity and viscosity, and hence pressure drop across the module. It is assumed that the power consumption is 0.2 kW m^{-2} for MF and UF, and 1 kW m^{-2} for RO with the plant operating 20 hours per day, 300 days per year and a unit cost of £0.05 per kWh. This corresponds to an energy requirement of 2 kWh m^{-3} of permeate for MF and UF and 10 kWh m^{-3} of permeate for RO if the permeation rate is 100 L m^{-2} h^{-1}. It is assumed that detergent cleaners with a cost of £0.01 per litre and a requirement of 15 litres per m^2 per day are required. Labour charges are taken to be £10 per man-hour. Maintenance costs are taken as 2% of the installed capital cost. With these assumptions and taking a simple view of process energy utilization, power consumption is seen to be a relatively small part of the overall cost for MF and UF, but much more substantial for RO due to the higher operating pressures.

It is worth comparing these guideline figures with recently published data for specific applications. Six examples are shown in Table 3. The data are generally based on the operation of pilot plants and the assumptions used in the calculations are rather different. Nevertheless, they confirm that the above estimates provide a good guide to at least the overall cost of running membrane plant.

Table 3   Costs of running membrane plant

Source	Type of process	Nature of feed	Annual cost per m^2 membrane area
Nielsen and Kristensen[4]	RO	Fermentation broth	£388*
Kristensen[5]	UF	Blood cell hydrolysate	£276
Le and Howell[6]	UF	Polysaccharide recovery	£590
Scott[7]	UF	Polysaccharide recovery	£817
Strathmann[8]	UF	Oil emulsion	£1000
Le et al[9]	MF	Cell harvesting	£311-£1147

*Energy use $\sim \frac{1}{2}$ of that assumed in Table 2.

Having established the overall running costs of membrane plant, it is then important to use the plant as efficiently as possible. That is, it should be operated so that the membrane permeation rate is as high as possible. Typically, permeation rates are in the range 10-100 L m^{-2}h^{-1}, being toward the upper end for the most porous membranes and for clean feeds. Unfortunately, it is not presently possible to predict the permeation rate for a given feed.

Up to a certain point the permeation rate can be increased by increasing the trans-membrane pressure, but at higher applied pressure, in the so-called 'gel-polarisation' region, the permeation rate becomes independent of applied pressure.   Increasing the cross-flow rate decreases the thickness of the boundary layer and increases the permeation rate.   However, the uncertainty is mostly due to the complex way in which solutes can interact with the membrane surface, either forming a dynamic membrane layer or simply by fouling the membrane surface.   These interactions are time dependent, the permeation rate often falling rather rapidly during the first hour of operation and more slowly afterwards.   The permeation rate also decreases as the solute concentration increases.   In practice, it is at present essential to run pilot plant studies for new feeds to determine permeation rates and the extent of membrane fouling. Variation of process parameters and modification of feed composition then allows optimisation of the process.   Membrane equipment manufacturers are often able to run such pilot studies.

With these difficulties in mind, the following sections will review some published studies of the costs of membrane processes compared to evaporation, centrifugation, and freeze-drying for specific separations.

3.2   Reverse-osmosis compared with evaporation

One of the most common operations in the food industry is concentration. Concentration is used to reduce the costs of drying, to induce crystallization, to reduce storage and transport costs, to reduce microbiological activity and to recover products from waste streams.   Concentration is an energy intensive process.   For example, it has been estimated that 50% of the energy input in the Dutch dairy industry is used for concentration[10].   Evaporation has been the standard means of concentrating, though there is now considerable interest in the use of reverse-osmosis or ultrafiltration.

Due to the energy intensive nature of these processes, and bearing in mind the present context, it is appropriate to look at the energy costs of reverse-osmosis and evaporation in the food industry.   A detailed study of costings in the Danish dairy industry has been published[10] and the present account is a modified version of these findings as they can be applied to the UK.   Multiple stage thermal vapour recompression (TVR) evaporators are widely used, with up to seven effects.   The number of effects is limited by the total temperature difference available.   In the food industries the maximum temperature is limited by the thermal stability of the product and the lower temperature by available cooling facilities and product viscosity.   Mechanical vapour recompression (MVR) is also used, again with several effects with regard to the vapour cycle.   The average energy requirements for these different types of evaporator are given in Table 4.

Table 4   Energy requirements and costs for evaporation and
            reverse-osmosis in the dairy industry

| Process | Direct energy input | | Energy cost |
	Steam t/t	Power kWh/t	£/t
TVR 3 effects	0.22	0.3	1.01
TVR 7 effects	0.10	0.3	0.47
MVR	0.02	20.0	1.09
RO	-	8.5	0.43

In these calculations the cost of electrical power has been assumed to be £0.05 per kWh and the cost of steam to be £4.50 per tonne. In terms of energy costs there is clearly very little difference between reverse-osmosis and thermal vapour recompression evaporators with seven effects. In both cases the energy costs will be a substantial part of the overall annual running costs. If only energy considerations are important the choice of process will very likely be determined by the local availability and relative costs of steam and electricity. Depending on the type of industrial location in which the plant is to be installed, there may be substantial differences in these costs. It is also worth noting that there may also be substantial differences between countries. For example, the relatively higher costs of steam in France and West Germany give the membrane process an economic advantage[10]. With the energy considerations so finely balanced, other factors may decide the choice. Thus the lower operating temperatures of membrane processes give rise to less deterioration of the product and allow retention of volatile aromatic components. Such factors are important in, for example, the concentration of fruit juices.

Reverse-osmosis is already widely used in the dairy industry. The main applications are concentration of milk or whey in order to save transport costs, pre-concentration in powder production, concentration of ultrafiltration permeate in lactose and alcohol production, concentration of skim-milk for ice-cream production and wash-water treatment. Ultrafiltration is also widely used. In conventional cheesemaking, the milk is first coagulated by adding rennet, concentration taking place when the whey is drained off the curd formed. This whey contains about 25% of the protein and about 10% of the fat content of the milk, which are therefore not utilised in cheese production. Concentration of this whey by ultrafiltration so that it may be used, with functional properties intact, in the food industry is the largest single use of membrane equipment in the dairy industry. Plants capable of treating several hundred tonnes of whey each day are characteristic. More recently, ultrafiltration has been used for concentration, either partial or total, before coagulation, so as to fully exploit the milk proteins and fats in cheese production. These latter processes are giving substantial financial benefit, with typical pay-back times on total investment of between 5 and 36 months[11].

3.3 Ultrafiltration compared with centrifugation

The recovery of proteins and enzymes from biological cells is an important aspect of the newer biotechnological industries. The recovery and purification is always a complex multi-stage procedure. The first stage is the separation of the cells from the fermentation broth or biological fluid. This is most often carried out by centrifugation, but can now equally well be achieved by microfiltration or ultrafiltration. If the enzyme or protein of interest is extracellular, that is, it has been excreted from the cell, one then has to further process the supernatant or permeate. However, many useful enzymes and proteins are intracellular, that is, they are retained inside the cell. In such cases, it is necessary to disrupt the collected cells so as to release the proteins to the solution. The ensuing stage is then the separation of the proteins in solution from the cell fragments. The conventional technique is again centrifugation, though this is not a simple matter as the fragments have dimensions of about a micron and densities close to that of the suspending medium. It is at this stage that the introduction of an ultrafiltration unit can show significant economic advantages.

A well-documented case is the separation of soluble proteins with molecular weights of 2000-4000 produced from enzymatic blood cell hydrolysis[5]. This case is of special interest in the present context as a direct comparison was made between ultrafiltration and centrifugation for identical scales of

operation.   The cell fraction from centrifugation of blood was first hemolysed
by adding about three times the volume of water.   An enzymatic hydrolysis was
then carried out, a: ter which the enzyme was inactivated by the addition of acid.
The separation of low molecular weight proteins could then be carried out, either
by centrifugation or ultrafiltration.   The equipment used was either two high
speed centrifuges with a capital cost of ∿ £300,000, or three membrane modules
with a total membrane area of 105 m^2, and a capital cost of ∿ £120,000.   The
annual running costs are given in Table 5, based on 8 hours operation per day,
200 days per year.

Table 5   Comparable data for ultrafiltration and centrifugation
of 25,000 L blood cell hydrolysate (ref.(5) modified)

Nature of cost	Cost (£)	Percentage of overall costs
(a) Centrifugation		
1. Capital charge	60,000	87
2. Power (35kW)	2,800	4
3. Labour	4,000	7
4. Maintenance	1,500	2
TOTAL	68,300	100
(b) Ultrafiltration		
1. Capital charge	24,000	58
2. Power (55kW)	4,400	11
3. Membrane replacement	10,500	25
4. Labour	2,000	5
5. Maintenance	500	1
TOTAL	41,400	100

Here the investment cost has been depreciated over 5 years and the power cost
is taken as £0.05 per kWh.   Clearly the membrane process has an economic
advantage, mainly due to the high investment costs of high speed centrifuges.
Differences in energy costs at this stage were relatively unimportant.
However, there were further advantages with the ultrafiltration process.
Firstly, whereas centrifugation gave 69% protein recovery with washing water
corresponding to 74% of the treated hydrolysate volume, the ultrafiltration
process allowed the recovery of 80% of the protein with addition of a 40%
volume of washing water.   The annual value of the extra protein collected was
£64,000.   Secondly, this reduction in the volume of washing water in the case
of ultrafiltration reduced the energy requirement in the subsequent
concentration stages of evaporation or reverse-osmosis and spray drying.

3.4   Ultrafiltration compared with freeze-drying

Freeze-drying is a process whereby a material is first frozen and then
dried by sublimation in a very high vacuum at a temperature in the range
240-260K.   It is thus suitable for the drying of heat sensitive biological
materials, especially as high local salt concentrations are not formed during

drying. Freeze-drying has been extensively used in the production of albumin
from blood plasma. A crude albumin is first produced by the addition of
alcohol and salts. The crude solution contains 3-10% protein and 10-20%
alcohol. To be of clinical use, the ethanol and some low molecular weight
materials must be removed and the solution concentrated up to about 20%
protein. Ultrafiltration is now being increasingly used for this purpose.

Some data for the comparison of ultrafiltration and freeze-drying for the
production of albumin are shown in Table 6. These data are extrapolated from
pilot plant experiences.

Table 6   Comparable data for freeze-drying and ultrafiltration in
the production of human plasma albumin, 50L batches

(ref. (12) modified)

Nature of cost	Freeze-drying	Ultrafiltration
1. Plant cost estimation	£45,000	£10,000
2. Time per operating cycle	48h	8h
3. Power	9kW	1kW
4. Pyrogen freewater requirement	50L	300L
5. Yield	93-95%	95-97.5%

The ultrafiltration process is run in two stages, first a washing stage at
constant volume to remove alcohol and low molecular weight materials, and then
a concentration stage. The ultrafiltration process has substantial
advantages in terms of capital cost, process time and energy requirements.
It also produces a product with better clinical properties,having lower salt
and alcohol content.

4.   FUTURE DEVELOPMENTS

The examples given have shown that the use of membrane separation
techniques in the processing of biological materials can give significant
advantages in terms of economics, process energy utilisation and product
quality. There are two major factors which will encourage and allow the more
widespread use of such techniques -- the ability to predict membrane permeation
rates for new process feeds and the development of means of improving such
rates. The ability to predict permeation rates is being studied in two ways.
Comparative studies are being made of membrane separations of many process
streams with different types of membrane equipment. In this way it is hoped
to discover correlations between permeation rates, feed properties and
membrane characteristics. This will supply the missing link in the economic
analysis. A more fundamental approach is to study in detail the nature of
the interactions between solutes and membranes, in effect to adopt a colloid
science approach to the problem[13]. This may lead to the development of
new types of membrane with increased permeation rates and improved fouling
resistance. Other means of improving permeation rates involve the
application of some further perturbation to the system where the main effect is
the in situ reduction or removal of membrane foulants. The use of pulsed
electric fields[14] for example, can substantially improve the overall rate of
membrane filtration with little increase in capital or running costs. This

is of paramount importance, for once the general costs of installing and running membrane plant have been incurred, the higher the throughput the greater are the profits.

References

1. Belfort, G. in 'Synthetic Membrane Processes', ed. G. Belfort, Academic Press, 1984, Chapter 1.

2. Strathman,H., J.Memb.Sci., 121, 9, (1981).

3. Beaton, N. C., I.Chem.E. Symposium Series No.51, 1977, p.59.

4. Nielsen, W. K. and Kristensen, S., Process Biochemistry, 8, 18(2), (1983).

5. Kristensen, S., Pro.Bio.Tech., October, 1985, p.vi.

6. Le, M. S. and Howell, J. A., in 'Comprehensive Biotechnology', ed. M. Moo-Young, Pergamon Press, 1986, Vol.3, Chapter 25.

7. Scott, J. A. in course notes for 'Short Course on Membrane Applications in Bioprocessing', Bath, 1986.

8. Strathman, H., in ref. (1), Chapter 9.

9. Le, M. S., Spark, L. B., Ward, P. S. and Ladwa, N. J., J.Memb.Sci. 307, 21 (1984).

10. Hansen, R., 'Evaporation, Membrane Filtration and Spray Drying', Northern European Dairy Journal, 1984, Chapter 8.

11. Ostergaard, B., 'Applications of membrane processing in the dairy industry', Kellogg Foundation Second International Food Research Symposium, Cork, 1985.

12. Schmitthauesler, R., Process Biochemistry, 13, 12(8), (1977).

13. Bowen, W. R. and Clark, R. A., J.Colloid Interface Sci. 401, 97 (1984).

14. Bowen, W. R., Proceedings of the International Congress on Membranes and Membrane Processes, Tokyo, 1987.

MULTIPLE BURNER STOICHIOMETRY CONTROL AND EFFICIENCY
OPTIMISATION BY MEANS OF A TELEOLOGICAL, HEURISTIC
ON-LINE EXPERT SYSTEM

J.T. Edmundson BSc, BA, PhD, FIAP*
D.P. Jenkins PhD, CEng, MIChemE, FInstE, MBCS*
J. Mortimer PhD, CChem, MRSC**

This paper describes a work programme in which
the relevant knowledge of a fuel
technologist is encapsulated in a programme on
a supermicro computer to form the rule base of
an expert system. Interfaced to the
combustion system's sensors and actuators, the
computer gives stoichiometric balance by means
of continuous experimentation on the
combustion system with the ultimate
teleological goal of maximum efficiency.

INTRODUCTION

Multiple burner systems in furnaces.

There are probably more examples of multiple burner installations
in furnaces and boiler plant than single burner installations.
The main difficulty associated with multiple burners is that the
waste gas analysis from the appliance reflects only the overall
air/fuel ratio, and not that of the individual burner.

If such a system is optimised in terms of excess air reduction,
one burner will clearly reach the point of sub-stoichiometry
before the others (shown by the appearance of CO in the flue gas)
whilst the others will remain with varying degrees of excess air.
In simple terms, this system would be rectified by a fuel
technologist by individual enriching of the burners until,
judging by the responses from the carbon monoxide and oxygen
analysers the burners achieve equalisation. This should be done
for every individual level of firing, and so this would clearly
be unrealistic in terms of manpower.

Even when inaccurate balance between burners has been accepted,
attempts at closed-loop stoichiometric control have nearly all
led to severe instability problems. This is due to the
interaction between the furnance temperature control loop and the
stoichiometry control loop.

*   Energy Research Department, BSC Welsh Laboratory,
    Port Talbot.
** Department of Mathematics and Computer Science,
    Bristol Polytechnic

Multiple small boiler plant systems.

An example of this type of application consists of six shell boilers of 12,000 lb/hour nominal steam capacity. Each boiler has twin burners and combustion chambers, and the flues are all trunked together to the stack.

To achieve feed-back control under these conditions, each boiler would need to have back-end brick wall separation so as to obtain anaylsis of the flue gas from each burner and a non-sampling gas analysis system for CO and O_2 (or CO_2) installed. Without consideration of the feedback control system, the analyser costs and flue separation would probably amount to over £60,000, which would render the system uneconomic. However, if the system is considered as a multiple burner installation, then the methods outlined above are applicable and the analyser costs are reduced to ¯£8,000. The control hardware costs in each case would not be very different from each other. Multiple burners in a single large boiler are even more amenable to treatment by this technique.

The capital cost of installing a sufficient number of analysis sensors to centrally monitor the performance of each individual burner is not economically viable. By using a small number of sensors with a control system using artificial intelligence techniques this dilemma can be resolved.

Expert systems.

This term is used to describe a computer program in which a data base of the knowledge, within a restricted subject domain, of a human expert(s) is incorporated. Such programs are generally used as an aid to decision making and have been successfully introduced into several areas of well defined expertise such as Spectroscopic Interpretation, fault finding in process plant, Medical Diagnostics and Financial Decision Making.

Expert Systems currently implemented are usually constructed in the form of a set of Horn clause rules,
i.e. IF condition(s) THEN conclusion(s).
Rules can be expressed either as absolute facts or in terms of probabilities of correctness. The logical connectives AND, OR, and NOT are used to combine conditions and rules to form further rules.

To simplify the task of programming, the rule base (i.e. the declarative part of the program) containing facts and rules about the problem, is generally separated from the interference engine (the procedural part of the program) that directs the flow of logic. This enables changes in the knowledge base (deletions, new rules, changed rules) to be made without causal effects.

It has been found in a number of cases that if this use of heuristic or rule of thumb methods is applied to process control applications, that system performance compares favourably with algorithmic methods. This allows considerable scope in developing control systems for plant which is difficult or impossible to model mathematically, or in cases where use of Expert System methods leads to more rapid convergence to set points. In addition, such systems are more easily understood and modified.

Work programme.

A program of work has been formulated and the initial phases implemented.

Phase 1 - Proving of prototype software on a 5 burner pilot plant. BSC Welsh Laboratory.

Phase 2 - Development of system on a 24 burner zone of the continuous annealing line at BSC Trostre Works.

Phase 3 - To extend the system to control all 6 zones of the C.A. line at BSC Trostre Works. To develop a boiler plant system.

Phase I has now been completed. Work on phase 2 has started and will continue to June 1987. This phase has attracted financial support from the Department of Energy through the Energy Technology Support Unit (E.T.S.U.) at Harwell. The third phase is currently at the planning stage and will incorporate modifications depending on the results of phase 2.

Control philosophy.

Physically, the control of the excess air to each burner is by means of a supplementary butterfly, or other valve, depending on combustion air supply pipe diameters. If electrical actuation is used, it is necessary to have an analog feedback position signal. Provision for failing safe in the fully open position is necessary no matter what the motive power of the valves.

The primary mode of control proposed is one of feed-forward signals to each air trimming butterfly valve. The level of these signals is determined by examination of a database whose compilation is discussed later. The real-time factors which determine the selection of the signal level from the data base are:-

(i)   Level of firing.

(ii)   Temperature, pressure and humidity of the air.

(iii)   Direction of movement of the level of firing.

(iv)   Fuel pressure and temperature in the case of a gas.

The firing level is the most important factor, since it is the parameter which is varied most frequently. Burners commonly have different excess air requirements at different firing levels and, additionally, the butterfly control valve characteristics will be different at varying flowrates.

The temperature, pressure and humidity of the air are important since they affect the mass of oxygen which a fan will pass.

The direction of the alteration is important mainly in terms of worn control linkages, which can thus be compensated for. The compensation also alters automatically with gradual wear.

The computer program.

The prototype program was not written in LISP or PROLOG (The usual languages employed in Expert System programming) for a number of important reasons. The British Steel team was fluent in the Alogorithmic Language FORTRAN and BASIC but would have to learn the new languages. In addition most implmentations of logical languages have a limited range of mathematical functions and are poor at input/output operations. This application comprised primarily numerical terms. The Supermicro-computer used in the project has a very fast and powerful BASIC compiler. Therefore for speed of development and ease of use this was employed.

The disadvantages of this approach were that as BASIC lacks certain functions such as recursion, it is difficult to write Expert System programs that are understandable and readily modified. The rules are embedded in the control, and difficult to isolate and generalise to similar applications.

The Centre for Advanced and Intelligent Systems at Bristol Polytechnic were contracted to assist British Steel in the development of the project. Their use of the IBM-PC based, very cheap Turbo Prolog by Borland overcomes many of the problems associated with logical languages.

Turbo Prolog has been criticised for it's lack of purity as an implementation of Prolog. It does however contain most of the features of Standard PROLOG and in addition contains a comprehensive range of built in mathematical, input/output, and device handling functions that are a positive advantage in designing this type of application. As a compiler implementation of Prolog it is fast in operation. In addition routines written in other languages such as "C" or in assembler are readily incorporated into the Code.

In addition the Bristol team are working on a transputer based high speed Prolog that will interface with the PC to enable more complex real time expert system applications to be built.

The Bristol team have converted the Intelligent Burners expert system controller into Turbo Prolog and are engaged in making the BSC specific program more generally applicable. They are developing the system into a self learning combustion control system which will be tested on different applications within BSC and then offered as a commercial product.

Operation of the development program.

The essential operation of the BASIC program is a simple loop:-

1. Scanning of the analog channels, and interpretation of the results in terms of percentage oxygen and parts per million carbon monoxide (or percentage carbon monoxide and parts per million oxygen if the system is applied to a non-oxidising or reducing furnace).

2. Determination of the conditions in terms of the acceptable ranges of oxygen and carbon monoxide.

3. Take action as dictated by these conditions.

4. Pause while the action takes effect.

5. Go back to step 1.

The scan of the analysers will give two of six possible results in terms of acceptable range (AR):-

RESULTS	TRUTH SYMBOL
Oxygen over AR	OO
Oxygen within AR	OW
Oxygen under AR	OU
Carbon monoxide over AR	MO
Carbon monoxide within AR	MW
Carbon monoxide under AR	MU

From this, the following inferences may be drawn:-

(these inferences and corresponding actions are shown diagramatically in figure 1.)

1. If MU and OU are true, then all the burners are capable of burning at lower excess air.

2. If MU and OW are true, then one or probably more burners are slightly too lean.

3. If MU and OO are true, then most, probably all, burners are too lean.

4. If MW and OU are true, then as 1., but to a lesser extent.

5. If MW and OW are true, then stoichiometry is possibly correct.

6. If MW and OO are true, then at least one, possibly
    more, of the burners is too lean.

7. If MO and OU are true, then at least one burner, and
    probably more, is too rich.

8. If MO and OW are true, then at least one burner,
    probably not more, is too rich.

9. If MO and OO are true, then at least one burner,
    possibly not more, is too rich, and at least one
    burner, possibly not more, is too lean.

The probabilities and possibilities may be quantified in terms of
how much the analyses are outside the AR's.

Figue 1 shows, for each of the 9 sectors: the inference, the
probable cause of the burner system giving rise to an analysis
within that sector, and also the action that should be taken to
return the burner system to "normal" operation.

This figure is an example of an expert system RULE BASE. The
particular rule, or course of action, is defined by the oxygen
and carbon monoxide concentrations.

Other factors.

One problem of optimisation of CO is that final combustion may be
delayed giving higher waste gas temperatures (although with the
theoretically correct analysis) and therefore, a final
improvement routine will be included.

Upon reaching the optimised balanced stoichiometry the system
will then attempt small increases in excess air. The purpose of
this is to see if the stack temperature can be reduced due to
slightly earlier burnout of the CO. An algorithm for calculation
of efficiency is included in the programme for this purpose which
will calculate the temperature and excess air effects.

There are further aspects to be covered in terms of inductive
logic. For example, if one burner is found to need adjustment
more than others, then it requires very little extra effort in
the programme to monitor this and report accordingly.

THE EXPERIMENTAL FURNACE

The combustion chamber and burners.

The existing hot blast air supply chamber at BSC Welsh
Laboratory was modified by the inclusion of three 'radiator'
panels to simulate a water tube boiler combustion chamber. The
length of the chamber was 1.7m and both breadth and height 0.9m.
A 150mm lining of stack-bonded (Kerlane 60) fibre was used to

insulate the chamber walls.   Three water cooled panels, 1.2m by
0.8m, were fitted inside the combustion chamber, one on each side
and one on the base. These were connected to the mains water
supply via a valve which was adjusted to maintain the exit water
temperature at a safe level to prevent boiling in the radiators.

The five burners used were Eddy Ray nozzle mixing type. For the
current work it was decided to derate the experimental furnace to
about   100 kW, using a propane fuel flow of up to 7 kg/hour.
The burners were thus operating at about 15% of their rated
capacity. This had no detrimental effects on the experimental
programme. The low flow was deliberately chosen as it would give
rise to higher carbon monoxide concentrations than at the full
rated capacity. This was necessary as the available Hartmann and
Braun analyser had a full scale range of 2% CO and so had poor
sensitivitiy below 100 ppm CO.

Figure 2 is a schematic diagram of the experimental furnace. A
butterfly valve and position controller was inserted in the air
supply line before each burner.   These valves were used to
simulate the air trim valves which were later installed on the
burners at Trostre. It is by adjusting these air trim valves on
the induced draught system burners that the individual burner air
to fuel ratio is controlled and hence the overall system
optimised.

Measurements and control.

The supplies of propane and combustion air to the burners were
monitored for temperature, absolute pressure and differential
pressure across orifice plates.   The transducers all gave
electrical signals and from these readings an NTP flowrate for
each gas was calculated in the on-line computer, and hence the
overall air : fuel ratio was determined.

The waste gas was analysed by a Hartmann and Braun analyser. The
analyser was checked for zero and span readings each day before
trials began. Although capable of analysing for H_2, O_2, CO, CO_2
and CH_4, only the CO and O_2 values were logged as only these two
gases are used by the rule base.

An Alpha Micro AM1000 microcomputer was used to monitor the state
of the experimental furnace, to perform the expert system
analysis and to send control signals to the firing level control
valves and the air trim butterfly valves. This is a multitasking,
68000 based, machine with a 10 Mbyte hard disk. There are two
terminals with keyboard and VDU, a printer and the THINKLAB
communications device containing 8 channels of DAC and 24
channels of ADC.

The thermocouples feed directly into the ADC whilst the flow
transducers and gas analysis readings are converted from mA
signals to 0-20 mV signals for input to the ADC.

The output from the DAC was 0 - 5 volts at about 1 mA, this was

not sufficient current to drive the firing level control valves
directly so a unity voltage gain current amplifier was used to
drive the electropneumatic converter which drives the valves.
The signal was sufficient to drive the position controllers for
the butterfly air trim valves.

Operation of the computer program.

The major function of the program is to determine, from the
oxygen and carbon monoxide concentrations in the waste gas, which
of the 9 rules apply. Specific action is then taken according to
whichever rule is the force. The program also calculates stack
loss, prints out data to the log file and resets various flags
and index pointers according to the firing level and air
condition status.

The action taken from rules 2, 3, 6, 7, 8 and 9 is intended to
correct the combustion system so that conditions are returned to
normal. From rules 1 and 4 the performance is better than
expected so the oxygen limits may be reduced.

The action of both rich and lean burner subroutines is to perform
a series of experiments to find and correct the individual
burners that are out of adjustment.

When rule 5 applies the system is working within its expected
performance limits. In this pilot plant work an "excess air
improvement subroutine" was called to attempt to minimise the
stack loss. In practice, this rule should be in force most of
the time and there would quickly become a time when further
adjustments have no practical effect. When this situation is
reached the only effect of rule 5 should be to confirm that it
still applies. Obviously, when rule 5 is invoked following a
change in conditions then the improvement routine should be
followed until again further adjustments have no measureable
effect.

Description of experiments.

The purpose of this series of experiments was to check the
control alogorithms and to debug the prototype BASIC program.

The sequence of events was to light the burners and allow the
furnace to warm-up under normal operating conditions (rule 5).
Then to adjust the firing level control valves and/or the air
trim valves to artificially create a non-standard operating
condition, i.e. rule 3, 6, 7, 8 or 9.(Oxygen and/or carbon
monoxide above the higher working limits.)

The expert system section of the program was then switched in and
this then took over the control of valve operation to attempt to
return the combustion conditions to normal, i.e. rule 5 (or 1, 2,
4). (both oxygen and carbon monoxide below the higher working
limits.)
The various measurements of gas analyses, flows, temperatures and

valve settings were printed to a log file for subsequent
analysis.

Results of the initial experiments.

A number of experiments were performed, firstly to debug the
software and check the operation of flowmeters and analysers and
secondly to confirm that the various correction subroutines
operated as expected. On the rule-base there are 5 rules where
oxygen and/or carbon monoxide are above their upper working
limits, i.e. rules 3, 6, 7, 8 and 9. A series of tests, two
starting in each of these 5 rules, are shown in Figure 3. The
initial position of the oxygen/carbon monoxide point is shown by
the test number (10-19). The 9 rules are shown by the circled
numbers. The final point of all tests is clearly within the
region of rule 5, the "oxygen within limits, carbon monoxide
within limits" sector. The starting and finishing points are
joined by straight lines for the sake of clarity, the actual path
from start to finish was obviusly not a simple straight line.
The tests were terminated soon after conditions entered the rule
5 sector.

Figures 4 and 5 show in more detail how the oxygen and carbon
monoxide concentrations, and stack loss vary with time for two of
these tests.

PHASE 2:   TROSTRE CA LINE

Application of the technique to the Trostre C.A. line has
attracted finance from the Department of Energy through ETSU at
Harwell and the SERC. An SERC/CASE studentship was awarded with
the BSC team. The student is working on the development of the
Prolog system which is undergoing parallel development with the
BSC program in order that it can be extended to operate over the
entire plant in phase 3 of the study.

Another member of the Bristol team is designing an advanced gas
sensor system using fibre optic technology to be used in
conjunction with the control system to increase the economic
benefit of using this system.

To date 24 burners, the two sets of 12 comprising zone 3 of the
CA furnace, have been fitted with pneumatically actuated control
valves and associated electropneumatic convertors. Signal cable
and sensor cables have been run back to a cabin.

The computer, interfaces and gas analysis systems are installed,
and working.

TROSTRE CA PHASE 3

The joint BSC/Bristol system will be extended to an IBM/AT based
system to enable a single microcomputer to handle the
input/output signals as well as expert system tasks. The

program will be adapted to deal with the increased number of burners. The Expert System will be extended to monitor its' own performance and improve the rule base using initially adaptive dialog evolutionary methods. In this technique the system checks the validity of its' suggested strategy with the human expert until it has learnt which are allowed new rules. This avoids the combinatorial explosion of rules often found in self organising controllers.

Other Applications.

Following the completion of Trostre phase 3 and the Boiler Plant systems the objective is that the software and hardware will be sufficiently well developed for general use in similar applications.

An industrial system based on this technology will be developed for stand-alone use. The software will be upgraded for engineers to program their own plant configurations, using a general burner technology data base .

CONCLUSIONS

The programme of work to-date has given confidence in the use of an expert system rule based program to control and optimise a multiple burner furnace. The development time of future work will be shortened and simplified through the knowledge gained from the initial test programme.

The emergence of new "supermicro" computers - with powerful processor, large RAM, rapid disk access and relatively sophisticated operating sytems - makes possible the type of control philosophy outlined here. In a works' situation such as a boiler plant, implementation of such a control system can be an add-on virtually independent control system in addition to the existing firing level system or, alternatively, the firing level control system may be easily incorporated into it.

Economic implications.

The savings accruing from implementation of such a scheme will entirely depend upon the circumstances, but experience of BSC's plant would give rise to an expected energy saving of 2-4% for boilers and 5-15% for furnaces. In extreme circumstances where there is no fuel department, or has been no R&D involvment, the savings could be substantially higher.

FIGURE 1

FUEL/AIR STOICHIOMETRY RULE BASE

Oxygen ← / Carbon monoxide →	LOW MU	EXPECTED MW	HIGH MO
HIGH OO	INFERENCE (3) — Most burners too lean. ACTION: Reduce air to all burners	INFERENCE (6) — One or more burners too lean. ACTION: Find and adjust lean burner(s)	INFERENCE (9) — One or more burners too lean + one or more burners too rich. ACTION: As (8) followed by (6)
EXPECTED OW	INFERENCE (2) — One or possibly more burners too lean. ACTION: Find and adjust lean burner(s)	INFERENCE (5) — Operating at expectation. ACTION: Lower oxygen limits	INFERENCE (8) — One or more burners too rich. ACTION: Find and adjust rich burner(s)
LOW OU	INFERENCE (1) — Performance better than expectations. ACTION: Lower oxygen limits slightly	INFERENCE (4) — Performance slightly better than expectations. ACTION: Lower oxygen limits slightly	INFERENCE (7) — Most burners too rich. ACTION: Turn all burners slightly leaner

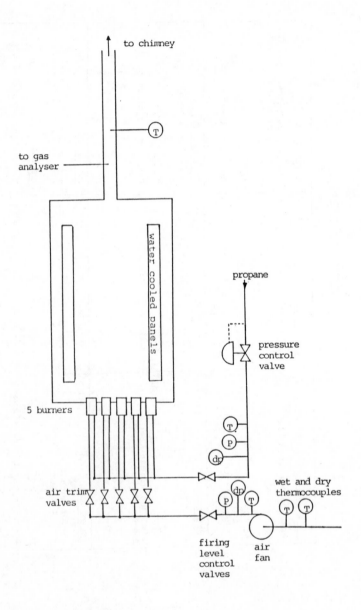

FIGURE 2   SCHEMATIC OF EXPERIMENTAL FURNACE

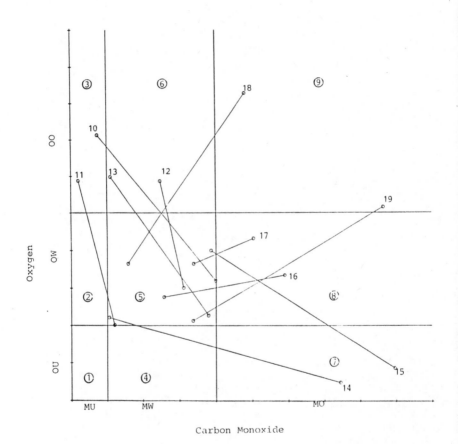

Carbon Monoxide

SUMMARY OF RESULTS

Showing the Starting Conditions for 10 Tests. All Tests
Finish in Sector 5.

FIGURE 3

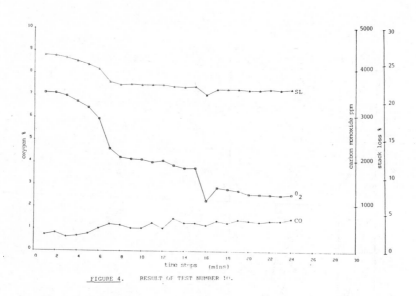

FIGURE 4.   RESULT OF TEST NUMBER 10.

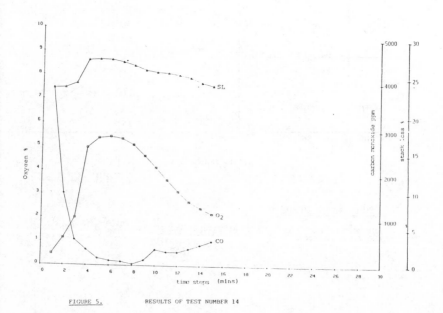

FIGURE 5.   RESULTS OF TEST NUMBER 14

TRANSPUTER TECHNIQUES FOR REAL TIME KNOWLEDGE BASED SYSTEMS IN PROCESS CONTROL

D.C. Bosley BSc*
J.T. Edmundson BSc BA PhD FIAP**
D.P. Jenkins PhD CEng MIChemE FInstE MBCS**
J. Mortimer PhD CChem MRSC*+

This paper describes a work program to improve the performance of Knowledge Based System Process Controllers by developing techniques of concurrency utilising the Inmos Transputer. The work is placed in the context of Energy Engineering Applications for which it is primarily being developed but has a far wider applicability.

INTRODUCTION

Control systems are traditionally based on a mathematical model of the response of the system to the control variables. These are often embedded in an algorithmic structure that contains implicit knowledge. In contrast, Knowledge Based Systems (KBS) explicitly deal with knowledge as a defined entity. Knowledge Based Systems have great potential in complex control engineering environments and in partnership with British Steel we have been applying such techniques to the control of multiple burners in a 24 burner zone of a BSC continuous strip annealing furnace (1).

In this project the relevant knowledge of the Fuel Technologist forms a KBS which is interfaced to the combustion system's sensors and actuators. The computer gives the stoichiometric balance by means of continuous experimentation on the combustion system with the goal of maximising the energy efficiency. As it is intended to develop this system to encompass the entire 108 burners in the CA line and also to develop a multiple fuel version of the system for use in large boiler plants we have found it necessary to investigate methods for accelerating the performance of KBS controllers.

*   Bristol Transputer Centre, Bristol Polytechnic

**   Energy Research Department, BSC Welsh Laboratory,
     Port Talbot

+   Lecture to be given by J Mortimer

KNOWLEDGE BASED SYSTEMS

Present KBS's comprise a knowledge base, which contains the facts and knowledge which the system knows about and an inference engine which is the process used to actively search the knowledge base. Communication with KBS's can be via human interaction or external sensors and controllers. Thus existing KBS's are essentially sequential systems. The techniques for searching rule based systems tend to be very time consuming as only one path is explored at a time. This is a consequence of the restrictions which are forced upon us by the underlying Von Neuman computer architecture.

Searching a KBS for a specific piece of information is usually demanding in terms of processor time. In a complex control system many tasks need to be processed within a small time period. Thus the extraction of information from a knowledge base within imposed time constraints may be impossible.

The use of a single processor as used in many real time systems exacerbates the problem by causing a bottleneck in throughput due to the sequential order in which tasks are time-shared.

In order to extend KBS's to a real time environment it is natural to consider developments in parallel computer architectures. This will enable the development of new models of KBS's involving distributed parallel inferencing and the utilisation of novel search techniques to enhance performance. The availability of the Inmos Transputer in commercial quantities provides a suitable vehicle for these investigations.

PARALLELISM

In virtually all sets of programmable tasks, certain operations may be carried out concurrently. The degree of concurrency exploited is known as the granularity. For example in the case of the Expert System Controller for Trostre Continuous Annealing Line described in the previous paper we have a set of 24 burners in a single zone controlled by a Knowledge Based System in a long sequential Prolog program (figure 1).

It is intended to extend this to the control of 9 zones all of which could be controlled independently using 9 implementations of the same program operating concurrently (figure 2). As the length of each program or process is large then this is known as Coarse Granularity.

Within each Prolog program are a number of related short clauses which could be searched in parallel and this would be Medium Granularity. This Granularity range is exploited by the Inmos Transputer.

Fine granularity looks for concurrency within a program line or down at instruction set level and will not be covered here.

TRANSPUTERS AND OCCAM

Transputers are designed to support parallel execution of processes and interprocess communication. The concepts of Communicating Sequential Processes as expressed in the Occam language are used to express and execute concurrent systems using transputers (2). Transputers are able to directly support real-time systems by means of timers, fast interrupt servicing and the prioritisation of processes. Increases in performance can mostly be achieved by building more transputers into the system.

The lowest level for programming the transputer is occam and in theory this language is designed to be architecturally equivalent to the transputer from the programmer's point of view. The programmer may define a procedure which is to be run in parallel with other procedures. If only one transputer is available, the occam compiler constructs occam code which will share the available processing time of the single transputer equally. If more transputers are available, the compiler will allocate one procedure per transputer if possible, thus executing them in parallel. All this is done by the occam compiler at compilation and so hidden from the programmer. The programmer does not have to modify his code at all. This enables the design of programs that can use the same concurrent techniques for both a single transputer and for networks of transputers. The logical framework can be designed on a single computer and then configured on a price/performance basis for the appropriate multiple transputer architecture.

In the occam language a system is described as a set of concurrent processes which communicate via channels (figure 3). Occam describes two forms of concurrency, PAR which is a set of processes executing in parallel all of which must complete before the next process in sequence may be executed and ALT which is a set of processes the first of which to complete execution allows the next in sequence to be executed. Processes are combined into constructs which in turn may become components of other constructs.

Occam is a simple language with limited syntax which nevertheless permits the construction of complex programs at a much higher level than programming at instruction set level. Indeed the occam/ transputer concept decrees that this is the lowest level at which the transputer may be programmed.

At present, however, programming tools are somewhat sparse and a full scale occam program is relatively hard to debug although the folding editor for program development simplifies design of concurrency in a one dimensional environment. Some manufacturers of transputer based systems have taken tentative steps towards remedying some of these problems.

For ease of programming development 'C', FORTRAN 77 and Pascal compilers have become available. Programs written using these compilers execute as single occam processes i.e. sequentially. Thus concurrent processes in these compilers must be linked using occam channels or physical transputer links.

OPTIONS FOR KBS IN A TRANSPUTER ENVIRONMENT

There are then several options for the development of KBS's in a transputer environment. The simplest is to transfer a KBS Shell written in another language to a transputer environment. Several well known Shells are written in the 'C' language, and whilst AI purists frown on the use of non declarative methods, these can be used in a control engineering environment if it is possible to interface user programs to handle complex I/O and algorithmic knowledge. Some very sophisticated real time development environments have been written in this language.

These programs can be ported as a single process within a transputer system and would be useful in taking advantage of the performance of a single transputer with perhaps further transputers being used to handle other functions of the system.

However, Knowledge Based Systems are generally written in a declarative language such as LISP or PROLOG. A PROLOG program for example comprises a basic structure of assertions and rules. It also contains a goal which it attempts to satisfy by matching it with the rules and assertions. The process of matching is known as Unification and the result of a match will be an instantiation of the variables involved in the matching (figure 4). If a match is not possible, the Unification algorithm will backtrack to the last successful match and try to resatisfy that match with an alternative rule or assertion.

Unless advantage is taken of concurrency this is little benefit to be gained from a Transputer environment. The widely available implementations of the logical languages do not support concurrency. Developments to introduce parallelism into the Prolog language have been published (3,4) and there are numerous opportunities to introduce parallelism into logic programs. These will be discussed shortly. If these could be compiled directly into a concurrent environment then major improvements in the performance of KBS's can be achieved.

There is, however, another problem in using Prolog that these implementations do not address, in that the language does not support real time operation. For addressing the problems of KBS Process Controllers this problem needs to be overcome.

Another major problem is that occam does not directly support certain features implicit to the execution of predicate languages, such as recursion. In the next phase of transputer development it is rumoured that the architecture is to be more open, thus permitting non standard software development. In the meantime we must solve these problems within the design constraints of the occam language.

A PARALLEL REAL TIME PROGRAMMING LANGUAGE IN LOGIC (PARTLOG)

We are developing a real time logic programming language suitable for control engineering purposes. This supports the logical structure of prolog and the principal features of the language but also is implicitly concurrent when mapped into a transputer environment, that is the compiler will look for inherent concurrency in the program during compilation.

The language in addition contains real time predicates not found in Prolog such that a KBS may have alternate strategies dependant on the processing time available to it.

PARALLEL ASPECTS OF PARTLOG

The procedural meaning of Prolog can be found in Graph Theory. Prolog is simply the searching of AND/ OR graphs and an implemented Prolog program gives an AND/ OR graph description of the problem. For example the AND/ OR Graph description of the Burner Controller described in the previous paper starts as shown in figure 5. We see that when an AND is met we carry out events in sequence but where an OR is met then all rules may be explored concurrently.

If we now extend this to Prolog itself and examine the following rules we can see the implicit parallelism.

    a(X,Y):- b(X,Y), c(X,Z).

    b(X,Y):- rule1(X,Y).

    b(X,Y):- rule2(X,Y).

    b(X,Y):- rule3(X,Y).

The b(X,Y) rules are OR parallel and can be searched without Backtracking under Concurrent operation.

In addition the b AND c rules are AND parallel in that whilst both b AND c must be matched the alternatives for b and c may be evaluated concurrently.

The problem of recursion can be solved by utilising a Frame based Structured sharing model as the implementation of the Unification algorithm for the compiler. A simplified unification algorithm is shown in figure 6.

REAL TIME ASPECTS OF PARTLOG

It is necessary, not only from the point of designing real time KBS controllers, but also to avoid deadlocks in concurrent operation, to introduce predicates for interrupts, and time dependent rules.

Predicates for interrupts include

```
read(Device,Location,State).
connect(printer,channel1).
```

Time dependant rules can be demonstrated with the following short example:

```
analyse_burner(Time_allowed,Condition,State):-

    time(Time_allowed,State),
    satisfy(Condition).
```

IF timeout occurs THEN rule fails. Thus

```
State:= success OR timeout.
```

Delays to enable time dependant actions can be carried out using for example:

```
service(Device1,Location,State):-
    delay(Time_interval),
    read(Device1,Location,State).
```

OPTIMISING PARALLELISM

Although the implicit parallelism within Prolog has been demonstrated programs may be written in such a way as to maximise or minimise concurrency and efficiency gains to be made.

For example the use of short clauses will reduce AND parallelism and the maximising of the number of clauses will increase OR parallelism. Even sized clauses will balance the hierarchy of the AND/ OR Graph.

There are however important operational considerations. The Transputer is a very fast processor and capable of extremely efficient sequential processing whilst having been designed for connection with other transputers for sequential operation. These links whilst operating at very high speed (20 MHz in the latest Transputers) are nevertheless much slower than the internal processing speed of the individual transputers. Experiments are showing that an array of transputers in a Coarse Grained set of concurrent processes appear to be more efficient than maximising parallelism as much as possible.

Thus the development of parallel logic programs offer considerable scope for optimisation of the target code.

To date the Prolog algorithm has been defined to provide a run time execution of PARTLOG programs. The algorithm is being implemented in Pascal on a VAX-750 Computer and provides for syntax and semantic phases, code generation (in Pascal at present) and a run time system. The next phase of code generation via occam is now underway and it is intended to port the compiler shortly to an Inmos ITEM 400 40 Transputer Development System.

FUTURE WORK

As soon as the compiler has been ported to the Transputer Development System and a number of benchmark tests made, it is intended to port the Intelligent Burners Program described in the previous paper (1) which is implemented under Turbo-Prolog on an IBM/PC over to the PARTLOG system running under TDS. This will allow the evaluation of a tried and simple KBS controller and enable practical experiments in optimisation to be made.

On the compiler development front we are exploring the possibility of using AI techniques themselves to determine whether it is possible for the compiler itself to produce the best implementation of the program as the more demands that can be placed on the Compiler mean less demands on the system designer. This leads to some interesting possibilities that the mapping from source to target code may require to be solved in a probabilistic manner.

Much work is underway to maximise the performance gains that are attainable by transputer systems. Our own group and others are developing new models of KBS's and novel search techniques through transputer based architectures such that the problems of real time KBS can be reduced in principle to adding more transputers to the system. However, even modest gains in processing speed in the execution of KBS control systems would be of great benefit at the present time and a stepwise move to these new techniques should be of considerable benefit.

CONCLUSIONS

A methodology for the implementation of a parallel real time logical language suitable for the development of knowledge based systems for use in Process Control has been described. The model of concurrency selected has allowed development for Systems based around the new Inmos Transputer.

The relevance and relationship of this paper to work by the group in developing expert systems in multiple burner stoichiometry control has been described and the general applicability to a wide range of topics demonstrated.

Future papers will describe applications of this work to signal processing, energy management systems and management information systems.

ACKNOWLEDGEMENTS

We wish to thank British Steel Corporation, The Department of Trade and Industry, Inmos PLC and the NAB Polytechnic Special Research Inititiative for their support of the project.

REFERENCES

1.   J T Edmundson, D P Jenkins, and J Mortimer, Paper to be presented at Inst Chem Eng Symposium Series No 105, "Innovation in Process Energy Utilisation", 16th-18th September 1987.

2.   I M Barron, Information Processing 86, pp 259 - 265, 1986

3.   J S Conery, ACM Comms, Vol 10, pp 163 - 170, 1981

4.   K Clark and S Gregory, ACM Transactions on Programming Languages and Systems, Vol 8, No 1, pp 1 - 49, 1986

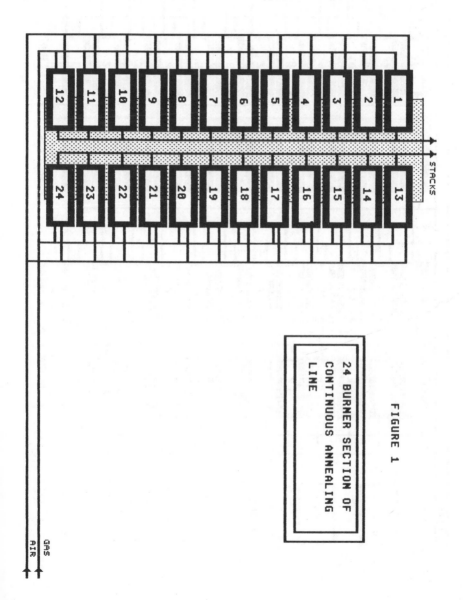

FIGURE 1

24 BURNER SECTION OF CONTINUOUS ANNEALING LINE

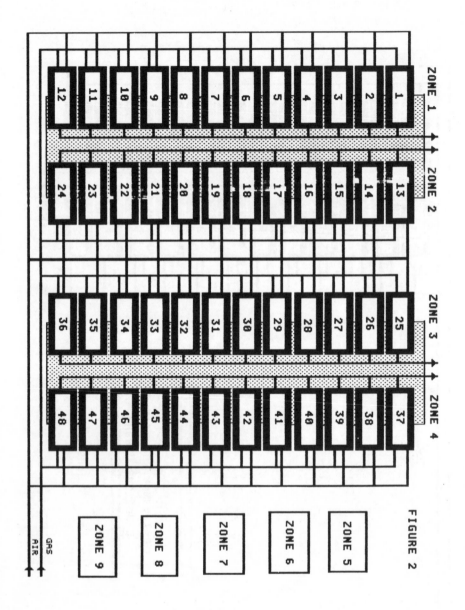

FIGURE 2

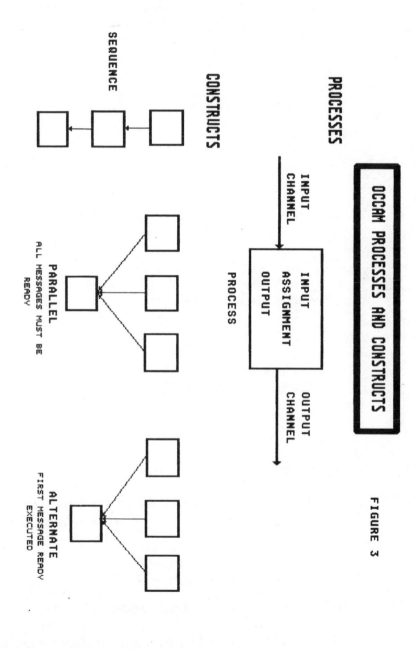

FIGURE 3

figure 4

PROLOG

```
optimise_burners(X,Y) :- oxygen(X,O),
                         temperature(X,T),
                         Y is O+T.

oxygen(burner_a, O):- ...
oxygen(burner_b, O):- ...



temperature(burner_a,T):- ...
temperature(burner_b,T):- ...
```

Run_time Enviroment :

Satisfy(goal)   Rulebase ───────→ Unification
                           ←─────── Algorithm

e.g. oxygen(X= burner_a,O) &
     temperature(X=burner_b,T).

No match possible, so Backtracking occurs.

PROLOG will attempt to find any possible matches.

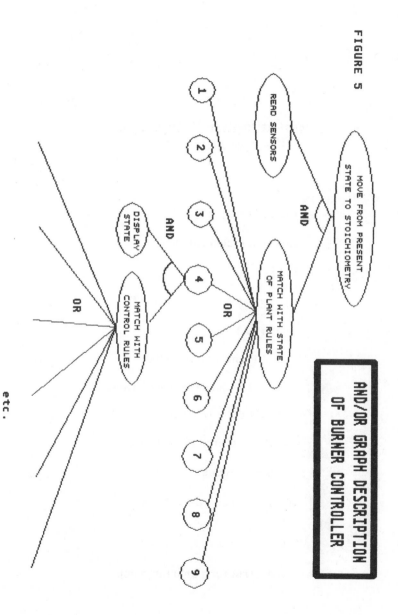

FIGURE 5

AND/OR GRAPH DESCRIPTION OF BURNER CONTROLLER

<u>figure 6</u>

<u>A Possible Implementation of the Unification Algorithm.</u>

```
BEGIN
        Initialise stack;

        Initialise current goal;

        REPEAT

            IF goal_not_empty THEN

            REPEAT

                IF unification THEN

                    create_new_frame

                ELSE retract_last_frame

            UNTIL finish OR failure

            ELSE success

        UNTIL success OR failure
    END;
```

Frame based Structure sharing model.

FURNACE DYNAMIC BEHAVIOUR : A MODELLING STUDY

R. G. Herapath[*] and S. Peskett[**]

Traditional methods of furnace control use temperature feedback to modulate fuel (or air) and a constant ratio link to modulate air (or fuel). This may be improved, giving better energy consumption and product quality, by replacing the constant ratio system by a second feedback loop using excess oxygen as the feedback variable. However, this dual loop control necessitates a detailed study of furnace dynamics. As a precursor to control design a model of the dynamic behaviour has been developed which indicates the types of responses the modified control system must accommodate. Complete validation of this model has, however, not yet been achieved.

INTRODUCTION

Although major developments have been made in the application of excess oxygen controllers to boilers there is little reported evidence of these controllers applied to furnaces although such control could give a reduction in furnace energy consumption and, in many cases, an improvement in product quality. However, the control of temperature is much more important in furnaces than in boilers and, to date, most furnaces operate with wall temperature control loops as in Fig.1, using a constant ratio system to maintain suitable excess oxygen levels. In many cases a digital computer schedules the temperature set-point to provide optimum heating strategies.

Any combustion control loop must act in parallel with the temperature controller thus modifying the furnace to a dual input/dual output control system as in Fig.2 with consequent increase in the complexity of the design of a suitably stable control system. Instinct tends to suggest that time constants associated with wall temperature changes are far longer than those associated with combustion and, if so, stability should not be a problem. However, in one case known to the authors the addition of combustion product control in a slab reheating furnace gave rise to an incurably unstable system and this confirms that an investigation of furnace dynamic behaviour is desirable to provide adequate data for control system design. Direct experimentation on a large furnace would be inconvenient without a basic understanding derived from a mathematical model of the system. This paper describes the work to produce such a model of furnace transient heat transfer to determine the dynamic relation between fuel and/or air flow changes and workpiece and wall temperatures.

*   *Department of Mechanical Engineering, University of Wales, Swansea.*
**   *Former research student, Energy Studies, University of Wales, Swansea, (now at Rutherford-Appleton Laboratory, BNSC Division)*

NOTATION

A	area m^2		r	radiative heat flux exchange kW
Cp	specific heat of gas kJ/kg K		T	temperature, K
c	specific heat of solid kJ/kg K		t	time, s
F	view factor		W	radiative flux per unit area
H	radiative flux per unit area incident upon a surface kW/m^2			leaving a surface kW/m^2
			x	distance through thickness of slab or wall, m
k	thermal conductivity kW/(mK)			
ℓ	slab thickness, m		α	thermal diffusivity m^2/s
M	dimensionless variable		ϵ	emissivity
m	mass flow kg/s		ρ	reflectivity
Q	radiative flux per unit area absorbed at a surface kW/m^2		ρ	density kg/m^3
			σ	Stefan Boltzmann constant $5.67 \times 10^{-11} \; kW/(m^2 K^4)$
q	rate of heat release within a model zone gas volume kW		τ	transmissivity

subscripts

a	absorbed		k	zone number
e	emitted		m	time increment
g	gas volume		n	distance increment
i,j	surface numbers		s	surface

LITERATURE REVIEW

A survey of the literature confirmed that much of the furnace control work to date has been to improve heating strategies with models of furnace behaviour playing an important role (1-6). Little reported work on excess oxygen control has been found. Record (7) used a flue mounted zirconia probe to modulate fuel flow and thus give excess oxygen control in an ingot soaking pit. He obtained significant improvements in fuel consumption. Amano (8) reported an optimum multivariable control system for a multizone reheating furnace in Japan. This involved the development of a fast response oxygen analyser and special valves to give good turn down of air flow. This is the only paper to indicate that problems of dynamic interaction of control loops might occur and Amano solved his problems by sampling combustion products close to the furnace wall using the fast response analyser. (Air infiltration may give uncharacteristic excess oxygen values near the furnace wall and the authors of this paper believe that better energy savings may be achieved by using a slower response system via a water cooled probe to sample gases from deep within the furnace).

Several exercises on furnace modelling have been reported (9,10,11) which are all based on the "zone" method developed by Hottel and Cohen (12). The furnace is discretised to give a number of surface and volume zones considered to be homogenous and isothermal. "Total exchange areas" are derived to describe the direct and indirect exchange of radiation between zones resulting in a set of simultaneous equations of the energy balance within the furnace. However, many of the models assume steady state operations with flow and temperature fields being time invariant. Although this is adequate for the study of steady state heat transfer, it is not suitable for a study of the dynamic behaviour necessary for control considerations.

It was not until the last stages of this work that a parallel study by Tucker and Lorton (13) on dynamic furnace modelling was noted. The purpose of their work was to investigate the thermal performance of reheating and heat treatment furnaces under transient conditions and is complementary to the work reported in this paper.

MODEL DEVELOPMENT

Assumptions

i) The model was to be based, for investigation purposes, on the five-zone slab reheating counterflow furnaces at the Port Talbot Works of the British Steel Corporation.

ii) Modelling is based on Hottel's "zone" method. However, as the furnaces are also divided into operating zones - preheat, tonnage, soaking - care must be taken not to confuse the terms used. In most cases the interpretation is obvious but where necessary, this paper will refer to model zone or furnace zone.

iii) The furnace may be split into a number of model zones as in Fig.3 each consisting of a number of real or imaginary surfaces and a gas volume. Walls and slab provide real surfaces, the interface of one zone with another is an imaginary surface. Each gas volume and surface is homogenous. For convenience, the model zone width has been taken as one slab width.

iv) Temperatures are constant over the length of each model zone, on each surface within a zone and in the gas volume. Temperatures are also constant over the furnace zone during each time-step of the computation.

v) Conductive heat transfer through the slab and furnace walls can be represented by a one dimensional model. All slabs are of the same size and material as changes in loading patterns are slow.

vi) Heat transfer between gas, furnace walls and slabs is by radiation and all surfaces are grey and opaque to radiation. Convective heat transfer was assumed to have a small effect. (This assumption was made on the advice of other workers in the field).

Model Equations

i) Slab and wall heat conduction equations

A one dimensional model was used, based on the assumption of uniform temperature distribution along the length of the slab (i.e. across the furnace) and along the width of the slab (i.e. within model zone). In the preheat and tonnage furnace zones the slab is heated from both sides and, initially, this heating rate is assumed to be symmetrical. With reference to the slab coordinates of Fig.3 the one-dimensional form of the heat conduction equation is:-

$$\rho c \frac{\partial T}{\partial t} = \frac{\partial}{\partial x} \left[k \frac{\partial T}{\partial x} \right] = k \frac{\partial_2 T}{\partial x^2} + \frac{\partial k}{\partial T} \left[\frac{\partial T}{\partial x} \right]^2 \qquad \ldots 1$$

The boundary condition are

$$-k \frac{\partial T}{\partial x} \left[\pm \ell/_2 , t \right] = Q_s = \begin{array}{l} \text{radiative heat flux absorbed} \\ \text{at the slab surface} \end{array} \qquad \ldots 2$$

These equations were solved by a finite difference technique. Expanding $T(x,t)$ by Taylor's Series leads to the explicit finite difference equation

$$T(x_n, t_{m+1}) = \frac{1}{M_n}\left[T(x_{n+1}, t_m) + T(x_{n-1}, t_m)\right] + \left[1 - \frac{2}{M_n}\right]\left[T(x_n, t_m)\right]$$

$$+ \frac{1}{4\,k_n M_n}\,\frac{\partial k\,(T(x_n))}{\partial T}\left[T(x_{n+1}, t_m) - T(x_{n-1}, t_m)\right]^2 \qquad \ldots 3$$

where $M_n = \dfrac{\Delta x^2}{\alpha_n \Delta t}$ and $\alpha_n = k_n/\rho c_n$

α_n is known as the thermal diffusivity and $\dfrac{1}{M_n}$ the Fourier number Fo.

Transforming the boundary conditions to finite difference form leads to

$$\frac{\partial T}{\partial x}\,(x_s, t_m) = \left[T(x_{s+1}, t_m) - T(x_{s-1}, t_m)\right]/2\Delta x \qquad \ldots 4$$

where x_{s+1} represents an imaginary slice beyond the slab surface which is necessary to enable the boundary conditions to be evaluated.

Thus on substitution into the equation 3

$$T(x_s, t_{m+1}) = \frac{2}{M_s}\left[Q_s\frac{\Delta x}{k_s} + T(x_{s-1}, t_m)\right] + \left[1 - \frac{2}{M_s}\right]T(x_s, t_m)$$

$$+ \frac{1}{k_s M_s}\,\frac{\partial k\,(T(x_s))}{\partial T}\left[\frac{\Delta x Q_s}{k_s}\right]^2 \qquad \ldots 5$$

At the centreline $T(x_{-1}, t_m) = T(x_{+1}, t_m)$ \qquad $\ldots 6$

Similar equations can also be written for heat conduction through the walls of the furnace.

ii) Radiative flux equations

A radiative flux heat balance on a grey and opaque surface is shown in Fig.4 assuming quasi-steady state conditions. The incident flux per unit area on surface i is H_i and the leaving flux per unit area is W_i which consists of reflected radiation, $\rho_i H_i$, and radiation emitted by the surface $\varepsilon_i \sigma T_i^4$

$$W_i = \rho_i H_i + \varepsilon_i \sigma T_i^4 \qquad \ldots 7$$

The heat flux absorbed per unit area of the surface is Q_i given by

$$Q_i = H_i - W_i \qquad \ldots 8$$

The radiation, $A_i H_i$, incident upon a surface within an enclosure containing a non-absorbing medium can be written in terms of the radiation leaving all surfaces within the same enclosure.

$$A_i H_i = \sum_j A_j W_j F_{ji} = A_i \sum_j W_j F_{ij} \qquad \ldots 9$$

Where F_{ji} is the view factor of surface i from surface j

However, the enclosure in this case contains an absorbing medium (i.e. the combustion gases) and hence the radiation leaving each surface will be attenuated by the medium. The medium also emits its own radiation and equation 9 must be modified to

$$H_i = \varepsilon_g \sigma T_g^4 + \sum_j W_j F_{ij} \tau_{ij} \qquad \qquad \dots \ 10$$

Substituting equation 10 into equation 7 gives for each zone k

$$W_{ki} = \rho_{ki} (\varepsilon_{gk} \sigma T_{gk}^4 + \sum_j W_{kj} F_{kij} \tau_{kij}) + \varepsilon_{ki} \sigma T_{ki}^4 \qquad \dots \ 11$$

An equation of this type is written for each real surface in each of the model zones.

In the case of an imaginary surface at a zone boundary the incident radiation is assumed to pass directly into the adjacent zone.

Hence, as shown in Fig.5

$$H_{k-1,4} = W_{k,2} \qquad \qquad \dots \ 12a$$

$$H_{k,2} = W_{k-1,4} \qquad \qquad \dots \ 12b$$

Substitution of equation 10 into equations 12a and 12b gives

$$W_{k,2} = \varepsilon_{gk-1} \sigma T_{gk-1}^4 + \sum_j W_{k-1,j} F_{k-1,4j} \tau_{k-1,4j} \qquad \dots \ 13a$$

$$W_{k-1,4} = \varepsilon_{gk} \sigma T_{gk}^4 + \sum_j W_{kj} F_{k2j} \tau_{k2j} \qquad \dots \ 13b$$

This pair of equations is written for each imaginary surface of the furnace model. Although there is a two-way exchange of radiation across the boundary, there is a one-way net flux, and it may easily be shown that

$$H_{k,2} - W_{k,2} = -(H_{k-1,4} - W_{k-1,4}) \qquad \dots \ 14$$

Assuming no recirculation of the products of combustion occurs, then with reference to Fig.6, an energy balance on the gas volume in zone k, assuming quasi steady-state conditions, gives

$$m_{gk-1} C_{pk-1} T_{gk-1} - m_{gk} C_{pk} T_{gk} + q_k + r_{ak} - r_{ek} = 0 \qquad \dots \ 15$$

The zonal heat release, q_k, is determined from a description of the distribution of combustion heat release within the furnace. The rates of radiative heat absorbed and emitted by the gas volume are represented by r_{ak} and r_{ek} respectively.

If radiation passing from surface j to surface i is partially absorbed by the gas, it follows that the rate of radiation absorption is given by

$$r_{ak} = \sum_i \sum_j A_{kj} W_{kj} F_{kji} (1 - \tau_{kji}) \qquad \dots \ 16$$

The rate of radiation emission from the gas volume is given as

$$r_{ek} = \Sigma A_{kj} \, \varepsilon_{gk} \, \sigma \, T_{gk}^4 \qquad\qquad \dots \; 17$$

The radiation equations, 11,13,15 (with 16 & 17) form a set of non-linear simultaneous equations of unknown gas temperatures and leaving fluxes. The solution leads to the boundary condition, Q_s, for the heat conduction Eqn.5.

Model Manipulation

Solution of the heat conduction equation is carried out by a "time-marching" method. However, within each time step the non-linear nature of the radiation equations and the fact that gas properties are temperature dependent means that the gas temperatures and leaving fluxes must be determined by an iterative procedure.

A steady-state flow field is established from data stores in file, describing the main flows entering and leaving each zone at time t = 0.

A double iteration procedure follows to calculate the gas temperature profile through the furnace and the leaving heat flux from each surface for this time step. An outer program loop calculates the values of emissivity and specific heat for each gas volume while an inner loop calculates the gas temperatures and heat fluxes using a Newton-Raphson iteration. When convergence is within the defined limits for this double iteration, the resulting solution is taken as the solution for this time step.

The net heat absorbed by each surface is calculated from equation 8 and is used in the heat conduction model to calculate the new temperature distribution through slabs and walls.

MODEL RESULTS AND FURNACE TESTING

The initial model development was made on a hypothetical furnace 4m long x 6m wide x 3m subdivided into 4 zones, 1m wide. Results are quoted for this model only. The model has been extended to cover the full slab-reheating furnace but results for this are, as yet, incomplete.

Transient Results on 4 zone model

Following the start up of a cold furnace with stationary slabs large temperature differences between gas, walls and slabs lead to high rates of heat transfer from hot gas to the cold slabs and the furnace fabric as shown in Fig.7. As the transient proceeds, radiative interaction leads to a further rise in gas temperature but now smaller temperature differences give lower rates of heat transfer to the slab surface. The effect of the conductive heat transfer from the slab surface to the centre is clearly shown in delaying the response of the slab centre temperature. Steady state conditions are reached after approximately 10 hours.

Fig.8 shows the continuing response, of the same model, as the slabs are pushed along the furnace moving one slab width at 10 minute intervals, commencing at the end of the heating transient (610 minutes). Air and fuel flows have been kept at the same values as for the start-up transient and hence a new steady condition is reached at much lower temperatures. Heat is now removed from the furnace by the slab passage. An interesting feature in this response is that for the first 2 pushes the slab centre temperatures are above the slab surface temperature, again demonstrating the delay in response due to heat condition but this time from the slab centre to the surface. The response is shown for zone 2 of the 4 model zone furnace, slabs entering at zone 4 and leaving at zone 1.

Similar information for each slab, rather than each model zone is shown in Figs.9 and 10. In this case the slab surface and centre temperatures are shown for each slab as it enters at zone 4 and leaves at zone 1. Because the furnace is initially at a relatively high temperature, the first new slab reaches a high drop out temperature, but, as more cold slabs are pushed, the furnace temperature is brought down and lower drop out temperatures result.

Note that in these transients the wall temperature is often lower than the slab surface temperature and this occurs because the radiative component of the gas is too high. Tests at different values of gas, wall and slab emissivity and reflectivity show that these do not affect the actual temperatures but have a significant effect on the length of the transient.

PRBS Disturbances

Because of problems faced in validating the model a set of model tests using a disturbance in the fuel flow (at constant air/fuel ratio) in the form of a pseudo random binary sequence was carried out. This is a recognised method of determining control transfer functions - particularly of linear systems - using statistical analysis by superimposing a small amplitude disturbance on the normal plant operating conditions (14). If the auto-correlation function of the input disturbance is an impulse then the cross-correlation function between output and input is the impulse response of the system. By using the statistical analysis of cross-correlation it is possible to eliminate the effects of other disturbances in the system. Thus a PRBS may form a useful method for validation of this model, particularly as it is primarily intended for control system design.

A 15 step repeatable sequence of +a, +a, +a, +a, -a, -a, -a, +a, -a, -a, +a, +a, -a. +a. -a, where a is the amplitude of the disturbance, was used in the model. The response of the 4 section model to a± 5% perturbation in fuel flow, is shown, with the disturbance, in Fig 11. The period of the sequence shown was 30 min (15 x 2 min) but the same sequence at periods of 90 min and 300 min were also used. The auto-correlation function of the fuel flow and the cross-correlation functions relating gas, wall, slab surface and temperatures to fuel flow are shown in Figs 12a, b and c. These results are rather confusing in that they indicate impulse responses which depend on sequence period and they are by no means convincing as a method of validation although they indicate some possibilities. There is little documented information on the use of PRBS methods to other linear and lumped parameter systems. The furnace is non-linear and, because of the heat conduction through the slab, is essentially a distributed parameter system. Further investigation is obviously required.

Model validation

Some attempts have been made to validate the model using data obtained from the slab reheating furnace at BSC Port Talbot. Validation is by no means complete but reports of this work will be presented at the conference.

CONCLUSIONS

A workable computer model of fuel fired furnaces has been developed which will be suitable for a study of the dynamic behaviour of furnaces but further work is needed for validation and parameter matching possibly using an experimental furnace.

Studies relating to control system design can now proceed, based on this model.

REFERENCES

(1) Ono, M et al. "The computer control system of billet reheating furnace". IFAC Control Science and Technology (8th Triennial World Congress), Kyoto, Japan, 1981, Vol 5, pp. 2613-2618.

(2) Iwahashi, Y. et al "Computer control system for continuous reheating furnaces". IFAC Control Science and Technology (8th Triennial World Congress), Kyoto, Japan, 1981, Vol 5, pp. 2625-2632.

(3) Hollander, F and Huisman, R L. "Computer controlled reheating furnaces directed to an optimum hot strip mill process". Proceedings of the ISI Conference 'Slab Reheating', Bournemouth, 1972, pp. 26-42.

(4) Ishida, R et al. "Temperature control system for reheating furnaces". Transactions ISIJ, Vol 23, 1983, p.3131.

(5) Fontana, P. et al. "An advanced computer control system for reheat furnaces". Iron and Steel Engineer, August 1983, pp. 55-62.

(6) Glatt, R D and Macedo, F X. "Computer control of reheating furnaces". Iron and Steel International, December 1977, pp. 381-396.

(7) Record, R G H. "Combustion control of a soaking pit, using a solid electrolyte oxygen probe". Iron and Steel, June 1972, pp. 317-323.

(8) Amano, H et al. "Development of O_2 control of furnaces in the iron and steel industry". IFAC Control Science and Technology (8th Triennial World Congress), Kyoto, Japan, 1981, pp. 2607-2612.

(9) Johnson, T K and Beer, J M. "The zone method analysis of radiant heat transfer: model for luminous radiation". Proceedings of the fourth symposium on flames and industry (predictive methods for industrial flames). British Flame Research Committee, Institute of Fuel, London, 1972, pp. 35-43.

(10) Johnson, T K et al. "Comparison of calculated temperatures and heat flux distributions with measurements in the Ijmuiden furnace". Proceedings of the fourth symposium on flames and industry (predictive methods for industrial flames.) British Flame Research Committee, Institute of Fuel, London, 1972, pp. 125-141.

(11) Mobsby, J A. "Thermal modelling of fossil-fuelled boilers". I Chem E Symposium series. No. 86, pp. 1013-1023.

(12) Hottel, H C and Cohen, E S. "Radiant heat exchange in a gas-filled enclosure: allowance for non-uniformity of gas temperature". American Institute of Chemical Engineering Journal, Vol 4, No 1, March 1958, pp..3-14.

(13) Tucker, R and Lorton. "Mathematical modelling of load-recuperative gas fired furnaces". MRS E418, May 1984, Midlands Research Station, British Gas, Wharf Lane, Solihull.

(14) Godfrey, K R. "The theory of the correlation method of analysis and its application to industrial processes and nuclear power plant". Measurement and Control, Vol 2, May 1969.

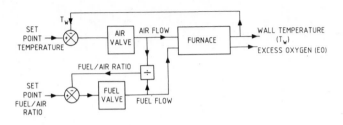

Figure 1   Existing control system

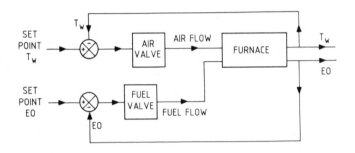

Figure 2   Possible excess oxygen control

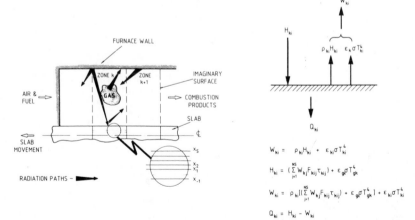

Figure 3   Representation of model zones and slab discretisation

Figure 4   Radiative heat balance on a real surface

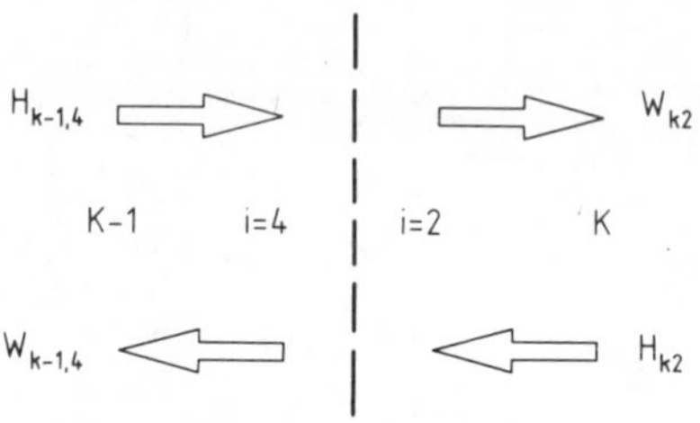

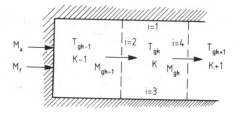

$$W_{k2} = H_{k-1,4}$$

$$H_{k-1,4} = (\sum_{j=1}^{NS} W_{k-1,j} F_{k-1,4j} \ \tau_{k-1,4j}) + \varepsilon_{gk} \sigma \ T_{gk-1}^{4}$$

$$W_{k2} = (\sum_{j=1}^{NS} W_{k-1,j} F_{k-1,4j} \ \tau_{k-1,4j}) + \varepsilon_{gk} \sigma \ T_{gk-1}^{4}$$

Figure 5   Radiative heat balance on an imaginary surface

$$M_{gk} C p_{gk} T_{gk} - M_{gk-1} C p_{gk-1} \ T_{gk-1} + E_k - S_k = \Phi_k (M_a C p_a T_a + M_f H_f)$$

$$E_k = \varepsilon_{gk} \sigma T_{gk}^4 \sum_i A_{ki}$$

$$S_k = \sum_j \sum_i A_{ki} W_{ki} F_{kij} (1 - \tau_{kij})$$

$$\sum_k \Phi_k = 1.0$$

Figure 6   Energy balance on a gas volume

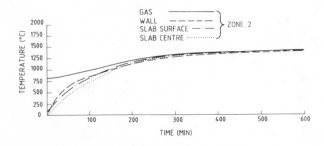

Figure 7   Furnace start-up transient

84

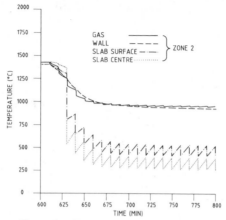

Figure 8   Transient with a slab "push"

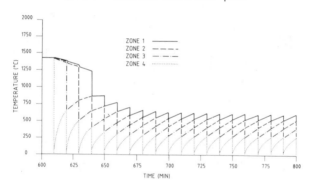

Figure 9   Slab surface temperature response (with "push")

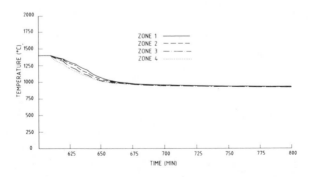

Figure 10   Roof temperature response (with "push")

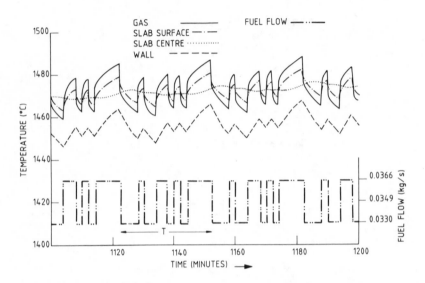

Figure 11   Response to a pseudo random binary sequence
(period T = 30 minutes)

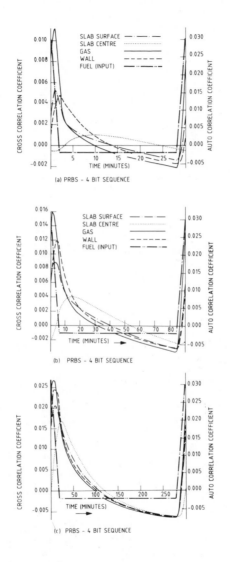

Figure 12   Furnace impulse response from PRBS

CONTROL OF THE HEAT INPUT TO A BATCH REACTOR

A. P. Wardle*, M. Alpbaz**, A. Mehta*, D. Vennor-Morris*
and D. Halksworth*

The operation of a pilot-scale, computer controlled batch
reactor involving a first order reversible reaction and run
on a minimum time per batch basis has been examined with a
view to reducing the energy input to the system.   Three
different primary level control strategies have been applied,
i.e. on-off, PI and self-tuning.   It was found that self-
tuning control of the heat input to the system resulted in
energy saving over the PI strategy which in turn was more
efficient than with on-off control.   An important
advantage in respect of the self-tuning mode over PI was
the facility with which it could be achieved.

INTRODUCTION

Batch processes offer considerable problems in modelling and control due to
their interesting dynamic nature.   Up to the present much of the work
carried out on batch reactors has been concerned with the problems of optimal
control (1,2,3).   Such optimal strategies have been involved with minimising
the time required to obtain a given degree of conversion and thus decreasing
the processing time per batch (4,5).   This type of optimal policy is
eminently desirable and will achieve a reduction in many of the associated
costs.   However, little interest has been shown in reducing the energy
consumption of a batch process operating under such an optimal strategy and
which may lead to further reductions in operating costs.   As the implementation
of this type of optimal control procedure involves both heating and cooling the
system (either consecutively or at the same time) it is possible to examine
different ways of combining the heating and cooling stages in order to minimise
the system energy consumption - whilst at the same time retaining the optimal
policy of minimum time per batch.   This paper describes the application of
on-off, PI and self-tuning control modes to a batch reactor operating on
minimum time per batch in an attempt to improve the energy consumption of the
system.

EXPERIMENTAL RIG

This is shown in Fig.1 and has been described in detail elsewhere (6).
It consists of a small batch reactor approximately 0.45 m high and 0.3 m in
diameter constructed of stainless steel.   An immersion heater is provided for
heating the reaction mixture and the power supplied in this way can be
measured by means of a wattmeter.   Cooling water is passed in parallel

* Department of Chemical Engineering, University College of Swansea,
  Singleton Park, Swansea, SA2 8PP, U.K.
**Department of Chemical Engineering, Ankara University, Fen F.Tandogan,
  Ankara, Turkey.

through 4 jackets welded to the curved wall of the reactor and has its flow-rate controlled by an on-off solenoid valve. The heating rate can be adjusted either manually by employing a variable transformer or automatically using a variety of control modes (see below). The reactor is equipped with an efficient stirrer to ensure good mixing and the whole is well lagged. The reaction involved is the fading of phenolphthalein using alkali (7), and the extent of the reaction is monitored by means of a spectrophotometer. The temperature of the reacting system is measured by a thermopile consisting of 18 Cr-Al thermocouples in series whilst the cooling water temperature is monitored by means of similar individual thermocouples.

A BBC model B microcomputer has been employed as the command and control module. This is interfaced with the rig as illustrated in Fig.2 via an IEEE filing system. Information from thermocouples and the spectrophotometer are accessed through an 8-bit A/D converter and filter at whatever sampling rate is required by the software. Signals from the BBC to the reactor are interfaced via the IEEE bus and a commercial D/A converter.

The voltage supplied to the reactor heating coil is adjusted by means of a thyristor drive unit constructed in the Department of Chemical Engineering. This consists essentially of a thyristor module mounted on a heat sink linked to a phase trigger device. The points at which triggering is carried out depend on a 0-5 V output applied from the computer via the D/A converter which in turn corresponds to a 0-120 V output to the heating element.

The control algorithms involved in the optimal time per batch strategy and those used to manipulate the energy supply to the system are all accessed by the microcomputer from disc. Plant data (viz. temperatures, concentration, heat input, etc.) are read back to disc via the BBC for off-line plotting and analysis. All programs have been written in a combination of BBC Basic and assembler.

THE OPTIMAL CONTROL ALGORITHM

(The Secondary Level Controller)

This has been implemented in the form of a supervisory control scheme in which the optimal path is predetermined according to *a priori* knowledge of the kinetics of the chemical reaction involved. Further work has been and is being carried out in which no knowledge of the system kinetics is assumed (6). This will be reported in another paper (8).

The task of the supervisory controller is to read the reactor concentration at consecutive intervals of time and to calculate the corresponding optimum operating temperature. This latter is derived using Pontryagin's Principle (see Appendix A) from the reaction kinetics as

$$T_{opt} = \frac{E_2 - E_1}{R \ln \frac{k_{02}E_2}{k_{01}E_1} + \ln \frac{x_3}{x_1 x_2}} \qquad \ldots \ldots (1)$$

The optimum temperature is employed as the set point for a primary level control loop which varies the reactor temperature by adjusting both the power supply to the heater and the cooling water flow-rate. The flow chart for the optimal controller is presented in Fig.3.

ENERGY CONSUMPTION - PRIMARY LEVEL CONTROL

Although the energy consumption of the system is partly governed by the time required to process a batch it is also dependent on the type of primary level control employed, i.e. by the means of adjusting the heating and cooling of the reactor in response to the overall optimal strategy. In this work three different primary level control schemes were examined, viz.

> a) On-off control
> b) PI control
> c) Self-tuning control

Common experimental details are listed in Table 1.

Table 1.   Experimental Details

Immersion Heater voltage range	0 - 120 V	Command and Control sampling interval	20 s
Maximum cooling water flow-rate	1.67×10^{-4} $m^3 s^{-1}$	k_{01}	1.4×10^4 $m^3 s^{-1}$ $Kmole^{-1}$
Initial concentration of Phenolphthalein	0.03 Kmole m^{-3}	k_{02}	$3.06 \times 10^9 s^{-1}$
Initial concentration of NaOH	0.09 Kmole m^{-3}	E_1	3.6×10^4 kJ $Kmole^{-1}$
Volume of reacting mixture	$7 \times 10^{-3} m^3$	E_2	7.74×10^4 kJ $Kmole^{-1}$

a) On-Off Control

This was provided by means of two relays, one being connected to the reactor heating element whilst the other was wired to the solenoid valve controlling the cooling water flow-rate. These relays are arranged such that a 1 V signal to the solenoid simultaneously causes the heater to be switched off and the cooling water to be switched on to full flow-rate (i.e. maximum cooling). Cutting off the 1 V signal produces the reverse mode, i.e. maximum heating (120 V across the heating element) and the cooling water shut off.

Fig. 4 shows the result obtained with this type of primary control action. The set point follows the optimal path produced by the secondary level controller. (The small sharp deviations in the latter were caused by air bubbles passing through the spectrophotometer cell). The control signal represents the voltage applied across the heater switching between 0 and 120 V and is a measure of the power consumption of the system.

b) PI Control

In this scheme the heat input to the reactor was controlled in a PI mode. The control action was programmed into the BBC using a standard finite difference algorithm. Procedures developed by Lopez et al (9) and Rovira et al (10) were employed to determine the optimum settings of the PI parameters. These were finely tuned by hand and gave a proportional gain of 2.0 and an integral time of 10 s. The control action was fed to the heater via the thyristor drive unit. The cooling water was still operated in the on-off mode - being turned fully on when the set point was exceeded.

The response of the reactor temperature and the level of heating required are presented in Fig.5.

c) Self-tuning Control

The self-tuning controller was a third order system based on the minimum variance procedure of Clarke and Gawthrop (11,12) which involves a system identifier based on a least squares recursive algorithm (see Appendix B).

The reactor temperature was initially stabilised using the PI feedback algorithm. Once stable control had been achieved the self-tuning controller was brought into operation. Start-up values of the diagonal elements of the associated covariance matrix were set to 10^3 whilst all initial parameter estimates were put at zero. Hence at the commencement of the run the pseudo-output polynomial λ (Appendix B) was set equal to zero in order to provide active control and to allow the parameter estimates to vary. After approximately 10 samples λ was increased to 0.01 and thereafter to 0.1 as required in order to stabilise the system. Once stable self-tuning control had been achieved the set point was continually evaluated using the supervisory control strategy. A typical run is presented in Fig.6 and the flow chart in Fig.3.

CONCLUSIONS

It can be seen by comparing the control signal in Figs. 4, 5 and 6 that the energy consumption employing the self-tuning controller is less than that occurring with PI action. The lack of resolution of the plotter attached to the microprocessor is responsible for the apparently continuous controlled variable signal, whereas in fact the control output changed in discrete steps at the 20 s sampling intervals employed. The energy consumption is proportional to $V^2 t$ and is therefore quite sensitive to small changes in the area under the control signal (V/t) plot. Direct comparisons are not entirely valid as the optimal trajectory in each run is somewhat different. Furthermore, it is difficult with a batch system to attain precisely the same starting point each time and consequently both set point and measured value appear first at slightly different values in Figs. 4, 5 and 6. As the intention of this paper is simply to emphasise the differences between the three primary control configurations reported the reaction is not shown to completion. As an indication the energy required by the reactor under self-tuning control was calculated over the first 85 sampling intervals after the reactor temperature had reached the set point and was found to be 66 kJ for the run shown in Fig.6. The energy consumption over the equivalent interval with the PI action indicated in Fig.5 amounted to 195 kJ. An additional advantage of employing self-tuning control is that no prior process testing is required to determine the controller parameters. If the PI action was perfectly tuned then its performance would approach that of the self-tuner more closely but would still not take the non-linearities of the system into account unless the PI parameters were updated during the batch cycle. On-off primary control action over the same time period required 335 kJ of heat

92

input (Fig.4) which is much less satisfactory. It is worth noting that a reduction in the sampling interval did not appear to improve the situation in the latter case due to the inherent lags in the system.

The investigation so far has been restricted by the limitation of the control of the cooling water flow-rate to an on-off mode. PI or self-tuning control of the latter could decrease the energy requirements of the process still further and work concerning this and other forms of control action will be reported at the meeting. However, it is clear already that it is possible to reduce the energy consumption of a batch reactor by improving the primary level of control whilst maintaining the optimal strategy of minimum time per batch.

REFERENCES

1.   A.A.T.G. Portugal, Computer Control of a Batch Process for Resin Manufacture, Ph.D. Thesis, Aston University, U.K. (1984).

2.   C. Kiparissides and S. R. Shaw, Automatica $\underline{19}$ (1983) 225.

3.   J. F. MacGregor, A. Penlidis, and A. Hamielec, Poly Proc. $\underline{2}$ (1984) 179.

4.   G. C. Goodwin, P. J. Ramadge and P. E. Caines, IEEE Trans. AC $\underline{AC-25}$ (1980) 449.

5.   K. S. Narendra and Yuan-Hao Lin, IEEE Trans. AC $\underline{AC-25}$ (1980)456.

6.   A. H. Baghaie, The Time-Optimal Control of a Batch Reactor, Ph.D. Thesis, University of Wales, U.K. (1978).

7.   M. D. Barnes and V. K. La Mer, J.Am.Chem.Soc. $\underline{64}$ (1942) 2312.

8.   A. P. Wardle, M. Alpbaz, A. H. Baghaie and A. Mehta, The Non-Supervisory Control of a Batch Reactor, in preparation.

9.   A. M. Lopez et al, Instrumentation Tech. $\underline{14(11)}$ (1967) 57.

10. A. A. Rovira et al, Instr. and Cont. Systems, $\underline{42(12)}$ (1969) 67.

11. D. W. Clarke and P. J. Gawthrop, Proc.IEE $\underline{122}$ (1975) 929.

12. D. W. Clarke and P. J. Gawthrop, Proc.IEE $\underline{126}$ (1979) 633.

NOMENCLATURE

E_1	activation energy of the forward reaction	J Kmole^{-1}
E_2	activation energy of the backward reaction	J Kmole^{-1}
k	integer number of system time delays	
k_{01}	Arrhenius coefficient for the forward reaction	Kmole^{-1}s^{-1}
k_{02}	Arrhenius coefficient for the backward reaction	s^{-1}
k_1	forward rate constant	
k_2	backward rate constant	
n_A, n_B etc.	order of A, B, etc. polynomials	

Nomenclature (contd.)

R	universal gas constant	$J \; Kmole^{-1} K^{-1}$
t	time	s
T	reactor temperature	K
T_{opt}	optimal temperature for maximum reaction rate	K
V	voltage	
x_1	concentration of (ph ph)$^=$	$Kmole \; m^{-3}$
x_2	concentration of (OH)$^-$	$Kmole \; m^{-3}$
x_3	concentration of (ph pH.OH)$^{\equiv}$	$Kmole \; m^{-3}$
z	z transform	
ζ	zero mean Gaussian white noise	

Subscripts

f	final value

APPENDIX A

When phenolphthalein (ph ph) is added to an alkaline solution it first undergoes a rapid irreversible conversion to a quinoid form (ph ph)$^=$, which has a very strong pink colour. The latter then slowly and reversibly reacts with further hydroxyl ions to form the non-resonant (hence colourless) carbinol form (ph ph.OH)$^{\equiv}$, viz.

$$(ph\ ph)^= + OH^- \quad \underset{k_2}{\overset{k_1}{\rightleftharpoons}} \quad (ph\ ph.OH)^{\equiv}$$

The rate of disappearance of the dye is given by:

$$\frac{dx_1}{dt} = -k_1(T) \; x_1 x_2 + k_2(T) x_3$$

with the reaction system also satisfying:

$$x_2 = x_1 + x_2(0) - x_1(0)$$
$$x_3 = -x_1 + x_3(0) + x_1(0)$$

where $x_1(0)$, $x_2(0)$ etc. are the initial
concentrations of components 1, 2, etc...

The optimal strategy is to choose a decision function which minimises

$$\theta = \int_0^{t_f} dt = x_4(t)$$

where $x_4(t)$ is an additional state variable required to solve the problem using Pontryagin's Principle. It follows that the performance equation for x_4

is

$$\frac{dx_4}{dt} = 1 \; ; \quad x_4(0) = 0$$

Hence the problem is transformed into that of minimising $\theta = x_4(t_f)$. The Hamiltonian function and the adjoint vectors for the latter case are respectively

$$H(z,x_1,T) = z_1 \left[-k_1(T) \, x_1 \, x_2 + k_2(T)x_3 \right] + z_2$$

$$\frac{dz_1}{dt} = -\frac{\partial H}{\partial x_1} \; ; \quad z_1(t_f) = 0$$

$$\frac{\partial z_2}{\partial t} = -\frac{\partial H}{\partial x_2} = 0; \quad z_2(t_f) = 1$$

According to Pontryagin's Principle, a necessary condition for $x_4(t)$ to be a minimum is that $\frac{\partial H}{\partial t} = 0$.

Hence

$$z_1 \frac{\partial}{\partial T} \left[-k_{01} \exp(-E_1/RT) \, x_1 \, x_2 + k_{02} \exp(-E_2/RT)x_3 \right] = 0.$$

Differentiating and solving for T gives equation (1).

APPENDIX B

It is assumed that the plant can be represented by the ARMA model:

$$A(z^{-1}) \, y(t) = B(z^{-1}) \, u(t-k) + C(z^{-1}) \, \zeta(t) + d$$

where A, B and C are polynomials in the backward difference operator acting on the system output, control signal and non-measurable disturbance respectively, and d is the system offset.

The self-tuning controller minimises a cost function

$$I = E\{\phi^2(t+k)\}$$

where $\phi(t+k)$ is the output of a pseudo-system defined by

$$\phi(t+k) = P(z^{-1}) \, y(t+k) + Q(z^{-1}) \, u(t) - R(z^{-1}) \, w(t)$$

P, Q and R are weighting polynomials in the backward difference operator whilst w(t) is the system set point.

Minimisation of the cost function leads to (11,12)

$$Gu(t-k) + Fy(t-k) + Qu(t-k) - Rw(t-k) + \delta = 0 \qquad (B1)$$

and

$$u(t) = \frac{Fy(t) + Hw(t) - \delta}{(CQ+G)} \qquad (B2)$$

where $\quad G = EB, \; h = CR$ and $\delta = E(1) \, d \quad$ (a scalar)

F, G and δ are identified recursively from (B1) using a least-squares technique, after which the control signal is evaluated from equation (B2). In order to restrict the number of parameters, each of n_A, n_B and k were set at 2 with $C(z^{-1}) = C_0 = 1$. This gave $n_E = 1$, $n_G = 3$ and $n_F = 1$ with pseudo-output polynomials as $P(z^{-1}) = R(z^{-1}) = 1$ and $Q(z^{-1}) = \lambda$.

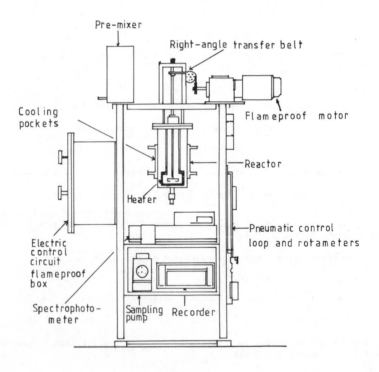

Fig.1   Major Details of the Experimental Rig
(Scale 1/16)

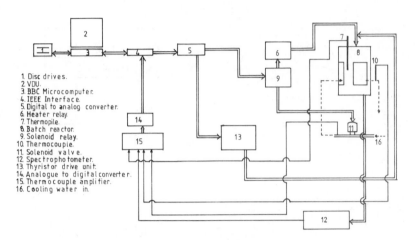

1. Disc drives.
2. VDU.
3. BBC Microcomputer.
4. IEEE Interface.
5. Digital to analog converter.
6. Heater relay.
7. Thermopile.
8. Batch reactor.
9. Solenoid relay.
10. Thermocouple.
11. Solenoid valve.
12. Spectrophotometer.
13. Thyristor drive unit.
14. Analogue to digital converter.
15. Thermocouple amplifier.
16. Cooling water in.

Fig. 2   Command and Control System

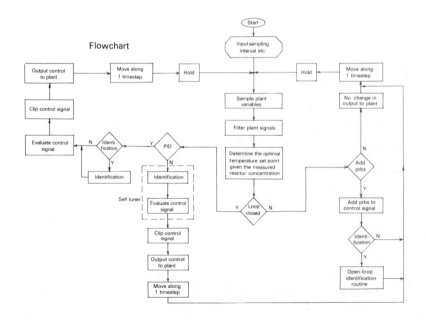

Fig. 3

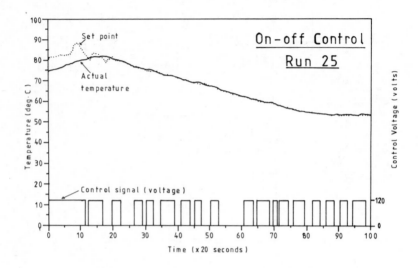

Fig. 4

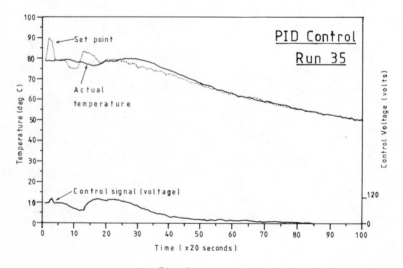

Fig. 5

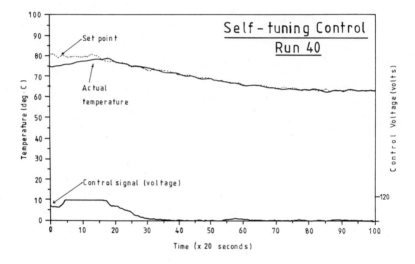

Fig. 6

CONTROL STRATEGY ANALYSIS OF ALUMINIUM REDUCTION CELLS
USING DYNAMIC PROCESS SIMULATION

A.R.Wright* and A.W.Wright*

Aluminium smelting is an energy intensive
electrochemical process. Most of the plant
variables cannot be measured directly hence
it is dificult to study the effect of control
strategy on the process. The paper describes
the development of a dynamic model of an
aluminium production cell and its use for
control strategy analysis for improving
energy utilisation.

Introduction

Primary aluminium smelting uses the most energy of all UK
non-ferrous industries. Aluminium is produced by the electrolytic
reduction of alumina dissolved in a cryolite melt using a carbon
anode and a carbon lined cell as a cathode. Aluminium is produced
at the cathode and oxygen at the anode.

$$2Al_2O_3 \longrightarrow 4Al + 6O$$

The oxygen reacts with the carbon anode to give CO_2. Electrolysis
of aluminium depletes the alumina content of the electrolyte
and, below a critical level, another reaction producing CF_4 gas
occurs at the anode causing the voltage to rise rapidly. This is
termed an anode effect.

Fig. 1 shows the cross section of a centre break pre-baked
anode cell. Alumina is fed into the bath by the operation of the
crust breaker and aluminium removed by 'tapping' the metal pad
periodically. A protective 'ledge' of frozen bath is formed
around the reactor walls. This serves to insulate the cell and
stabilise the metal pad which rotates under the influence of the
magnetic field induced by the high operating current.

There are two types of centre break cell, the half break and
the point feed. In a half break cell, two crust breakers, each
half the length of the cell, operate alternately and large
amounts of alumina are fed at each feedtime, typically about 90
kg every 70 minutes. In a point feed cell only a small area of
crust is broken and alumina fed in small quantities more
frequently, about 9 kg every 7 minutes.

* Dept. Chem. and Proc. Eng.,University of Newcastle upon Tyne

The energy consumption of a cell is given by

$$298.1 \ \times \ \frac{\text{cell volts}}{\text{current efficiency } \%} \qquad \text{DC kWh per kg}$$

   Clearly the way to improve energy consumption is to decrease the cell voltage and/or increase the current efficiency of the process. Such improvements may be made through better cell design and instrumentation[1,2] but for a typical smelter operating some 300 cells, the capital requirements would be prohibitive. Although cell designs are being improved, extended cell life through better process operation and design means that it takes several years for complete pot change over. A reduction in energy use may therefore be more readily achieved by improvements in cell operation and in particular by the development and improvement of control software.
   The cells are constantly operating in unsteady state. For optimal control it is neccessary to approach steady state conditions in the heat and mass balance loops of the process. It is not possible however to measure directly any process variables apart from cell voltage hence it is difficult to study the effects and interactions of control strategies on the plant. This paper concentrates on the development of a dynamic model simulation for the purpose of developing an improved control strategy. This work is part of a collaborative project between Anglesey Aluminium and the Department of Chemical and Process Engineering, Newcastle University, aimed at improving pot control and energy utilisation.

Dynamic Model

   The model is shown conceptually as a block diagram in fig 2. The interconnections between the functional blocks relate the flow of mass, energy and information. Each block represents a set of equations describing the relationships between the inputs and outputs of that block. These relationships take the form of differential and algebraic expressions derived from analysis of published work, plant data and mathematical modelling techniques.

Process inputs
   The process has three primary inputs:-

1. The breaker action. This regulates the amount and frequency of alumina fed to the pot. It affects both the heat and mass balances in that immediately following a breaker action there is a sudden increase in the mass of alumina in the bath followed by an increase in the weight percent of alumina. A loss of heat also results both from heating the fed alumina to the bath temperature and from the exposed bath where the breaker has broken through the crust.

2. The anode beam position. This governs the anode cathode distance and hence the voltage drop across the bath. The anode beam position is the distance from the bottom of the anodes to the top of the cathode and the ACD determined from the difference between this value and the depth of the metal pad.

3. The line current. In this model line current is assumed to be constant.

Stochastic Disturbances

Two stochastic disturbances are assumed to enter the system. D_1 represents the uncertainty in the amount feed added by the breaker action. D_2 represents the uncertainty in the anode beam position.

Block 1   Anode Change and Sludge Formation

During an anode change an amount of frozen crust consisting largely of an alumina bath agglomerate enters the bath. This addition is likely to be a major cause of sludge formation as the agglomerate would not dissolve before sinking to the bottom of the pot and forming sludge[4,5].

Block 2   Dissolution Rate Routine.

Alumina in the bath dissolves from alumina held in suspension. Although the mechanism for dissolution is not completely understood the dissolution rate appears to obey a first order rate law.

$$\text{Rate of dissolution} = K.A_{Al2O3}.(C_{sat} - C_{Al2O3})$$

According to Kachanovskaya[3] the rate constant is dependent upon both the alumina content and temperature of the bath.

Block 3   Electrolyte Mass Balance

The primary outputs from this block are the aluminium production and the alumina concentration but dynamic mass balances for all ionic species are required. The general mass balance equation is

rate of change   =   mass flow - mass flow - rate of consumption
of species   in   out   by reaction

where the rate of consumption of a species i by an electrochemical reaction is given by

$$r_i = I.CE_i/z_iF$$

CE_i is the current efficiency specific to the reaction of species i and is dependent upon the dominance of the reaction over competing reactions.

For Al^{3+} ions in the bath

In: Al^{3+} ions from dissolution of alumina, $2*molerate*27$
Addition of Al^{3+} ions as bagged AlF_3, M_{Albag}

Reaction: Electrolysis to Aluminium at cathode, $I/z_{Al}F*CE_{Al2O3}*27$
Formation of Al^{3+} ions by reoxidation of aluminium, $r_{back}*27$

which gives

$$dM_{A13}/dt = (2*molerate-I/z_{A1}F*CE_{A12O3}+r_{back})*27+M_{Albag}$$

For O^{2-} ions in the bath

In: O^{2-} ions from dissolution of alumina, $3*molerate*16$

Reaction Electrolysis to CO_2 at the anode, I/z_0F*CE_0*16
Formation of O^{2-} ions by reoxidation of aluminium, $3/2*r_{back}*16$

which gives

$$dM_0/dt = (3*molerate-I/z_0F*CE_0+3/2*r_{back})*16$$

Dynamic mass balances are also derived for alumina in suspension in the bath, the aluminium in the bath and in the cathode pad and the fluoride in the bath. The weight percent of alumina in the bath is calculated from the O^{2-} content. The NaF/AlF_3 wt. ratio is also determined. Loss of overall current efficiency is assumed to be due to the chemical reaction between aluminium in the electrolyte and dissolved carbon dioxide[6].

$$2Al + 3CO_2 \longrightarrow Al_2O_3 + 3CO$$

Block 4   Electrolyte Conductivity

The specific conductivity of the bath is dependent upon the composition and temperature of the bath and is calculated from an empirical relationship.
The IR drop across the bath is calculated from specific conductivity, line current, area for current flow and the anode-cathode distance.

Block 5   Electrolysis Charge Balance

The total current density of an electrode is the sum of the partial current densities of the various reactions occuring on that electrode.

$$i_{tot} = \sum i_i$$

The partial current density for a species i may be related to the bulk concentration of the species, $C_{b,i}$, the mass transfer coefficient of the species, k_1 and the polarisation voltage, E.

$$i_i = \cfrac{C_{b,i}}{\cfrac{1}{z_i.F.k_1} + \cfrac{1}{K_i.exp(-b_i.E)}}$$

For a fixed line current, using these two equations, it is possible to calculate the polarisation voltage, E for each electrode and the current efficiency of each reaction. The bath voltage is then given by the sum of the two polarisation voltages and the Ohmic drop across the electrolyte.

$$V_{bath} = E_{anode} + E_{cathode} + V_{drop}$$

Under normal operating conditions, Al^{3+} and Na^+ ions are reduced at the cathode while O^{2-} ions are oxidised at the anode. At anode effect F^- ions are oxidised at the anode, rapidly suppressing the other anodic reactions. In this model the process goes into an anode effect when the partial current density of the oxide reaction falls below the line current density.

Block 6 Bath Resistance and Electrical Energy

From the bath voltage and line current, the bath resistance and electrical energy supplied to the cell are calculated. The model therefore simulates the dynamic changes in bath resistance due to electrolysis conditions and bath composition. It provides a means of analysing feed strategy based on the practice of monitoring cell resistance.

Block 7   Heat Losses

The primary factors affecting heat loss from the pot are the pot design and insulating materials. The side walls of the pot are purposely made of materials with poor insulation properties so as to encourage a layer of frozen bath to form. Losses occur through the crust, hood, bottom of the pot and through the collector bars. Although the crust actually behaves in a similair way to the freeze crust thickness is assumed to be constant. Other losses occur intermittently, changing anodes and crust breaking when feeding allows heat to be lost from the exposed surface of the bath. A new anode will also absorb heat causing localised freezing. Heat lost due to breaks in the crust either by breaker action or at anode change is made up of convective and radiative parts and is given by:-

$$Q = h_{break} \cdot A_{break} \cdot (T_b - T_a) + e \cdot A_{break} \cdot \sigma \cdot T_b^{4}$$
$$\phantom{Q = h_{break} \cdot A_{break} \cdot } \text{convection} \text{radiation}$$

Heat losses through the bottom and the collector bars, are assumed constant and are average values measured on a number of pots.

Block 8   Electrolysis Heat Balance

This block calculates the net rate of heat generated by the chemical reactions during electrolysis.

Block 9   Pot Heat Balance

A dynamic heat balance on the pot is formulated from the net heat due to electrolysis (from block 8), the heat losses (block 7) and the heat generated due to the resistance of the electrolyte (block 6). From this heat balance the temperature of the bath is determined.
The rate of heat change in the pot in the model is given by:

$Q_{pot}=$
     Electrical Power in
     - heat of electrolysis
     - heat from bath to freeze
     - heat from pad to freeze

- heat from pad to cathode
- heat lost through top of pot
- heat lost by breaker action at feed
- heat lost at anode change
- heat lost in heating feed to bath temperature
- heat of dissolution of alumina
- heat content of tapped metal

Block 10   Freeze

A mass and heat balance on the freeze is considered in two parts, at the bath-freeze boundary and at the metal-freeze boundary. The layer of freeze at each boundary is assumed to have the same composition as the molten bath. Heat which is not conducted away by the freeze affects the freeze melt rate. For the bath this gives

$$dM_{freeze}/dt = \frac{U_f \cdot A_{bf} \cdot (T_{liq} - T_a) - h_{bf} \cdot A_{bf} \cdot (T_b - T_{liq})}{H_{latent}}$$

where the resistance to heat transfer, $1/U_f$, is given by

$$\frac{1}{U_f} = \frac{X_f}{k_f} + \frac{X_w}{k_w} + \frac{1}{h_s}$$

The freeze melt rate at the metal-freeze interface is similarly calculated.

Model Validation and Tuning

The simulation has been used to model a variety of cell operating conditions. The results have shown that the simulation exhibits very similar behaviour to measurable parameters both in terms of dynamic response and interactions. A real time version of the model running on an IBM PC has been interfaced with the existing control system on site and has enabled further testing and tuning.

Simulation of Simple Feed Strategies

At present there is no easy method for the direct measurement of alumina concentration in the bath. Control is therefore attempted through alumina feed strategies. These strategies may be broadly split into two categories, open loop and closed loop. The open loop strategies make no attempt at determining the alumina concentration but simply feed alumina at constant intervals. The feed time and amount of alumina fed are chosen to maintain the alumina content at either a low (2-3 wt%) or high (6-7 wt%) level. In the latter case periodic anode effects are scheduled to prevent excessive sludge formation.

The closed loop strategies try to infer alumina concentration changes in the bath from resistance measurements. As a cell approaches anode effect the cell resistance increases more rapidly and by detecting this change the cell may be fed and anode effect prevented. However, resistance slope changes may also occur for other reasons and to safeguard against overfeeding this strategy is windowed about a base feedtime. This makes it unsuitable for use with point feed cells, instead a strategy based on overfeeding and underfeeding may be used. The cell is underfed until an approaching anode is detected when a period of overfeeding is started. This continues until resistance changes indicate that the alumina concentration has moved outside the desired band and the cycle is repeated. Both these strategies try to control the cell within a broad deadband rather than to a set point.

Using the model it is possible to examine feed strategies for both half break and point feeder cells under a variety of operating conditions. The model predicts the dynamic variations in pot composition and electrolysis products, temperature, freeze thickness, current efficiency, polarisation potentials, cell voltage and resistance. Figures 3 and 4 shows the results of a simulation of a 150 kA half break cell. The simulation used a simple open loop feed strategy with 90 kg of alumina being fed every 70 minutes. The voltage is controlled within a dead band by automatic movement of the anode beam position. The simulation clearly shows the disturbances caused by this type of strategy and demonstrates some of the problems associated with anode effects.

This simulation provides a simple demonstration of the use and predictions of the dynamic model. Clearly for the rigorous testing of different strategies the simulated pot operational time would have to be days or weeks. At present the simulation is being used to investigate feed strategies and energy utilisation through cell voltage regulation.

Symbols

A area, cm^2
C_{sat} saturation concentration, mol cm^{-3}
C_b bulk concentration, mol cm^{-3}
dM/dt rate of change of mass with time, g s^{-1}
e emissivity
E polarisation voltage, V
F Faraday constant, C mol^{-1}
h heat transfer coefficient, W $cm^{-2} K^{-1}$
H_{latent} latent heat of fusion, J g^{-1}
i current density, A cm^{-2}
I current, A
k thermal conductivity, W $cm^{-1} K^{-1}$
k_l mass transfer coefficient, cm s^{-1}
K rate constant for alumina dissolution, cm s^{-1}
K_i,b_i tafel coefficients for species i,
molerate   rate of dissolution of alumina, mol s^{-1}
M_{Albag} mass of bagged AlF_3, g
r_{back} rate of back reduction of Al, mol s^{-1}
T temperature, K
T_{liq} liquidus temperature of bath, K
U overall heat transfer coefficient, W $cm^{-2} K^{-1}$
V voltage, V
x thickness, cm
z number of electrons passed

suffixes
Al2O3   alumina
Al3   aluminium 3+ ions

b   bath
bf   bath freeze interface
f   freeze
i   species i
0   oxygen 2- ions
sat   saturation
tot   total
w   sidewall

References.

1. G.C.Bushell, "Energy Efficiency in the Aluminium Industry, Some Case Studies", Aluminium and Energy Conference, Bangor, 19th September 1986
2. D.J.Salt, G.Guelfo, A.J.Sargent, "Improvements in Cell Energy Efficiency at Anglesey Aluminium", Aluminium and Energy Conference, Bangor, 19th September 1986
3. Asbjornsen et al, "Kinetics and Transport Processes in the Dissolution of Aluminium Oxide in Cryolite Melts", Light Metals 1977, vol 1, pp137-15
4. Rudolf Keller, "Alumina Dissolution and Sludge Formation", Light Metals 1984, pp513-51
5. J.Thonstad, P.Johansen, E.W.Kristensen, "Some Properties of Alumina Sludge", Light Metals 1980, pp227-239
6. B.Lillebuen and Th.Mellerud, "Current Efficiency and Alumina Concentration", Light Metals 1985, pp637-646

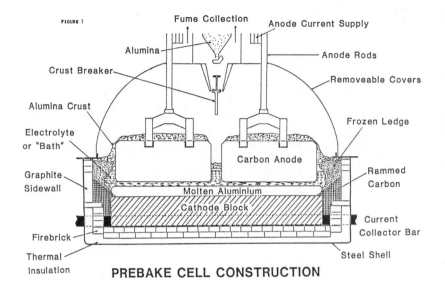

PREBAKE CELL CONSTRUCTION

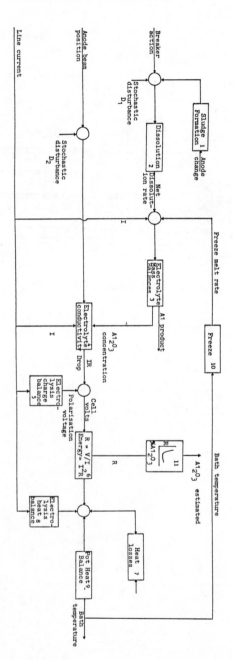

Figure 2. Conceptual Model.

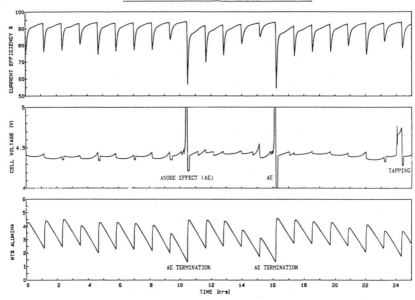

FIGURE 3   DYNAMIC SIMULATION OF AN ALUMINIUM PRODUCTION CELL

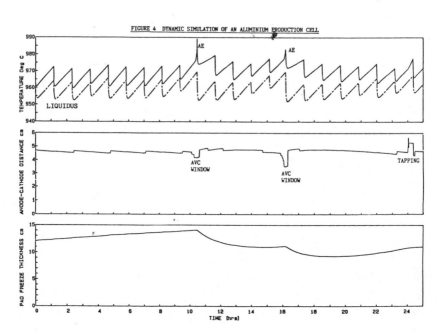

FIGURE 4   DYNAMIC SIMULATION OF AN ALUMINIUM PRODUCTION CELL

A NEW CO2 RECOVERY SYSTEM FOR BREWERIES

A R Coleman MA MSc CEng MInstE MInstR*

This paper describes a new system for liquefying carbon dioxide (CO_2) gas recovered from brewing processes. The plant produces a high yield of recovered liquid CO_2 and uses less power relative to conventional techniques.

The new process uses the refrigerating effect from vapourising liquid CO_2 to condense the recovered material. Liquid stripping is then used to remove oxygen, and other volatile impurities, to provide the quality of gas required for brewing purposes.

The first full scale plant is operational at a brewery in Ireland and the second plant is being installed in one of the largest UK breweries.

INTRODUCTION - THE USE OF INERT GAS IN BREWING

Having largely gained control of microbial infection, the brewers' drive to improve product quality has recently centred on reducing the level of dissolved oxygen in the beer. Even small quantities of oxygen can oxidise flavour components thus reducing customer acceptability. Levels of oxygen down to 0.15 mg/litre, or below, are being sought, as compared to levels of around 11 mg/litre in tap water.

Breweries have, in the past, only used inert gas in critical processes such as packaging and filtration where turbulence or residence time caused oxygen pick up from headspace air. Air was widely used to provide top pressure in cold conditioning and bright beer tanks. Now inert gas is being used at all stages. This has led to rising cost to the extent that a large bright beer brewery may be paying as much as £700,000 per annum for supply of inert gases.

Traditionally, the inert gas used has been CO_2 though nitrogen (N_2) has gained particular acceptance in keg racking of ale beers since it forms smaller and more stable bubbles when dispensed and thus gives a longer lasting head in the glass. This, however, leads to problems of stablefob formation, reduced efficiency of racking and, on occasion dispensing difficulties in the public house. Control of the nitrogenated level is more critical than with carbonation.

* Of WS Atkins Management Consultants

CO_2 will always be used for injection into beer (carbonation) since it makes the beer 'sparkle' on the palate, whilst N_2 does not. The decision as to which gas to use for blanketing and counter pressure in processing is largely a matter of cost, although considerations of the partitions of gases between solution in the beer and in the vapour space are also important.

CO_2 bulk liquid costs around £72/tonne. When vapourised this is equivalent to around £144/1,000 m^3 at STP. Correspondingly, the price for nitrogen works out to around £65/1,000m^3 at STP. Understandably, N_2 is being used for more of the processing requirements though there is resistance to its use from some brewers. Any gas dissolved in the beer will tend to come into equilibrium with the gas in the vapour space above it. Under an N_2 atmosphere, CO_2 comes out of the solution while N_2 is taken into solution. Thus the balance of gases in solution can move away from specification. Many breweries are using only CO_2 for their processing needs.

THE BREWING PROCESS

In order to understand the importance of CO_2 in brewing it is worth recapping on the brewing process. Figure 1 is a simplified block diagram showing the main flow path of beer in a bright beer brewery. Bright beer is the name given to filtered ale and lager beers as dispensed from kegs or sold in bottles or cans. It is distinct from cask conditioned ale which is packaged into casks directly from fermentaion and requires significantly less processing. Carbon dioxide is central to brewing operations; any interruption in supply can halt production.

The Brewhouse

The first stage in brewing is to use hot water to extract the maltose and other sugars from malted barley. This sugary solution, called wort, is then boiled, cooled and passed to fermentaion.

Fermentation

Yeast is added to the wort and encouraged to grow, feeding on the sugars. The products of the yeast action are alcohol and CO_2. Some of the CO_2 is retained in solution in the beer, but the majority is evolved and vented from the vessel. Some 25-35 kg of CO_2 is evolved for every 10 hl of beer fermented.

Conditioning

The beer is passed to cold conditioning vessels where it is held for up to three weeks at temperatures down to -1°C. A number of complex and slow reactions take place with protein material being precipitated. CO_2 is used to flush the tank before filling and is vented as filling proceeds.

Carbonation

The beer is then carbonated by controlled injection of CO_2 to a specified concentration. Different beer qualities require different concentrations varying from 2 to 5 kg/10 hl. The beer now contains more CO_2 than equilibrium at atmospheric pressure and it must be held under top pressure at all times to prevent breakout of the dissolved gas.

Filtration

The beer is then filtered to clarify it. The flow rates and pressures through the filter are critical and buffer tanks are used at inlet and outlet to regulate the conditions. These tanks are supplied with CO_2 which is let in or vented to maintain the top pressure as the levels in the tanks fluctuate.

Bright Beer Tanks

These tanks provide the buffer between filtration (and other processing) and packaging. A typical filling cycle is:

1. Counter pressure empty tank with CO_2 to one bar gauge.

2. Fill against counter pressure, exhausting CO_2 to maintain one bar gauge.

3. Close for checking beer specifications

4. Raise top pressure to 1.5 bar gauge and pump out, filling with CO_2 to maintain top pressure.

5. Release pressure down to one bar gauge to receive next batch.

Thus 4.5 kg of CO_2 is used to transfer 10 hl of beer through the bright beer tanks before consideration for cleaning - which can as much as double this quantity, depending on the detergents and cycles used.

Pasteurisation

Before packaging into kegs or bulk delivery tanks, most breweries pasteurise their beer in a standard regenerative flash pasteuriser. Control of the beer flow is again critical and, as with filtration, inlet and outlet buffer tanks are used to take up fluctuations in supply and demand rates. CO_2 is again used to maintain top pressure in these vessels, being let in or vented under pressure control as the levels fall or rise.

Keg Racking

Kegs must again be counter-pressured with CO_2 before filling to prevent gas breakout. Higher pressures are often used than in the bright beer tanks due to the speed of filling required and consequent turbulence. Simplified filling cycle is:

1.   Blow out sterilising steam using CO_2.

2.   Counterpressure to 1.75 bar gauge.

3.   Fill against counterpressure exhausting CO_2 to maintain pressure.

4.   Recover beer left in the filling head by scavenging with CO_2.

Other Controlled Uses

CO_2 is also used for top pressure in the provision of deaerated water, beer tankering, bottling and canning.   Many breweries also fill CO_2 cylinders with liquid for use in dispensing at the public houses.

Uncontrolled Uses

When a tank has been opened for inspection or cleaned with caustic detergent it will be full of air.   Prior to filling, the air is flushed out with CO_2 in an often inefficient and uncontrolled manner.   Similarly if a tank of beer is found to have too much dissolved oxygen, it is often removed by bubbling an uncontrolled flow of CO_2 through it.

The mass balance of CO_2 in a hypothetical bright beer brewery producing around 2,500,000 hl of beer a year is illustrated in Table 1.

At the same time, around 8,000 tonnes per annum will be evolved from the fermenters.

It is not difficult to see that, given an adequate system for collection of vented CO_2 and CO_2 from fermentation, the brewery should be able to provide all its own requirements.

CONVENTIONAL CO2 RECOVERY SYSTEM

Many breweries, particularly the larger and older established ones, have CO_2 collection equipment installed.   Of these, the majority collect from fermentation only.   The usual plant arrangment is shown in Figure 2.

This equipment is quite standard and is offered by a number of international contracting companies.   The pattern is followed by every one of the twenty plants the author has visited in the UK and Europe.

CO_2 from the fermenters (and/or other sources) is collected into a balloon.   When it has filled, the compressor is started on a level switch and continues to run until empty.   The balloon is protected against overpressurising by a hydraulic seal.

The gas is then water washed in a scrubbing tower and passes through a knock out pot to prevent any water droplets being entrained in the suction of the compressor.

Compression is normally by a two stage, reciprocating, oil free compressor delivering at around 21 barg.   This is the limit that can be achieved with two stages due to the temperature limit to on the PTFE rings (around 130°C).

The compressed gas is then cooled in a standard water cooled aftercooler, and any condensation removed in a separator. It is important to cool the gas as much as possible and to achieve good separation - the gas will later pass into a dessicant drier. A shortfall of only 10°C in the aftercooling will double the load on the drier.

Deodourising and Drying

The gas now passes into a twin column, carbon granule deodouriser where trace components such as alcohols, ketones etc are removed. The CO_2 will eventually be injected into beer at the carbonation stage, and it is important to remove any compounds that would affect flavour.

The gas now passes into a twin column dessicant dier and is dried to a dewpoint of around -40°C or sometimes lower. This is essential, since the liquefier operates at around -28°C and any moisture in the gas will cause tube frosting and loss of capacity. Correct drier operation is critical to the operation of the plant.

Liquefication

The dry gas is now passed into the liquefication plant. This is usually a package set comprising a shell and tube liquefier with R22 or R502 evaporating in U tubes under TEV control; a reciprocating compressor; and a water cooled shell and tube condenser. The compressor is started and stopped on the CO_2 pressure in the liquefier, as sensed by pressure switch P2. The condensing temperature of pure CO_2 at 21 barg is -17°C, refrigerant evaporating temperatures of around -28°C are usually used.

The gas passes along the liquefier towards the purge point and CO_2 is condensed and falls under gravity into the liquid tank below. The residual gas containing the non condensibles is purged to atmosphere. The control of the purge flow can be manual, or by a timer interlocked to the compressor, or most effectively, by an analyser that checks the purity and opens an automatic valve when the CO_2 concentration is below a set level (normally around 94% by volume).

Defrosting

Defrosting is usually required at intervals depending on the reliability of the drier. This is normally done by steam and can take a significant time since the liquefier must be depressurised, purged of air and repressurised before liquefication can continue.

PERFORMANCE OF CONVENTIONAL PLANT

The typical yield from this system is however quite poor. While 25-35 kg of CO_2 is evolved from fermenters for every 10 hl of beer fermented (depending on the strength of the particular brew), typical recovery rates are only 8-14 kg/10 hl as liquid.

At best, then, less than 50% and often only 30% of available CO_2 is liquified. The main reason for this is that the plant cannot produce CO_2 liquid with an acceptable level of dissolved oxygen (less than around 10 vppm as re-vapourised gas) from feedstocks of lower than 99.5% v/v CO_2.

This means that the brewing operator cannot initiate collection from the fermenter at the time of filling. He must wait until the fermentation has become established, the air in the headspace has been fully purged away and the evolving gas has reached the required purity. This can take some time since the first gas to be evolved strips nitrogen out of the solution in the wort. Substantial quantities of CO_2 can be lost during purging. Furthermore, since most breweries have a manual changeover system, additional losses are often incurred through late connection - or even no connection at all - due to the pressures on operators in a busy brewery.

THE NEW LIQUEFIER

It was observed that CO_2 liquid from bulk storage was being vapourised for site use with a steam vapouriser at the same time as mechanical refrigeration was being used to liquefy recovered gas. Furthermore, it was observed that the vapourised gas was being reduced from 21 bar gauge to 6.5 bar gauge for distribution immediately downstream of the vapoursiser. Thus the idea was conceived that by flashing the liquid from the storage tank down to 7 bar gauge before vapourising, it could be used in a heat exchanger to liquefy recovered CO_2.

The basic flow pattern is illustrated in Figure 3. At 7 bar gauge the saturation temperature is -46°C. Thus after allowing a 5°C temperature approach, the recovered gas can be cooled to -41°C, equivalent to a purity of only 45% CO_2 by volume in non condensibles (with a total pressure of 21 bar gauge). This means that CO_2 can be condensed from feedstocks with purities of 80% or even below, and the yield from feedstocks of all purities is enhanced.

The plant was originally conceived as a vent condenser for the purge gas from an R22 liqufier. However, it was quickly realised that for the majority of the week it could be used to liquefy the bulk of the recovered gas, thus reducing the energy cost both for liquefication and vapourisation.

Oxygen Removal

The liquid produced in the new liquefier contains an unacceptable level of dissolved oxygen (up to 2,000 mole ppm). A stripping column is used to clean up the liquied by contacting it with a counter current of reboiled gas from the collecting tank.

PLANT CONFIGURATION IN PRACTICE

The practical configuration of the plant is illustrated in Figure 4. The gas compression path is standard as per a conventional plant except that a lower water dewpoint (-55°C) is required from the drier. The high pressure, dry, recovered gas (at about 28°C) is passed through the reboiler coil and is cooled (to around -10°C), so providing the flow of reboiled gas for the stripping column. The precooled, recovered gas is then passed through the refrigerated liquefier into the new liquefier. Here it contacts the cold tubes and CO_2 is condensed and falls to the stripping column. The non-condensibles travel along the heat exchanger to the purge point where they are released through the purge valve under control of a purity analyser and at around 50% CO_2 by volume. Stainless steel 304L has be found to be a suitable material for all vessels.

High pressure liquid from store is expanded into the surge drum. The level in the surge drum is regulated by a float control system. Safety protection is provided should the surge drum pressure approach the dry ice point. The liquid now at 7 bar gauge (-46°C) circulates by natural convection through the tubes of the shell and tube heat exchanger. The two phase returning mixture is separated in the surge drum, with the vapour being distributed to the site.

Electric heaters are also installed in the collection tank (under control of a purity analyser) to boost the rate of purification if required. Liquid from the refrigerated liquefier can also be passed through the column. Addition of such a column and reboiler to a conventional plant would allow it to accept feedstocks down to around 95% purity, by volume, a tenfold increase in non condensibles.

Defrosting is performed with hot CO_2 gas. Defrosting can thus be performed without depressurising the plant and is thus quicker and easier than the conventional steam defrost.

PLANT OPERATION IN PRACTICE

Naturally, the supply of recovered gas and the demand for vapourised pure gas is not always in balance on the site. Figure 5 is a graph of typical recovered gas flows from a large brewery operating a three shift pattern. The chain dotted line shows the recovery profile of CO_2 from fermenters. Due to five day working in the brewhouse this profile tends to peak at the weekend and drop away at the start of the week when the previous week's production has been fully fermented and the present week's production is yet to start. The dotted line shows the available recovery from brewing operations, as discussed earlier in the paper. This recovery tends to complement the profile of fermenter recovery. Breweries recovering from all sources can achieve high utilisation of the installed compressor and liquefication plant.

Figure 6 shows the balance of recovered gas with the demand for pure vapourised gas. The dashed line shows the combined recovered gas flow, while the full line shows the demand for the vapourised gas. This latter flow peaks during the weekday shifts and falls away to a very low flow at the weekends.

Thus for the majority of the working week the rate of vapourisation is generally adequate to liquefy most of the recovered gas and the refrigerated liquefier is seldom called to run. A conventional steam vapouriser can be used to supplement during the peaks in gas demand. At the weekend the rate of gas demand has fallen away and refrigerated liquefier is called to perform all the duty. A bleed flow of gas from the surge drum of approximately 15% of the recovered gas flow is required to maintain the purging of the system. The let down line is opened under the control of pressure switch P4.

The profile of operation of the fermenters does however mean that the recovered gas at the weekend is nearly 100% pure since fermentation of the week's production is well established. Thus the plant does not normally have to handle large quantities of air at the time when the vapourisation of liquid has ceased.

ECONOMICS

Running cost of the New Plant

The cost of producing liquid CO_2 from a conventional plant is around £15/tonne including power, water and maintenance cost. The new plant reduces this cost by around 10-20% by reducing the running on the liquefier and by reducing the steam demand for vapourisation.

Capital Cost of New Plant

The plant can be economically fitted to an existing recovery plant or incorporated into a new system.

The retrofitting of a 2 tonne per hour system onto an existing fermenter recovery plant can cost around £150,000 depending on site circumstances. Savings of around £80,000 pa can be expected over a well run conventional recovery pant. The marginal additional cost of recovery from other brewery operations can be around £250,000 and yield savings of around £200,000 pa.

CONCLUSIONS

The science of CO_2 recovery in breweries has not advanced significantly in many years. Low yield from conventional plants have been accepted as a fact of life. This development is seen as the way ahead for CO_2 recovery: not only in the UK but also abroad, where CO_2 pricese can be substantially higher.

TABLE 1   CO2 MASS BALANCE FOR A TYPICAL LARGE UK BREWERY

OPERATION	USAGE (TE/YR)
Cold Conditioning	900
Carbonation	800
Filter Buffering	300
Bright Beer Tanks	2,000
Pasteuriser Buffering	400
Kegging	1,500
Tankering	600
Bottling/Canning	500
Deaerated Water	400
Cylinder Filling	900
Unaccounted	1,700
TOTAL	10,000

FIGURE 1   SIMPLIFIED FLOW DIAGRAM OF MAIN BREWERY PROCESSES INVOLVING CO2 USE OR VENTING

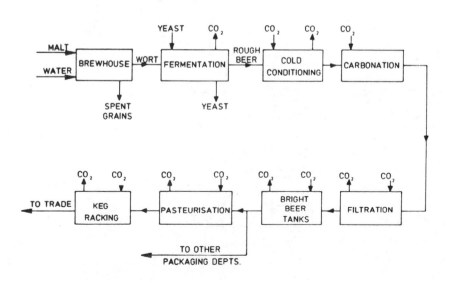

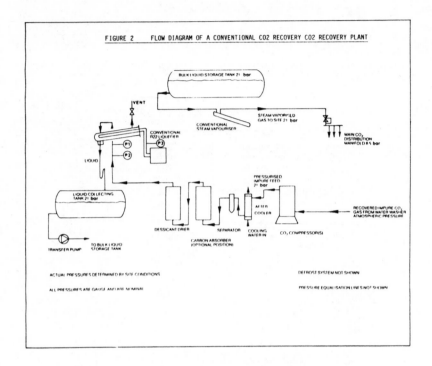

FIGURE 2   FLOW DIAGRAM OF A CONVENTIONAL CO2 RECOVERY CO2 RECOVERY PLANT

FIGURE 3   SIMPLIFIED FLOW DIAGRAM OF NEW LIQUEFIER SYSTEM

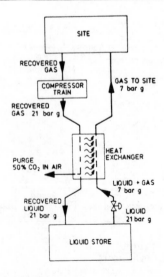

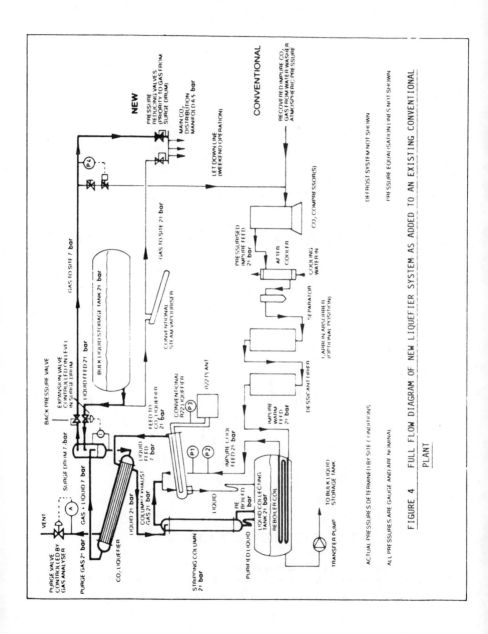

FIGURE 4   FULL FLOW DIAGRAM OF NEW LIQUEFIER SYSTEM AS ADDED TO AN EXISTING CONVENTIONAL PLANT

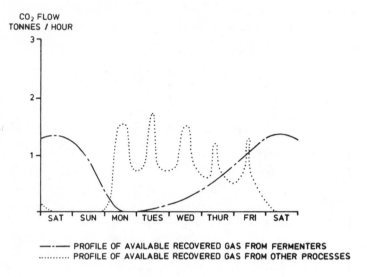

————— PROFILE OF AVAILABLE RECOVERED GAS FROM FERMENTERS
.... PROFILE OF AVAILABLE RECOVERED GAS FROM OTHER PROCESSES

FIGURE 5   FLOW PROFILES FOR CO2 RECOVERY IN A TYPICAL LARGE UK BREWERY

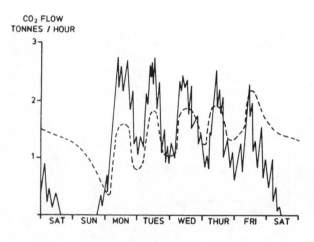

————— PROFILE OF CO_2 GAS DEMAND ON SITE
- - - - PROFILE OF TOTAL FLOW OF RECOVERED GAS FROM ALL SOURCES

FIGURE 6   FLOW PROFILES FOR CO2 RECOVERY AND SUPPLY IN A TYPICAL LARGE UK BREWERY

STUDIES ON A SUPPLEMENTARY FIRED AIR HEATER COMBINED CYCLE
FOR INDUSTRIAL CHP

M. St.J. Arnold, P.J.I. Cross, J.E. Davison, R.C. Green, T.J. Murphy *

This paper is concerned with the technical and economic
evaluation of an advanced coal-fired combined heat and
power cycle. The cycle is configured around a
circulating atmospheric fluidised bed combustor which
supplies heat for both a back pressure steam turbine
cycle and a gas turbine cycle. The gas turbine inlet
temperature is increased by supplementary firing with
fuel gas derived from partial gasification of coal in a
fluidised bed gasifier. Performance estimates for a
number of process variations have shown that attractive
power to heat ratios may be achieved at high overall
efficiency and the flexibility in terms of power to heat
ratio which accrues from the supplementary firing feature
has also been demonstrated. A preliminary assessment of
the economics of the subject cycle design is presented.

INTRODUCTION

The use of advanced coal fired options for industrial combined heat and power
generation is likely to be most favoured for technologies which give a high
power to heat ratio, because in general the value of electric power is higher
than that of heat.

One such technology of potential interest in this context is the atmospheric
pressure circulating fluidised bed boiler system with a combined air and
steam cycle which uses an air heater in the external heat exchanger. However,
conventional designs which have been proposed for this type of cycle are
normally limited to air inlet temperatures to the turbine of around 800°C, by
the availability of suitable materials of construction for the hot end of the
air heater. This constraint means that the turbine is operated in a
substantially derated condition and that the power/heat ratio of the cycle is
not optimised.

The present paper is concerned with studies of an air heater cycle,
configured around a coal-fired atmospheric circulating fluid bed boiler, in
which the turbine inlet temperature is increased by supplementary firing of
the gas turbine with fuel gas derived from partial gasification of some of
the coal feed. The gasifier considered is based on British Coal technology
and is air blown. The char from the partial gasification process is fired in
the circulating bed boiler.

The cycle is described in detail in the following section. MCR performance
estimates are presented for a 'base case' cycle and a number of process
options. The performance calculations were made using the ARACHNE

* British Coal, Coal Research Establishment, Stoke Orchard, Cheltenham

flowsheeting package developed by British Coal. The calculations are based on design assumptions appropriate to an industrial system producing about 80 MW$_{th}$ of steam with variable electrical generation. The studies consider the sensitivity of cycle performance to gas turbine inlet temperature, steam turbine inlet conditions and the use of alternative methods of fuel gas clean-up. The results of the performance calculations for the base option have been used to develop a preliminary evaluation of the economics of the system.

CYCLE DESCRIPTION

Overall definition

The cycle concept is illustrated in Figures 1 and 2 which show the gas and steam elements of the overall system respectively. The major subsystems of the cycle are as follows:-

- gas turbine generator package

- partial gasifier unit

- gas cleanup plant

- circulating bed boiler system

- steam/water circuit and back pressure steam turbine

In the 'base case' process option inlet air is compressed to ca. 13 bar A in the gas turbine compressor and heated to 800°C in air heater tubes in the external heat exchanger of the circulating bed boiler system. The heated air is forwarded to the gas turbine combustion chambers where fuel gas is fired to raise the combustion gas temperature to 1000°C before expansion through the turbine. Fuel gas required for this duty is generated from coal in an atmospheric fluid bed air blown partial gasifier with a cold gas cleanup system. Turbine exhaust gas passes to the combustor and external heat exchanger of the circulating bed boiler system where most of it is used for combustion of the char produced in the gasifier along with some additional coal. The balance of the turbine exhaust gas is ducted to the convective boiler in the circulating bed system. In the base case option the turbine exhaust flow split and the additional coal firing rate are fixed by the design assumptions of a scheme size of 80 MW$_{th}$ back pressure steam output and 18% excess air in the circulating bed unit.

The steam cycle consists of deaerator, feed pump, economiser, assisted circulation evaporator, superheater and back pressure steam turbine as indicated on the schematic, Figure 2.

Partial gasification plant

The partial gasification plant (Figure 3) is based on the British Coal gasification process (refs 1, 2) which involves fluidising a bed of char on a mixture of air and steam. A major proportion of the fluidising gas is introduced as a jet at the apex of the conical base. This promotes rapid recirculation within the bed enabling caking coals to be processed without encountering agglomeration problems. The coal is fed overbed from a lock hopper system. Limestone is added with the coal to retain sulphur which would otherwise be released in the fuel gas. The addition of limestone at the level of 10% by weight of the coal feed can retain up to 90% of the sulphur in the coal.

For the purposes of this study it is assumed that the gasifier is operated at 980°C with a bed height of 2m and produces a fuel gas with a calorific value of about 4MJ/Nm³. The raw gas contains a high proportion of carbonaceous fines which are collected in a high temperature cyclone and may be injected into the base of the gasifier using a steam driven ejector in order to improve carbon conversion. Fines present in the gas leaving the high temperature cyclone are captured in a second cyclone downstream and transferred to the combustor. A small stream of char is continuously withdrawn from the bottom of the gasifier to remove any stone or debris. This stream is also transferred to the combustor in order to achieve maximum carbon utilisation.

Circulating bed boiler system

For the purposes of the present study, design assumptions appropriate for Battelle Laboratories Multisolids Fluidised Bed (MSFB) process (ref 3), which is licensed in the UK by Foster Wheeler Power Products, have been used. The MSFB boiler, which is shown in Figure 4, consists of four main components:

- combustor vessel

- high efficiency hot gas cyclones

- external heat exchanger

- convective boiler and economiser

Virtually all combustion of the fuel takes place within the combustor, which is a vertical refractory lined vessel. In the present concept the inputs to the combustor are coal, gasifier char, gas turbine exhaust gas and limestone for sulphur retention. The operating temperature within the combustor is controlled by recycling solids which have been cooled in an external heat exchanger (EHE). Fluidising air carries the entrained bed solids through the combustor and into high efficiency hot gas cyclones. Entrained bed solids separated in the cyclones are fed to the EHE, which is a refractory lined vessel composed, in the present scheme, of three compartments separated by refractory weirs. Each compartment contains a low velocity fluidised bed and two compartments contain heat transfer tubes, which perform evaporative duty for the steam cycle and preheat the turbine inlet air to 800°C. The very low fluidising velocity and benign environment in the EHE result in very low heat transfer tube erosion and corrosion.

The flue gas stream leaving the high efficiency cyclones is cooled in a convective waste heat boiler which raises superheated steam. The coal ash, which is generally very fine soft material, remains in the flue gas stream, passes through the boiler, and is captured by bag filters before the induced draught fan and the stack.

Gas cleanup

In the base case system a cold gas cleanup system is used to remove entrained solids from the raw fuel gas before it is fired in the gas turbine. Gas leaving the partial gasifier process area is cooled to around 165°C by generating first high pressure saturated steam for the steam cycle and subsequently low pressure superheated steam for the gasifier process requirement and for deaeration. The cooled gas passes to a venturi scrubber system and the scrubbed gas is forwarded to wash vessels in which the separation of clean gas from the solids-laden water is effected. The clean gas is compressed to 16.1 bar A in a motor driven compressor and forwarded to the gas turbine area.

Gas turbine

The gas turbine assumptions used for the base case evaluation are for a conceptual machine with a pressure ratio of about 13:1, an inlet temperature of 1000°C and a compressor mass flow of about 60 kg/s, giving a full load electrical output of about 13 MW. These assumptions were selected because the air mass flow aligns reasonably with the scheme size selected for evaluation (80 MW_{th} steam output) and because the inlet temperature of 1000°C provides significant scope for supplementary firing. Selection of an actual turbine for this type of system would require a machine which could be modified relatively easily to enable air to be extracted to and returned from the AFBC air heater. Most modern gas turbines are designed to be very compact, particularly at relatively small mass throughputs, so making it very difficult to extract and return the air. Machines which could be considered for this application include the Mitsui SB60 (ref. 4) or the John Brown Frame 3 (ref. 5).

Steam cycle

In the steam cycle plant (Figure 2), process condensate returned from outside battery limits is initially deaerated at about 3 bar. Steam for deaeration is generated in the waste heat recovery train of the gas cleanup package. Deaerated feedwater passes to the electrically driven main boiler feed pump and thence to the steam drum via economiser surface located in the convective pass of the circulating bed boiler.

In the assisted circulation system, feedwater from the drum is pumped through parallel flow paths including evaporative surface in the external heat exchanger and in the gas cleanup train and screen tubes and waterwall tubes in the convective pass. Dry steam exiting the drum is advanced through the superheater surface located in the convective pass and reaches the back pressure steam turbine inlet flange at 87 bar A/410°C at MCR load. The turbine, which drives a generator through a reduction gearbox, exhausts at 14 bar/217°C and the full load steam flow of 34.2 kg/s is sufficient to develop 9980 kW at the generator terminals. The turbine exhaust steam is available at battery limits for process heating.

PERFORMANCE PREDICTIONS

Options evaluated and methodology

In order to assess the basic performance of the supplementary-fired air heater cycle described above and to explore the sensitivity of the performance to changes in process parameters, estimates of cycle efficiency have been developed for a suite of six cases comprising a 'base case' and five variations. The main features of the base case are summarised as follows:-

1. The gas turbine is operated at its design point inlet conditions of 13 bar/1000°C.

2. The partial gasification plant operates at just above atmospheric pressure and a cold clean-up system with a venturi scrubber is used to clean the fuel gas product.

3. The net process heat output, based on enthalpy flows of back pressure turbine exhaust steam and of returned process condensate, is 80 MW$_{th}$

4. The circulating bed boiler system is operated at 18% excess oxygen.

5. A consequence of (3) and (4) is that the turbine exhaust gas flow is in excess of the requirements for combustion air in the circulating bed boiler. Part of the turbine exhaust gas is therefore ducted directly to the convective waste heat boiler in the circulating bed system.

6. The steam turbine inlet conditions are 87 bar A/410°C

The design efficiency of the cycle defined in this way was determined by thermodynamic modelling of the cycle using the ARACHNE process flowsheeting package developed by British Coal (ref 6). With this package the system to be modelled is represented as a network of modules representing reactors, compressors, turbines, heat exchangers, stream splitters, etc. and the compositions of input streams and the values of parameters such as turbine efficiencies are defined. The model then determines the heat and material balance for the complete system and enables the predicted MCR performance to be derived.

The model of the cycle set up to represent the base case was readily modified to enable five step-off cases to be investigated. These cases are identified as follows:-

Case 1: In this case the gas turbine inlet temperature is reduced to 800°C, assumed to be the maximum outlet temperature achievable with the air heater in the circulating bed boiler system. In this extreme situation supplementary firing is reduced to zero and hence the partial gasification plant is no longer required.

Case 2: In this case the gas turbine inlet temperature is increased to 1200°C, resulting in an increase in the gas turbine fuel gas requirement and hence the quantity of coal fed to the gasifier.

Case 3: In this case the steam turbine inlet conditions are increased to 140 bar A/540°C.

Case 4: In this case the 80MW$_{th}$ scheme size limitation is removed and all of the gas turbine exhaust gas is used for combustion in the circulating bed boiler system. Excess air in the combustor is maintained at 18% and hence the coal feed and steam turbine output are increased.

Case 5: In this case a hot gas cleanup system is used to remove solids

from the gasifier product gas and the gasifier is operated at a
pressure of about 17 bar, to align with the gas turbine fuel gas
supply pressure requirement. Air for the partial gasifier is
sourced from the gas turbine compressor.

The results of the performance predictions for the base case and five options
are discussed below.

Performance estimates

Performance estimates for the base case and five step-off cases are presented
in Table 1, which shows the coal inputs to the gasifier and combustor and the
net outputs of process heat, gas turbine power and steam turbine power. The
efficiencies from coal to electric power and process heat are also indicated.

The base case cycle has a net process heat output of 80 MW$_{th}$ as steam at
14 bar/217°C with conjoint gross electrical generation of 23.4 MW. Net
electrical generation, after deduction of auxiliary power consumption
including the gasifier air fan, the fuel gas compressor, the combustor
induced draught fan and the main boiler feed pump, is estimated to be
19.4 MW. The base case cycle therefore, has a power to heat ratio of 0.24
and is predicted to have an overall efficiency of about 81.4%, based on the
higher heating value of the coal feed. An energy flow diagram for the base
cycle is presented in Figure 5, and illustrates the major energy exchanges
between the principal plant subsystems. The diagram indicates that the major
system loss is via the boiler stack (13% of the coal feed). Small but
significant cycle parasitics are represented by heat loss and char loss from
the partial gasifier package and by the auxiliary power consumption of the
plant.

The results for cases 1 and 2 taken in conjunction with the base case
illustrate the benefits of the supplementary fired cycle in terms of the
variations in power to heat ratio that can be achieved. Case 1 represents an
extreme with turbine inlet temperature 800°C and zero supplementary firing
and is predicted to have a power to heat ratio of 0.18 with an overall
efficiency of 84.8%. Case 2, with the turbine inlet temperature increased to
1200°C by feeding a high proportion of the coal to partial gasification, is
predicted to have a power to heat ratio of 0.31. With typical relative
values of power and heat, this increase in power to heat ratio is likely to
be sufficient to offset the reduced overall efficiency of 77.9%. Figure 6
shows these results in terms of a plot of power/heat ratio against the
percentage of the total coal feed which is processed in the partial gasifier.
Overall efficiency is also indicated on the plot.

The result for case 3 compared with that for the base case indicates the
potential for improving power to heat ratio by adopting more advanced steam
conditions. Overall cycle efficiency is predicted to increase slightly, to
81.6%, and power to heat ratio increases to 0.29.

The result for case 4 demonstrates the effect of relaxing the 80 MW scheme
size constraint adopted for the other cases. In this case the net process
heat output increases to around 118MW. The power to heat ratio of 0.20 is
significantly lower than that of the base case, but the overall efficiency is
significantly higher at 84.7%.

The result for Case 5 demonstrates that the use of hot gas cleanup in
conjunction with a pressurised gasifier enables a power to heat ratio of

0.25, similar to that of the base case, to be achieved but with an efficiency advantage of around 3 percentage points over the base case.

ECONOMICS

In order to provide a preliminary assessment of the operating economics of the supplementary fired air heater cycle with partial gasification, capital and operating cost estimates have been developed for the base case system described above. This information has then been used to estimate payback times for the system, as a function of coal price, when viewed as a competitor to a steam-only circulating fluidised bed coal fired boiler.

Capital cost

The capital investment for the base case system with 80 MW thermal output and 19.4 MW net electrical generation is estimated to be about £25 million (±30%) at November 1986 price levels. This cost is for a turnkey plant on a greenfield UK site, assuming overnight construction. It includes coal reception and storage, gasification, gas cleanup and fuel gas compression, circulating bed boiler system and ancilliary plant, gas turbine, steam turbine, civil works, engineering, construction and commissioning and a 15% contingency. A partial breakdown of the cost is presented in Table 2. It is emphasised that this is a preliminary estimate which would undoubtedly be subject to revision in the event of progressing to a detailed design. In addition, it should be noted that site-specific factors and the client's design standards can have a major impact on capital cost.

Operating costs

The annual non-fuel operating costs of the base case plant are estimated to be about £2.0 million, assuming operation of the plant for 8,000 hours per annum. The operating costs, which are shown in Table 3, include operating labour, maintenance labour and materials, overheads, limestone and ash disposal.

Discussion

The overall economic performance of a coal-fired CHP plant is clearly a function of the coal price, the capital payback time and the values of the steam and electricity produced. For the purposes of the preliminary evaluation presented here the value of electricity is taken as a variable and the value of steam is assumed to be equal to the cost of steam generated in a new coal fired circulating bed boiler with the same payback time as the coal fired CHP plant. The assumptions used for the coal fired boiler are shown in Table 4. With these assumptions, and using the operating performance, capital cost and operating cost data described above for the base case scheme, the capital payback time for the scheme can be presented as a function of coal price for various electricity values as indicated in Figure 7. Thus, with coal at £1.5/GJ and electricity valued at 3p/kWh a payback time of less than 5 years is projected for the base case scheme.

CONCLUSIONS

A preliminary study of the technical performance and economics of an advanced coal fired industrial cogeneration scheme based on an atmospheric circulating fluid bed airheater cycle with supplementary firing has been completed.

Technical performance information has been developed for a base case cycle and a number of process variations and has illustrated that attractive power to heat ratios may be achieved at high overall efficiency. The flexibility in terms of variable power to heat ratio which accrues from the supplementary firing feature has been investigated and the system has been found to be potentially capable of meeting a wide range of end user requirements in respect of power to heat ratio.

The overall economics of the type of scheme described will clearly be strongly dependent on site-specific factors and on the client's criteria for investment in energy projects. However, a preliminary economic assessment indicates that payback times of less than 5 years may be achievable for the base case cycle investigated, assuming a coal price of £1.5/GJ and an electricity value of 3p/kWh.

REFERENCES

1. R.C. Green and J.C. Whitehead, 'Gasification for the industrial market', paper presented to Coaltech 85 Conference, London, December 1985.

2. R.C. Green, N.P. Paterson and I.R. Summerfield, 'Demonstration of fluidised bed gasification for industrial applications', Energy World, July 1984, pp7-10.

3. A. Nack et al, 'Battelle MSFBC process'. Proceedings of the 5th International Conference on FBC, Volume III, 1977.

4. Mitsui Engineering and Shipbuilding Company Ltd, 'Mitsui SB60C-M gas turbine for industrial use'.

5. John Brown Engineering Gas Turbines Ltd 'Industrial gas turbines'.

6. J. Holmes et al., 'The development and application of the ARACHNE package to predict the efficiency and cost of processes to convert coal to synthetic fuels'. NCB final report on EEC research project no EHC-51-018-UK to the Commission of the European Communities, February 1984.

TABLE 1   CYCLE PERFORMANCE PREDICTIONS

Case	Base	1	2	3	4	5
INPUTS (MWHHV)						
Coal to gasifier	47.23	0	107.03	47.23	47.23	33.11
Coal to boiler	74.79	111.52	27.96	78.84	120.56	85.47
Total	122.02	111.52	134.99	126.07	167.79	118.58
OUTPUTS (MW)						
Net process heat	80.00	80.00	80.00	80.00	118.29	80.00
Gas turbine power	13.41	5.79	22.77	13.41	13.41	11.49
Steam turbine power	9.98	9.98	9.98	13.83	14.75	9.98
Gross power output	23.39	15.77	32.75	27.24	28.16	21.47
Ancillaries power	-4.03	-1.25	-7.59	-4.32	-4.29	-1.56
Net power output	19.36	14.52	25.16	22.92	23.87	19.91
EFFICIENCIES (%)						
Net electric power	15.87	13.02	18.64	18.18	14.23	16.79
Net process heat	65.56	71.74	59.26	63.46	70.50	67.47
Overall	81.43	84.76	77.90	81.64	84.73	84.26
POWER:HEAT RATIO	0.242	0.181	0.315	0.286	0.202	0.249

TABLE 2   CAPITAL COST ESTIMATE FOR BASE CASE CYCLE

Item	Cost, £M
Coal reception and storage, gasification, gas cleanup and fuel gas compression	3.2
Circulating bed boiler system and ancilliary plant	10.9
Gas turbine generator package	3.8
Steam turbine generator package and steam pipework	1.2
Civil works	2.1
Spares	0.5
Contingency	3.2
Total	24.9
Say	£25M, ±30%

TABLE 3   ANNUAL NON-FUEL OPERATING COSTS FOR BASE CASE CYCLE

Item Annual cost,	£M
Maintenance (labour and materials)	1.0
Operating labour	0.2
Overheads	0.5
Limestone	0.2
Ash disposal	0.1
Total	2.0

Note: Variable cost elements are based on 8,000 hours operation per
         year.

Table 4   EFFICIENCY AND COST ASSUMPTIONS FOR COAL FIRED BOILER

Steam output (MW_{th})	80
Capital cost (£M)	11
Thermal efficiency (%, HHV basis)	85
Maintenance, labour, overheads, limestone and ash disposal costs (£M/yr)	1
Electricity consumption (MW)	1.4

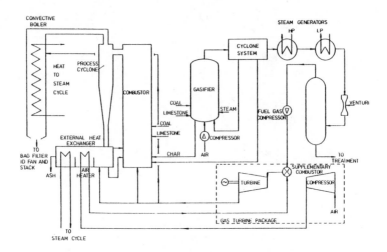

FIGURE 1. SIMPLIFIED FLOWSHEET OF GAS CYCLE

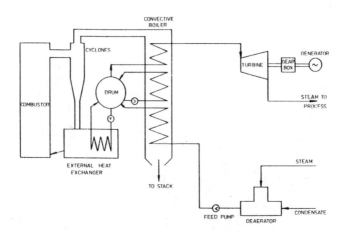

FIGURE 2. SIMPLIFIED FLOWSHEET OF STEAM CYCLE

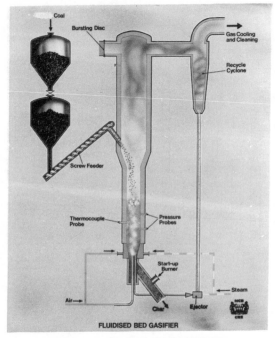

FIGURE 3. BRITISH COAL GASIFIER

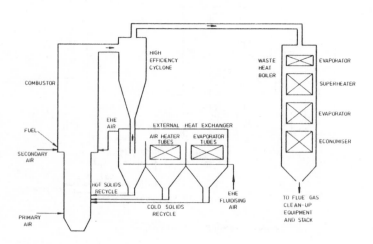

FIGURE 4 CIRCULATING BED BOILER SCHEMATIC

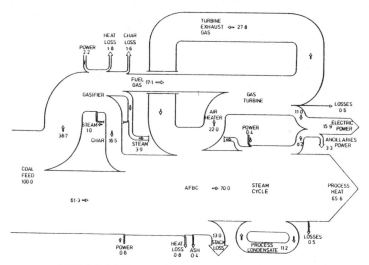

FIGURE 5. SIMPLIFIED ENERGY FLOW DIAGRAM FOR BASE CASE CYCLE

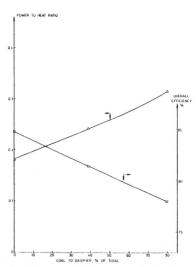

FIGURE 6 POWER TO HEAT RATIO AND OVERALL EFFICIENCY vs
% COAL TO GASIFIER (BASE CASE, CASE 1, CASE 2)

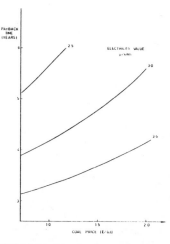

FIGURE 7 PAYBACK TIME vs COAL PRICE FOR BASE CASE SCHEME

A NEW APPROACH TO DISTILLATION SEQUENCE SYNTHESIS

A J Finn *

Previous approaches to designing optimal distillation column sequences have only considered sequences of simple columns. A procedure is presented which allows the comparative performance of different sequences to be assessed. Sequences may include simple or complex columns. The bases used for comparison are the thermodynamic efficiency at minimum column flow conditions and the corresponding lost work of separation, both determined via exergy analysis. An Underwood-based procedure is presented for determining the minimum column flow conditions for complex systems consisting of a main column linked to a side-column. The energy consumption and thermodynamic efficiency of these arrangements are compared to more conventional column arrangements.

INTRODUCTION

The problem of separating a multicomponent mixture into a number of products is solved by using a sequence of separators. Considerable effort has been expended over recent years in developing systematic methods for synthesising 'optimal' sequences i.e. those of minimum venture cost. Most synthesis studies have concentrated on separation via distillation. One reason for the enduring attraction of distillation is that it is potentially thermodynamically reversible whereas other forms of separation such as absorption or via membranes are not. Thus, despite the high energy consumption of distillation systems they can offer scope for considerable energy savings (1).

Most of the investigations into distillation column sequencing reviewed in the literature (1 - 3) have concentrated on the use of 'simple' columns which have one feed, two products and recover components to high purity levels. 'Complex column' arrangements such as columns with multiple feeds, intermediate heaters/coolers or sidedraws have been largely ignored even though they may well be attractive in terms of energy consumption. The potential for integration of the various energy loads within the column sequence has usually been disregarded. Undoubtedly, the viewpoint taken has been that the determination of the optimal column configuration is a difficult enough problem without considering the use of complex columns or energy-integration between columns.

*Costain Petrocarbon Ltd., Gateway House, Styal Rd., Manchester M22 5WN, England

The most common approach for tackling the synthesis of distillation column sequences has been by the use of heuristics – rules of thumb developed from experience. These guide the sequence designer towards the best sequences by eliminating options. This approach has been coupled with evolutionary methods to critically examine sequences for potential improvement (4 – 7). Algorithmic methods have attempted to isolate the best sequences by use of analysis and rigorous mathematical optimisation.

The heuristics were developed with the objective of minimising both energy consumption and capital cost. Their success is based on minimising the flowrates of non–key components throughout the column sequence if this is justified with regard to other concerns such as reduction of overall flowrates. It has been shown (8) that because of this basis the sequences resulting from the application of a set of heuristics not only have low internal flowrates and energy loads but less extreme condenser and reboiler temperatures. This increases the potential for energy–integration of the condensers and reboilers of the sequence. Greater energy–integration leads to lower energy consumption for the sequence.

The heuristic set used was:

1) Remove the predominant component (molar basis) first.

2) Leave the most difficult separations (in terms of relative volatility and product recovery level) till last.

3) When neither relative volatility nor molar percentage in the feed varies widely, favour the direct sequence (remove components one–by–one in the top product in order of decreasing volatility).

4) Favour equimolar splits between the top and bottom product.

The use of the heuristic approach for evaluation purposes means that a significant number of potential sequences can be readily dismissed, so isolating those that need more detailed examination. Thus, the use of heuristics assists in reducing the scope of the problem but does not necessarily identify the best single sequence.

A different approach to the synthesis problem was proposed by Tedder and Rudd (9) who undertook a comprehensive 'parametric' study of eight distillation sequences for separating ternary feeds of light hydrocarbons. Included in the eight sequences were a number of complex column arrangements including the 'thermally coupled' arrangement of Petlyuk et. al (10 – 12) which is more thermodynamically reversible than a simple column arrangement, as discussed later under 'Thermally Coupled Columns'. Tedder and Rudd concluded that the optimal sequence based on venture cost is defined by feed composition, distribution of products and the 'Ease of Separation Index' (ESI) which is a simple function of component relative volatilities. Two triangular diagrams were constructed giving the regions for which sequences would be optimal as a function of feed composition, one for ESI less than 1.6 and one for ESI greater than 1.6. These diagrams provide a valuable guide for sequence selection but their use for separations dissimilar to those considered for their development may give uncertain conclusions.

Both the heuristic approach and the parametric approach are relatively quick and easy to use but rather subjective and approximate. A more formalised procedure, which may be used with one or both methods, is needed for quantitative evaluation of distillation column sequences. This method should not exclude the possibility of using complex columns or energy–integration, must be reasonably rigorous but not difficult to use. Exergy analysis, for assessing thermodynamic efficiency, has been widely used as a guide in optimisation of processing systems including distillation. It allows the comparative performance of different column arrangements to be quantified, unlike the other methods. It is particularly valuable where energy costs dominate the overall venture cost, such as in low temperature processing where the heuristics and the parametric approach are unproven. It must be stressed that there is no simple correlation between thermodynamic efficiency and economic optimality. However, by identifying areas of significant inefficiency, exergy analysis is a major tool for improving process designs and consequently minimising venture cost.

EXERGY ANALYSIS FOR EVALUATION OF THERMODYNAMIC EFFICIENCY

As noted in a recent review (13), exergy analysis is based on evaluating thermodynamic losses using both the First and Second Laws of Thermodynamics. The convertability of energy between different forms is limited by restrictions defined by the Second Law and this enables assessment to be made of different forms and grades of energy. Exergy is the maximum potential of a system to do work or of a particular form of energy to be transformed into any other form of energy within a given environment. Exergy is often termed 'available work' or 'availability'. The environment is considered as a reservoir of zero-grade energy and a reference environment needs to be defined in terms of temperature and pressure conditions at which energy has no value.

Exergy distinguishes between energy as such and energy that is available for useful work. The exergy of a process stream at the state (T,p) is given by

$$Ex\,(T,p) = H\,(T,p) - H\,(T_o,p_o) - T_o\,[S(T,p) - S(T_o,p_o)] \qquad -- \quad (1)$$

and is the maximum amount of energy or work that can be obtained from the process stream in bringing it from the state (T,p) to equilibrium with the environment at (T_o,p_o).

The Second Law only allows energy transformations in which entropy is not reduced. In a real (irreversible) process exergy must be lost overall so that

$$\triangle Ex_{irr} = \Sigma Ex_{in} - \Sigma Ex_{out} \qquad -- \quad (2)$$

where $\triangle Ex_{irr}$ is the irreversible exergy loss in the process and

$$\triangle Ex_{irr} = \sigma\,T_o \qquad -- \quad (3)$$

The term on the right of equation 3 is commonly referred to as the 'lost work' as irreversible losses mean the equivalent potential for doing work is lost. Only in a reversible process is exergy conserved and the quantity of lost work is consequently zero. The reversible process may be considered as a 'yardstick' against which real processes can be compared, by use of exergy analysis, to quantify the irreversibility within the real process. Clearly lost work should be minimised to improve process efficiency as it represents wastage either of work input to the process or work that could have been utilised from the process.

EXERGY ANALYSIS IN EVALUATION OF DISTILLATION SYSTEMS

Real, irreversible distillation systems may be considered to effect separations by degradation of energy. Consider a simple column with a reboiler which consumes a quantity of energy Q_R and a condenser which rejects a quantity of energy Q_C.

The exergy or work value of a quantity of energy Q_i at temperature T_i is given by

$$W_i \quad = \quad Ex_i = Q_i \left[1 - \frac{T_o}{T_i}\right] \qquad -- \quad (4)$$

Clearly

$$W_R \quad = \quad Q_R \left[1 - \frac{T_o}{T_R}\right] \qquad -- \quad (5a)$$

and

$$W_C \quad = \quad Q_C \left[1 - \frac{T_o}{T_C}\right] \qquad -- \quad (5b)$$

Therefore to achieve a specified separation an amount of exergy is required as given by

$$W_{act} \quad = \quad W_R - W_C \qquad -- \quad (6)$$

The actual work consumption of the distillation system is defined as the difference between the work that could have been obtained with a reversible heat engine exploiting the energy entering the system (Carnot cycle work) and the work that could have been obtained with a reversible heat engine exploiting the energy leaving the system (1). The temperature differences in the condenser and the reboiler are ignored as these are process losses and do not influence the efficiency of the distillation system itself. For ease of evaluation, pressure drop between column feed and products is also ignored.

W_{act} is the actual work expended in achieving the separation and is equal to the minimum work requirement for the separation, W_{min} plus the irreversible exergy loss, ΔEx_{irr}

$$W_{act} \quad = \quad W_{min} + \Delta Ex_{irr} \qquad -- \quad (7)$$

The minimum work or exergy requirement to achieve the separation is given by the difference in exergy between the feed and product streams

$$W_{min} \quad = \Sigma Ex_{prod} - \Sigma Ex_{feed} \qquad -- \quad (8)$$

and this is the work or exergy requirement of the distillation process if performed reversibly.

The thermodynamic efficiency of the separation could thus be defined as

$$\eta \quad = \frac{W_{min}}{W_{act}} \times 100\% \qquad -- \quad (9)$$

Linnhoff and Smith (14) discussed this method of assessing distillation systems. By using equation 8 assessment is made relative to an 'ideal distillation' process rather than relative to an unspecified 'ideal' separation process. This is because equation 8 accounts for the temperature differences between products and feed that are unavoidable in distillation. However, the definition of minimum work of separation given in equation 8 can also result in W_{min} being negative. This shows that a reversible ideal distillation system is in fact capable of exporting useful work, depending on the boundary conditions used for the exergy analysis and especially on the exergy of the feed to the distillation system. If the minimum work of separation, W_{min} is negative then the concept of thermodynamic efficiency becomes redundant and assessment must be based on the lost work of separation, found from re-arrangement of equation 7. Lost work is in general a more useful parameter than thermodynamic efficiency.

Linnhoff and Smith considered the thermodynamic efficiency of distillation systems in detail and also account for separation of non-ideal mixtures which are not rigorously considered in our procedure.

For distillation sequences which have identical feeds to the sequence and identical product streams from the sequence, the minimum work of separation, W_{min} is identical. Therefore, by equation 7, by evaluating the actual work of separation, W_{act} for each sequence, comparison between the irreversible exergy loss (lost work) of each sequence may be made directly. The sequence with the lowest value of W_{act} has the lowest irreversible exergy loss and is also the most thermodynamically efficient. Although the thermodynamic efficiency cannot be quantified using this approach, as the value of W_{min} for each sequence is not actually known, no stream exergy values are needed. The most efficient sequence can be found from energy loads and column top and bottom temperatures alone, by use of equations 5a, 5b and 6.

A number of workers (1,2,14 - 17) have shown how reversible distillation may be achieved theoretically for a binary mixture. They have stressed that the two major contributions to irreversibility in real columns are energy degradation through the column and irreversible mixing of liquid and vapour phases which are not in equilibrium. Figure 1 shows the McCabe-Thiele diagram for reversible distillation of a binary mixture. To achieve co-incident equilibrium and operating lines the column required would need an infinite number of stages. Each stage below the feed point would have a heater and each stage above the feed point would have a cooler so that the driving forces for both heat and material transfer would be negligible throughout the column. Distillation at minimum reflux also requires an infinite number of stages but minimum reflux conditions result in significant column irreversibility so the required work of separation is not the minimum. This is due to the significant driving forces in the column arising from supplying energy solely at the bottom of the column and withdrawing it solely at the top. However, the actual energy loads required at minimum reflux are identical to those for reversible distillation (15).

Many methods have been utilised to reduce column energy degradation and mixing irreversibilities. These include intermediate column heaters or coolers, heat pumping schemes and others. The benefits from using such methods are best evaluated by exergy analysis.

Umeda et al (18) and Naka et al (19) both used thermodynamic efficiency to examine distillation column sequences. Both approaches could be useful in identifying optimal column sequences with energy-integration, particularly if combined with a heuristic-based procedure (8). However, neither considered the potential for energy saving via complex columns, though use of these is not precluded.

'THERMALLY COUPLED COLUMNS'

A more nearly reversible separation of a multicomponent mixture into its constituent components is possible when each column section removes only one component from the product of that section (1, 10-12, 20). This is the basis of the 'thermal coupling' arrangement advocated by Petlyuk (10) for ternary mixtures and shown in Figure 2. In the prefractionator a crude split is performed so that component B is distributed between the top and bottom of the column. The upper section of the prefractionator separates AB from C whilst the lower section separates BC from A. Thus, both sections remove only one component from the product of the section and this is also true for all four sections of the main column. Irreversible mixing and heat transfer effects are minimised and therefore the 'Petlyuk arrangement' is potentially a more efficient method of separating ternary mixtures than the conventional schemes – the direct sequence and the indirect sequence shown in Figures 3 and 4.

These two sequences both suffer from a similar deficiency. The first column of the sequence isolates a single component as a product which is in conflict with the above proposal for increasing the reversibility of the overall separation. With both sequences the composition of component B increases as the composition of the isolated component diminishes but then it must decrease as the composition of the other component, with which it forms the feed to the second column, increases. In the direct sequence the composition of component B increases down the column only to reach a maximum below the feed as the liquid enriches in component C. In the indirect sequence the composition of component B increases up the column only to reach a maximum above the feed as the vapour enriches in component A.

It may be questioned as to why thermally coupled columns have not been favoured in actual processing systems considering their low energy consumption and relatively high thermodynamic efficiency. Control, inflexibility to changes in operating conditions, uncertainty of benefits, uncertainty in design approach and a general fear of unproven technology are possibly some of the answers. However, the use of side-columns (Figures 5 and 6) for the separation of ternary mixtures is proven technology and it is known that these also reduce energy consumption by removing the intermediate component at approximately that point in the column where its composition is maximised. This is where this component is 'pinched' i.e. its composition in the vapour phase is equal to its composition in the liquid phase. Whereas the Petlyuk system ensures that all components are always moving in the correct direction for separation and thus mixing irreversibility is minimised, the side-column arrangement prevents the flow of the intermediate component from going in the wrong direction. Because of this the side-column arrangement may be more thermodynamically efficient than the conventional schemes.

In comparing Figures 3 and 5, whereas for the direct sequence of Figure 3 energy is supplied via reboilers at the bottom of each column, in the corresponding side-column arrangement of Figure 5, energy is supplied for both columns via the single reboiler at the bottom of the upstream column. Even if the sum of the energy loads for the two reboilers will exceed the energy load for the single one, the single reboiler supplies energy at the same temperature as the reboiler on the bottom of the downstream column in Figure 3. This is a more extreme temperature level than for the upstream column. In terms of thermodynamic efficiency, by equation 4, it can be seen that although we are reducing the value of Q_i, we are increasing the value of $(1 - T_0/T_i)$ and thus it cannot be certain whether the side-column arrangement will be significantly more thermodynamically efficient than the conventional sequence, unless exergy analysis is employed. An analogous argument applies with the condensers for Figures 4 and 6.

For a ternary feed the preferred column configuration between the direct sequence of Figure 3 or the indirect sequence of Figure 4 may be selected by use of the heuristics. However, unless the choice is obvious i.e. all the relevant heuristics indicate the same sequence as being the better one, exergy analysis should be employed to evaluate between them. Furthermore, by the ESI method of Tedder and Rudd, the thermally coupled side-stream rectifier arrangement of Figure 5 may appear more attractive than the direct sequence of Figure 3 and similarly the thermally coupled side-stream stripper arrangement of Figure 6 may appear more attractive than the indirect sequence of Figure 4. Exergy analysis helps to quantify the potential benefits. Tedder and Rudd state that the thermally coupled side-column arrangements should be evaluated if less than half of the feed to the sequence is intermediate component.

For the separation of more than three components a similar strategy to that for a ternary separation is desirable. It would be convenient if the heuristics could be used to isolate sequences of simple columns which included the most thermodynamically efficient sequences. These could then be studied further.

We earlier identified non-zero driving forces for heat and material transfer as a source of irreversibility in distillation. Allied with this, the flow of components in the opposite direction to which they should be flowing for separation is also a major reason for irreversibility, as discussed above for a ternary system. This 'reverse fractionation', as termed by King (1), is more significant for multicomponent distillation the greater the quantity of non-key components present. Therefore, the irreversibility of multicomponent distillation increases with increase in non-key component flows. Minimisation of non-key flows is the major objective of the heuristic approach and thus the heuristics clearly have a basis in improvement of the thermodynamic efficiency of column sequences. The heuristics have previously been found to give good energy-integrated sequences (8,21) without realisation as to why they did so. The link between the heuristics and thermodynamic efficiency was not appreciated. Future development work on synthesis procedures for multicomponent problems should exploit this link and use the heuristics to identify the most thermodynamically effcent sequences of simple columns. Energy saving can then be achieved by appropriate integration (21,22) with exergy analysis playing an important role in identifying where integration will be of most benefit and whether thermally-coupled arrangements will be of use.

EVALUATION PROCEDURE

A convenient and consistent basis for assessing the thermodynamic efficiency of a column system is to use the minimum internal column flows. Flows at minimum reflux can usually be determined relatively easily. The minimum flows at each stage may be determined without much more effort (20) but even this is not worthwhile for the level of evaluation proposed here.

From known feed conditions of temperature, pressure, flowrate and composition a material balance may be defined by setting product specifications based on desired component recovery levels. As the procedure is being used for the suitability or otherwise of a column arrangement it is not essential that calculations are exact and for identification of minimum column flowrates the Underwood procedure may be used or a simpler method such as that proposed by Grunberg (11). We employed a computerised procedure for determining minimum column flows (23) which also provided the column splits. Exergy analysis was carried out by performing energy balances and obtaining stream exergy values from 'in-house' software.

The conventional sequences of Figures 3 and 4 may easily be evaluated by use of the Underwood procedure. However, for evaluation of the side-columns of Figures 5 and 6, the Underwood procedure cannot be used directly as it is only applicable to single feed, two section columns. By 'visualising' the function of the different column sections in terms of the key components of the particular section, it is possible to modify the thermally coupled arrangements into two 'transformed' columns each of two sections as shown in Figures 7 and 8. The numbered column sections are analogous to those of Figures 5 and 6. Gikas (23) shows that the 'net feed' to a column can be used in the solution of the Underwood equations in identical fashion to a single feed. Considering Figure 7, the net feed to the downstream column is $(Ls_{min} - Vs_{min})$ and the liquid fraction of the feed is given by q = $\dfrac{\text{Liquid entering downstream column}}{\text{Net feed to downstream column}}$

= $\dfrac{Ls_{min}}{Ls_{min} - Vs_{min}}$ — — (9)

The flowrate and composition of $(Ls_{min} - Vs_{min})$ is obtained from the material balance around the upstream column. From the feed condition, determined by use of equation 9, the minimum internal flowrates of the downstream column can be subsequently determined. By modifying the sequence of Figure 7 back to the actual thermally coupled sequence of Figure 5 a feasible material and energy balance around the actual columns may be fully defined. A similar treatment applies for Figure 6.

The computerised procedure we employed could handle subcooled and superheated feeds which aided considerably in definition of the minimum flows for the 'transformed' columns.

EXAMPLE AND DISCUSSION

The method outlined above has been employed several times at Costain Petrocarbon and has provided some enlightening results. The minimum flow calculation method has been found to predict material and energy balances around thermally coupled sequences with sufficient accuracy for meaningful exergy analysis.

The sequence evaluation procedure and example calculations are detailed in the Appendix. The example considered is discussed in detail by Tedder and Rudd who use it to stress the energy saving and overall cost benefits of the side-column rectifier, Figure 5 over the direct sequence , Figure 3. The calculations show that reduction in energy loads for the side-column rectifier are significant but the actual work requirement of the side-column rectifier is marginally higher than that of the direct sequence. Even allowing for any slight inaccuracies in the calculations due to non-constant molal overflow and relative volatilities the thermodynamic efficiency of the two sequences are similar. As discussed earlier, energy loads are reduced by thermal coupling at the expense of greater use of extreme energy levels. For this example at least, one counter-balances the other.

For the example considered the ESI method and exergy analysis give differing conclusions but the two methods do have different objectives. The ESI method aims to select the column arrangement with the lower venture cost. As Tedder and Rudd show, this decision depends greatly upon the energy consumption of the sequences. Exergy analysis examines not only the effect of energy consumption but also the temperature level at which energy is supplied or obtained. For the example in the Appendix the fact that more energy must be supplied at the higher temperature level for the side-column rectifier arrangement is not considered by the ESI method. This is the major reason why the ESI method finds the side-column rectifier arrangement is optimal. The ESI method also assumes optimal system pressure and feed conditions which were not considered for the exergy analysis.

For some sequences the magnitude of the energy consumed by the sequence is significant and energy temperature levels are not so important. The proposed evaluation procedure can still be used to find the energy load savings obtainable from the use of thermally coupled side-column arrangements, as shown in the Appendix. However, for many sequences the temperature level at which energy is supplied or obtained is significant, particularly in low temperature processing where thermodynamic efficiency is closely linked to overall capital cost or where opportunities exist to integrate condensers and reboilers with each other or other process energy loads (22).

It would be inadvisable to derive any major conclusion from the example given or other problems we have studied other than that the use of thermally coupled side-columns can reduce energy loads compared to conventional distillation sequences. More work is needed to show whether or not significant improvement in thermodynamic efficiency or saving in lost work is possible by use of such arrangements. The evaluation procedure presented should enable easier assessment to be made. For all the problems we have investigated the heuristics and the ESI method gave a good guide as to which sequences should be compared via the evaluation procedure.

CONCLUSION

In the comparison of distillation sequences for separating multicomponent mixtures, exergy analysis has a significant role to play in aiding designers identify optimal configurations. It is especially valuable where energy costs are a high proportion of overall venture cost as it enables identification of the most thermodynamically efficient configurations and areas of inefficiency which may be improved upon. It complements the heuristic and parametric approaches to sequence selection.

Thermally coupled side-column arrangements have been shown to have low energy consumption compared with conventional sequences. Their attraction in terms of relative thermodynamic efficiency and lost work saving needs more investigation. A simple method of performing an exergy analysis to compare such arrangements with conventional sequences has been shown.

ACKNOWLEDGEMENTS

The author would like to thank a number of people for their help in developing the ideas presented, notably David Limb at Costain Petrocarbon who reviewed the manuscript and made a major contribution on the use of exergy analysis in particular, Spiros Gikas for development and programming of the short-cut distillation procedure, his supervisor, John Flower of the University of Leeds and the management of Costain Petrocarbon for allowing this project to develop and be published, particularly Mel Duckett. Thanks to the Word Processing Department for their sterling efforts in typing the manuscript. Finally thanks to Robin Smith at UMIST for initiating several of the ideas discussed.

REFERENCES

1. King, C J, 'Separation Processes', 2nd edition, McGraw-Hill, New York (1980).

2. Henley, E J and J D Seader, 'Equilibrium – Stage Separation Operations in Chemical Engineering', Wiley, New York (1981).

3. Nishida, N, G Stephanopoulos and A W Westerberg, 'A Review of Process Synthesis', AIChE J, 27, 321 (1981).

4. Seader, J D and A W Westerberg, 'A Combined Heuristic and Evolutionary Strategy for Synthesis of Simple Separation Sequences', AIChE J, 23, 951 (1977).

5. Nath, R and R L Motard, 'Evolutionary Synthesis of Separation Processes', AIChE J, 27,578 (1981).

6. Lu, C D and R L Motard, 'A Strategy for the Synthesis of Separation Sequences', I Chem.E Symp. Ser. 74 (1982).

7. Nadgir, V M and Y A Liu, 'A Simple Heuristic Method for Systematic Synthesis of Initial Sequences for Multicomponent Separations', AIChE J, 29,926 (1983).

8. Stephanopoulos, G, B Linnhoff and A G Sophos, 'Synthesis of Heat Integrated Distillation Sequences', I Chem.E Symp. Ser. 74 (1982).

9. Tedder, D W and D F Rudd, 'Parametric Studies in Industrial Distillation', AIChE J, 24, 303 (1978).

10. Petlyuk, F B, V M Platonov and D M Slavinskii, 'Thermodynamically Optimal Method for Separating Multicomponent Mixtures', Int. Chem. Eng., 5, 555 (1965).

11. Grunberg, J F, 'The Reversible Separation of Multicomponent Mixtures', Proc 4th Cryogen. Eng. Conf. (1956).

12. Stupin, W J and F J Lockhart, 'Thermally Coupled Distillation - A Case History', Chem. Eng. Prog., 68, (10), 71 (1972).

13. Kotas, T J, 'Exergy Method of Thermal and Chemical Plant Analysis', Chem. Eng. Res. Des., 64, 212 (1986).

14. Linnhoff, B and R Smith, 'The Thermodynamic Efficiency of Distillation', I Chem.E Symp. Ser. 56 (1979).

15. Fitzmorris, R E and R S H Mah, 'Improving Distillation Column Design Using Thermodynamic Availability Analysis', AIChE J, 26, 265 (1980).

16. Kayihan, F, 'Optimum Distribution of Heat Load in Distillation Columns Using Intermediate Condensers and Reboilers', AIChE Symp. Ser. 192, Vol. 76 (1980).

17. Stephenson, R M and T F Anderson, 'Energy Conservation in Distillation', Chem. Eng. Prog., 76, (8), 68 (1980).

18. Umeda, T, K Niida and K Shiroko, 'A Thermodynamc Approach to Heat Integration in Distillation Systems', AIChE J, 25, 423 (1979).

19. Naka, Y, M Terashita and T Takamatsu, 'A Thermodynamic Approach to Multicomponent Distillation System Synthesis', AIChE J, 28, 812 (1982).

20. Franklin, N L and M B Wilkinson, 'Reversibility in the Separation of Multicomponent Mixtures', Trans. I Chem. E, 60,276 (1982).

21. Minderman, P A and D W Tedder, 'Comparisons of Distillation Networks: Extensively State Optimised vs. Extensively Energy Integrated', AIChE Symp. Ser. 214, Vol. 78 (1982).

22. Linnhoff, B, H Dunford and R Smith, 'Heat Integration of Distillation Columns into Overall Processes', Chem. Eng. Sci., 38, 1175 (1983).

23. Gikas, S, 'Optimal Sequences of Distillation Columns' M.Sc. Dissertation, University of Leeds (1986).

NOMENCLATURE

A	Light component of ternary mixture	
B	Intermediate component of ternary mixture	
C	Heavy component of ternary mixture	
Ex	Exergy	kW
H	Enthalpy	kW
Lr	Liquid flowrate between 'transformed' columns, Figure 8	kmol/s
Ls	Liquid flowrate between 'transformed' columns, Figure 7	kmol/s
p	Pressure	kPa
Q	Energy load	kW
q	Feed molar liquid fraction	
S	Entropy	kW/K
T	Temperature	K
Vr	Vapour flowrate between 'transformed' columns, Figure 8	kmol/s
Vs	Vapour flowrate between 'transformed' columns, Figure 7	kmol/s
W	Work equivalent	kW
ΔEx	Difference in exergy	kW
ΣEx	Net exergy	kW
η	Thermodynamic efficiency	%
σ	Entropy change due to a process taking place	kW/K

Subscripts

act	actual
C	condenser
feed	feed stream
i	system i
in	flow into a system
irr	irreversible
min	minimum

o   reference conditions of temperature and pressure

out   flow out of a system

prod   product stream

R   reboiler

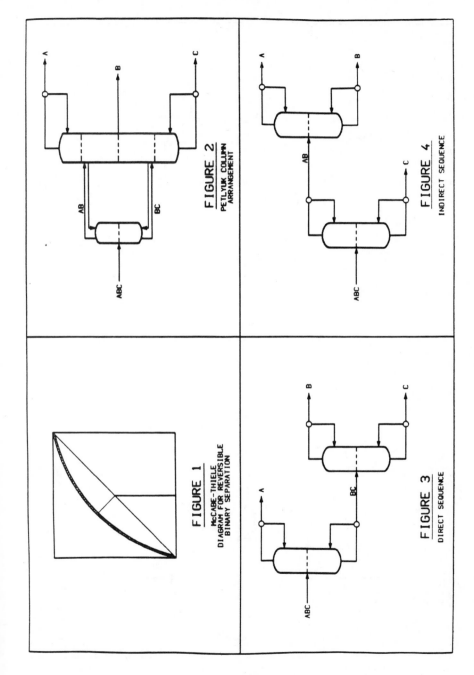

FIGURE 1
McCABE-THIELE DIAGRAM FOR REVERSIBLE BINARY SEPARATION

FIGURE 2
PETLYUK COLUMN ARRANGEMENT

FIGURE 3
DIRECT SEQUENCE

FIGURE 4
INDIRECT SEQUENCE

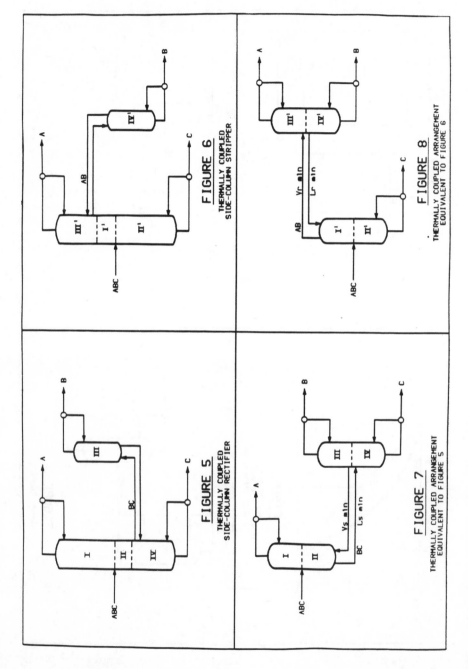

FIGURE 6
THERMALLY COUPLED
SIDE-COLUMN STRIPPER

FIGURE 8
THERMALLY COUPLED ARRANGEMENT
EQUIVALENT TO FIGURE 6

FIGURE 5
THERMALLY COUPLED
SIDE-COLUMN RECTIFIER

FIGURE 7
THERMALLY COUPLED ARRANGEMENT
EQUIVALENT TO FIGURE 5

APPENDIX

The evaluation procedure for assessment of thermally coupled side-column arrangements versus conventional sequences is illustrated by example. All vapour-liquid equilibrium and thermodynamic data were obtained from 'in-house' software. Equations referred to are from the paper. Reference temperature for calculation of work, T_0 = 293.15K.

Example

Separate into three high purity liquid products a saturated liquid feed of 80% propane, 10% i-butane and 10% n-butane.

For both sequences the same overall component recovery was used. This was at least 95%. System pressure was defined as 1200 kPa which is high enough to allow propane vapour overheads to be condensed by cooling water. A basis of 0.02778 kmol/s of feed was used.

(a) Considering the direct sequence, Figure 3:

(i) From definition of component relative volatilities and recoveries the computerised procedure predicted product flowrates and compositions for the upstream column. The key components were propane and i-butane. A minimum reflux ratio of 1.1 was calculated.

(ii) From the calculated minimum reflux ratio value the energy and material balances around the upstream column were completed to give the condenser and reboiler energy loads.

(iii) As the flowrate and composition of the feed to the downstream column were now known a similar procedure to that for the upstream column was used. The key components were i-butane and n-butane. A minimum reflux ratio of 7.8 was calculated and from this the condenser and reboiler energy loads were determined.

(iv) From the calculated energy loads and column product temperatures the actual work requirement was found for each column by use of equations 5a, 5b and 6.

		Upstream Column	Downstream Column
Condenser load	kW	655.6	400.7
Condenser temperature	K	307.6	343.9
Reboiler load	kW	670.8	401.3
Reboiler temperature	K	351.7	360.8
Actual work consumption	kW	80.8	16.1

The total actual work requirement for the sequence is therefore:

$$W_{act} = 80.8 + 16.1$$

$$= 96.9 \, kW$$

(v)   From the column feed and product compositions and flowrates the minimum work requirement could be found for the columns by use of equation 8. By use of equations 7 and 9 the lost work of the sequence and the thermodynamic efficiency of the sequence could be calculated. However, as explained in the paper, as the two sequences being compared have an identical feed and identical products the minimum work of separation is identical and comparison between the sequences can be made simply by comparison of the total actual work consumption of each sequence.

(b)   Considering the side-column rectifier, Figure 5 'transformed' to the arrangement of Figure 7:

    (i)   As with the direct sequence (a), the computerised procedure predicted product flowrates and compositions for the upstream column.

    (ii)   From the calculated minimum reflux ratio value the energy and material balances around the upstream column were completed. The magnitude of the 'net feed' to the downstream column was found by material balance around the upstream column. The liquid flow to the downstream column, Ls_{min} was found from the reflux liquid flow and by assuming constant molal overflow in the upstream column. By equation 9 the value of q was calculated as 8.5.

    (iii)   As the flowrate and composition of the feed to the downstream column were now known the computerised procedure was able to predict product flowrates and compositions. A minimum reflux ratio of 4.1 was calculated and from this the condenser and reboiler energy loads were determined.

    (iv)   As with the direct sequence (a), from the calculated energy loads and column product temperatures the actual work requirement was found for each column.

		Upstream Column	Downstream Column
Condenser load	kW	655.6	228.1
Condenser temperature	K	307.6	343.9
Reboiler load	kW	–	899.5
Reboiler temperature	K	–	360.8
Actual work consumption	kW	–30.8	135.0

The total actual work requirement for the sequence is therefore:

$$W_{act} = -30.8 + 135.0$$

$$= 104.2 \text{ kW}$$

(c)   The results given were verified by use of a rigorous equilibrium-stage model. Although calculated values of energy load and work requirement differed slightly from those presented here the actual work requirement of each sequence was as the results given within a few percent.

FLEXIBLE HEAT EXCHANGER NETWORK DESIGN: COMMENTS ON THE PROBLEM DEFINITION
AND ON SUITABLE SOLUTION TECHNIQUES

E. Kotjabasakis* and B. Linnhoff*

The design of flexible heat exchanger networks is
currently attracting a lot of research work. Most of this
work is based on complex mathematical methods to solve an
unrealistic problem, ignoring the governing energy/capital
trade-off. In this paper we show that Pinch Design Method
when applied to this kind of problem is simpler to use and
yields cheaper designs. A realistic problem definition is
then suggested which takes into account the three-way
trade-off between energy, capital and flexibility. It is
shown that the combined use of two practical design tools,
Pinch Technology and Sensitivity Tables, takes on board
these trade-offs and results in significant project cost
savings.

INTRODUCTION

Many chemical plant operators are faced with plant flexibility problems.
These range from handling different feedstocks to coping with changes in
product demand and plant throughput. De-bottlenecking can be considered as a
flexibility problem.

Most of the academic research on flexible process design has focused mainly on
heat exchanger networks. This is probably so because the overall processes are
considered too complex. In addition, temperature control is a major factor
determining overall process operability. Thus the design of flexible heat
exchanger networks can yield valuable results at less than the full
complexity.

Generally, the objective of such research (1-9) has been to find heat exchanger
networks which are "rigorously" flexible. One of the criteria used, for
example, is that network operation should be guaranteed throughout all ranges
and combinations of deviations with minimum energy requirements based on an
assumed allowable temperature approach (ΔTmin) (1). This is a challenging
objective. However it is clear that it cannot be a practical one. It cannot
be cost-effective to guarantee maximum energy recovery *for all* circumstances.
The solution of this problem, whilst being very flexible, will also require an
excessively high level of investment (10).

* Centre for Process Integration, Chemical Engineering Department, U.M.I.S.T.

In addition, little attempt is made to economically evaluate the level of flexibility demanded. This often results in designs which are too flexible from an overall cost point of view. In reality a three-way trade-off between flexibility, operating costs and capital costs exists. The cheapest design overall is that with appropriately adjusted trade-offs in these three dimensions. Thus methods aimed at flexible process design should firstly be able to evaluate the energy/capital trade-off and secondly permit easy repetition to allow the study of various flexibility levels required.

In this paper we consider the problem of designing a "flexible" heat exchanger network. More specifically we consider a plant that must operate under different specific conditions for identified periods of the year. That is, we consider a "multiperiod operation" problem.

A MULTIPERIOD OPERATION PROBLEM

In a recent publication Floudas and Grossmann (12) solved a multiperiod operation problem using a version of the mixed integer linear programming (MILP) transshipment model. In this problem a plant was required to operate under three different conditions. The problem data consisted of three individual sets of related hot and cold stream data and the operating period for each condition (Table la-c).

Floudas and Grossmann defined their objective in a similar manner to that reported above. They sought the design which gave feasible operation for each condition whilst featuring minimum capital cost given the constraint of:

* minimum utility cost for each period of operation, based on a $\Delta Tmin$ of 10 deg C.

The cost data used are shown in Table 2.

The selection of a $\Delta Tmin$ of 10 deg C was arbitrary. As a result, the energy recovery level is completely defined *regardless* of consequences (i.e. capital). No attempt was made to determine the energy/capital trade-off associated with the problem. No attempt was made to evaluate possible advantages of using different $\Delta Tmin$'s in each operating condition. This problem approach is representative of a wide research field, as mentioned in the introduction.

SOLUTION USING MILP

Floudas and Grossmann (12) extended the linear programming (LP) and mixed integer linear programming (MILP) transshipment models proposed by Papoulias and Grossmann (13) and the automatic synthesis of heat exchanger networks for one period of operation proposed by Floudas et al (14) to synthesise, automatically, flexible heat exchanger networks for multiperiod operation. Their solution to the problem presented above is shown in Figure 1. Details of the heat exchangers proposed are given in Table 3.

The solution involves the use of seven heat exchangers. Around three process-to-process exchangers there are bypasses installed. It is through the adjustment of the flows through these bypasses and the steam and cooling water flowrates, that the required flexibility is obtained.

The energy cost for this solution is 8.418 $/hr during period 1, 27.465 $/hr during period 2 and 11.033 $/hr during period 3. This is the minimum utility cost for a $\Delta Tmin$ of 10 deg C for each period of operation. The required investment cost is $269,400.

ALTERNATIVE SOLUTION USING
PINCH TECHNOLOGY AND SENSITIVITY TABLES

Two complex computer programs (LINDO and MINOS) had to be used in order to obtain the above result (12). It is also probable that a significant amount of expertise had to be employed both in the use of these programs and in the interpretation of the intermediate results obtained.

Designing for flexibility is clearly not a simple problem. That may be the reason why most of the research to date tries to solve the problem by complex mathematical formulations. However, before adopting such a technology we should ask ourselves what we can achieve using other, perhaps simpler, established techniques.

Application of Pinch Design Method

The Pinch Design Method for heat exchanger networks (15) is now extensively developed and widely used. It is generally applied to problems in which the plant can be considered to operate at, or close to, a single condition. The question raised here is "Can this method be used to solve a multiperiod operation problem?".

Let us try. Let us start by seeking to solve the problem posed by Floudas and Grossmann.

Using the Pinch Design Method we can approach the problem as follows:

First, the problem specifies that an energy recovery level associated with a ΔTmin of 10 deg C should be achieved in each operating condition. So, using this temperature approach, we determine the pinch location (15) and energy consumption for each operating case. Using this information and the pinch design rules (15) we can determine an efficient network design for each period.

Generally, in heat exchanger network designs the vital design elements are those around the pinch, where the number of options is restricted. Away from this location the designer has substantial freedom of design options.

The Pinch Design Method lays open all these design options. The engineer is given all the information he requires to choose those which will produce one design suitable for all operating conditions. Therefore, when the Pinch Design Method is used for multiperiod operation problems, the designs are not conducted in isolation of each other and there may be no need to take any individual period design to completion. The obvious thing to do is to conduct the pinch region designs for each period, looking for common structures. A simple examination of the "stream populations" in each location could prove to be all that is required in order to obtain a design which will be efficient for all operating conditions.

When we apply this approach to the example problem we obtain the design shown in Figure 2. The required degree of flexibility is achieved by installing bypasses around all process-to-process heat exchangers and by adjusting the steam and cooling water flowrates. Operating details of each period for all the exchangers are presented in Table 4.

Large parts of the network are used for all three operating conditions. Heater and coolers heat loads vary from period to period and are adjusted in the normal ways. In period 1 all the installed area on the process-to-process exchangers is utilised. During periods 2 and 3 bypasses are used, with

157

suitably adjusted flowrates, in order that the duties specified in Table 4 are obtained.

We kept ΔTmin of 10 deg C across all periods of operation in order to have a valid comparison with the solution obtained using MILP on the basis of identical energy costs, see Table 5. The required investment cost is $223,300. This is 17% lower than that achieved by Floudas and Grossmann.

A More Realistic Problem

We mentioned in the introduction that most research appears to seek to answer an incomplete problem. The object of a heat exchanger network is energy recovery. The question that the plant operator is likely to be asking is: "How do I make the best trade-off between the capital I invest and the energy recovery I make?". He is seeking to minimise the "total annualised cost" of his plant and *not* the investment cost for a fixed energy recovery level (i.e. ΔTmin). However, this requires the determination of annualised costs of energy and capital.

In order to convert the multiperiod operation problem posed above into this more appropriate one we must add two factors to the cost data in Table 2: total number of hours of operation per annum; assumed plant lifetime. A total annual operating time of 8500 hours is assumed and for the purpose of simplicity we assume that the capital cost given in the original problem is the annualised capital cost.

In the rest of this paper we aim to make two points. Firstly, we seek to demonstrate how such a problem can be handled. Then we aim to demonstrate that the application of Pinch Technology alone is not the best way!

Application of Pinch Technology and Sensitivity Tables

Above we applied the Pinch Design Method to a flexibility problem. However, we never truly utilised its power. To be comparable we pre-specified the energy recovery level. Pinch Technology however has reached a stage of development in which targets can be set for both energy consumption and capital cost (16,17). We can use it to evaluate the trade-off between capital and energy and so minimise the total annualised cost. "Can we use it to answer our 'more realistic' problem?".

Where only one operating case is involved Pinch Technology can be used to accurately determine the energy/capital trade-off for a process ahead of any design activity (16,17). Unfortunately, where more than one operating case is to be considered the determination of this trade-off becomes more difficult for it becomes a function of network design.

We need a tool which allows us to evaluate the energy/capital trade-offs associated with operating different cases on a fixed network structure. The recently introduced Sensitivity Tables (11) provide such a tool.

A combination of Pinch Design Methodology and Sensitivity Tables allows us to answer our problem. The strategy is:

1. Use Pinch Technology to obtain alternative efficient designs
2. Use Sensitivity Tables to guarantee the required flexibility and to minimise the total annualised cost of the plant.

Pinch Technology is very versatile. It can be used in a number of ways.
Several design strategies can be envisaged. In the solution of this problem
we have tried just one, one of the simplest.

A pinch analysis of the three cases indicates that the most difficult problem
is that associated with period 1. Here we have both the more complex design
problem and the highest investment cost. Consequently, we choose to design a
network for this case. The starting point for the design is the optimal
energy-capital trade-off for period 1.

Having obtained an optimal design for period 1 we use Sensitivity Tables to
make it suitable for the other operating conditions. We seek to maximise the
use of installed capital. Details of how Sensitivity Tables are used can be
found elsewhere (11).

The result is shown in Figure 3. This design requires six heat exchangers,
one less than the two previous solutions. The operating details of all
exchangers proposed are shown in Table 6. Examination of Table 6 shows that
there is a better area utilisation than that for the solution obtained using
the Pinch Design Method (compare Table 4). To obtain the duties specified,
we need to install bypasses around all process-to-process exchangers, but
their use is moderate compared to that in Table 4. This is the result of
using Sensitivity Tables for maximum area utilisation between different
operating conditions.

The total annualised cost for this design is $327,000. This is made up of
$144,100/annum for energy and $182,900/annum for capital costs. There is a
reduction of 8% in total cost relative to the solution obtained using Pinch
Design Method on its own (see Table 7).

DISCUSSION

The figures presented in Table 5 clearly indicate that the already
established heat exchanger network technology can be used on the type of
problem currently being tackled by researchers developing techniques for the
solution of flexibility problems. It is simpler to apply, easier to
understand and yields acceptable solutions.

The real problem facing the plant operator is not being addressed by most of
the current research workers. The problem perceived by the practical designer
involves the determination of the minimum total cost per year, not just
capital investment or energy cost. The solution we found (i.e. Solution 3)
clearly demonstrates the benefits of treating the problem this way. In this
solution more energy is consumed but a lower capital investment is required.
The result is a lower annualised total cost.

In obtaining this solution we have made use of two practical process
engineering tools: Pinch Technology; and Sensitivity Tables. The result is
a practical methodology and a practical design. We cannot claim "global
optimality". But therein lies the challenge - "Can you do better?". The
example is fully documented for the reader to attempt his own solution.

For the solution of this multiperiod operation problem we considered that the
degree of flexibility demanded by the plant has been stated in the problem.
This implies that we must have that amount of flexibility at "any price". Is
this realistic? If the cost of the flexibility was found to be high wouldn't
the question be asked "Do we really require this amount of flexibility?".

We shouldn't just ask this question when forced to. We should recognise that at all times we have a "three-way trade-off" between operating costs, capital costs and flexibility requirements. In some cases it has been found that just a small reduction in flexibility requirement has resulted in significant cost savings (11).

The combined use of Pinch Technology and Sensitivity Tables is sufficient for the practical evaluation of this complex trade-off.

ACKNOWLEDGEMENTS

Thanks are due to the Bodossakis Foundation, The Onassis Foundation and the Process Integration Research Consortium, who supported Mr Kotjabasakis's research at UMIST. BASF, Union Carbide, and Dow helped this work greatly by providing meaningful industrial case studies. Special thanks are due to Dr G T Polley for his help and constructive criticism in preparing the manuscript.

REFERENCES

1.   Marselle E. F., Morari M. & Rudd D. F., 1982, Chem. Eng. Sci., 37(2):259

2.   Saboo A. K. & Morari M., 1984, Chem. Eng. Sci., 39(3):579

3.   Townsend D. W. & Morari M., 1984, AIChE National Meeting, San Francisco, California, USA.

4.   Grossmann I. E. & Floudas C., 1985, paper no. 51, Process Systems Engineering Conference, PSE '85 I Chem E Symposium Series No. 92, 619

5.   Swaney R. E. & Grossmann I.E., 1985, AIChEJ 31(4):621

6.   Stephanopoulos G., 1985, AIChE Spring Meeting, Session 88, Houston, Texas, USA

7.   Beautyman A. C. & Cornish A.R.H., 1984, 1st. U.K. National Heat Transfer Conference, IChemE Symp. Ser. No. 86, Vol. 1:547

8.   Parkinson A. R., Liebman J. S., Pederson C. O. & Templeman A. B., 1982, AIChE Symp. Ser. No. 214, Vol. 78:85

9.   Halemane K. P. & Grossmann I.E., 1983, AIChEJ, 29:425

10. Linnhoff B. & Smith R., 1985, paper no. 24f, AIChE Spring Meeting, Houston, Texas, USA

11. Kotjabasakis E. & Linnhoff B., 1986, Chem. Eng. Res. Des., 64:197

12. Floudas C.A. & Grossmann I.E., 1986, ASME Winter Meeting, Anaheim, California, USA. ASME AES Vol. 2-1:75

13. Papoulias S.A. & Grossmann I.E., 1983, Computers and Chemical Engineering 7: 707-721

14. Floudas C. A., Ciric A. R., Grossmann I.E., 1986, AIChEJ, 32: 276-290

15. Linnhoff B. & Hindmarsh E., 1983, Chem. Eng. Sci., 38(5):745

16.   Linnhoff B. & Ahmad S., 1986, ASME Winter Meeting, Anaheim, California,
      USA.   ASME AES-Vol. 2-1:1

17.   Ahmad S. & Linnhoff B., 1986, ASME Winter Meeting, Anaheim, California,
      USA.   ASME AES-Vol. 2-1:15.

TABLE 1

The stream data for the example Multiperiod operation
problem as proposed in ref (12).

PERIOD 1 (1/3 of total time)				
Stream No	TS (°C)	TT(°C)	CP (kW/°C)	
1	249	100	10.55	(a)
2	259	128	12.66	
3	96	170	9.144	
4	106	270	15.00	

PERIOD 2 (1/3 of total time)				
Stream No	TS (°C)	TT (°C)	CP (kW/°C)	
1	229	120	7.032	(b)
2	239	148	8.44	
3	96	170	9.144	
4	106	270	15.00	

PERIOD 3 (1/3 of total time)				
Stream No	TS (°C)	TT (°C)	CP (kW/°C)	
1	249	100	10.55	(c)
2	259	128	12.66	
3	116	150	6.096	
4	126	250	10.00	

TABLE 2

Cost data for the example problem as proposed in ref (12).

Cost of Heat Transfer Area

Installed cost (\$) : 4,333 (A)$^{0.6}$

A : area of exchanger in m^2

Cost of Utilities

*   Steam, 300°C

      Cost : 171.428 x 10^{-4} \$/kWhr

*   Cooling water : inlet temperature : 30°C

                        outlet temperature : 50°C

      Cost : 60.576 x 10^{-4} \$/kWhr

Film Heat Transfer Coefficients

*   Process Streams : 2.0 kW/m^2 °C

*   Steam   : 1.33 kW/m^2 °C

*   Cooling water   : 0.5 kW/m^2 °C (*)

                      : 0.353 kW/m^2 °C (**)

(*)   if matched with stream No 1

(**)   if matched with stream No 2

TABLE 3

Surface area installed and utilised for each heat exchanger for SOLUTION 1 (Figure 1).

Exchanger	PERIOD 1		PERIOD 2		PERIOD 3		Area Installed (m²)
	Duty (kW)	Area Utilised (m²)	Duty (kW)	Area Utilised (m²)	Duty (kW)	Area Utilised (m²)	
1	126.6	11.765	0.0	0.0	0.0	0.0	11.765
2	676.7	20.15	676.7	20.15	207.3	15.8	20.15
3	817.9	123.44	89.8	61.8	200.0	123.44	123.44
4	1177.1	54.84	768.0	13.1	1030.0	41.0	54.84
Heater	338.4	10.6	1602.1	28.45	10.0	0.3	28.45
Cooler 1	77.4	3.1	0.0	0.0	1164.7	26.7	26.7
Cooler 2	354.8	11.6	0.0	0.0	628.5	18.67	18.67
Total Area		235.5		123.5		225.9	
Total Area Installed							284.02

TABLE 4

Surface area installed and utilised for each heat exchanger for SOLUTION 2 (Figure 2).

Exchanger	PERIOD 1		PERIOD 2		PERIOD 3		Area Installed (m^2)
	Duty (kW)	Area Utilised (m^2)	Duty (kW)	Area Utilised (m^2)	Duty (kW)	Area Utilised (m^2)	
1	126.6	11.8	51.3	0.7	100.0	9.1	11.8
2	1197.0	66.4	766.5	32.0	678.0	24.6	66.4
3	798.0	23.7	40.0	0.3	452.0	11.7	23.7
4	676.7	27.2	676.7	12.3	207.3	2.8	27.2
Heater	338.4	10.6	1602.1	28.6	10.0	0.3	28.6
Cooler 1	375.0	12.1	0.0	0.0	894.0	22.6	22.6
Cooler 2	57.6	2.1	0.0	0.0	899.0	24.6	24.6
Total Area Utilised		153.9		73.9		95.7	
Total Area Installed							204.9

TABLE 5

Comparison between SOLUTION 1 as given in ref. (12) and SOLUTION 2
as obtained using the Pinch Design Method.

	Energy Cost	Total Area Installed (m^2)	Capital Investment ($)
SOLUTION 1	Minimum for $\Delta T_{min} = 10°$	282.02	269,400
SOLUTION 2	Minimum for $\Delta T_{min} = 10°$	204.9	223,300
SOLUTION 2 as a percentage of SOLUTION 1	—	72%	83%

TABLE 6

Surface area installed and utilised for each heat exchanger for SOLUTION 3 (Figure 3).

Exchanger	PERIOD 1		PERIOD 2		PERIOD 3		Area Installed (m^2)
	Duty (kW)	Area Utilised	Duty (kW)	Area Utilised	Duty (kW)	Area Utilised (m^2)	
1	0.0	0.0	0.0	0.0	0.0	0.0	0.0
2	1120.9	42.2	766.5	32.0	722.2	42.2	42.2
3	819.9	19.2	91.5	0.8	518.3	19.2	19.2
4	676.7	20.2	676.7	12.3	207.3	2.7	20.2
Heater	519.3	14.5	1602.1	28.6	0.0	0.0	28.6
Cooler 1	451.1	14.0	0.0	0.0	849.9	21.9	21.9
Cooler 2	161.9	5.7	0.0	0.0	932.9	25.3	25.3
Total Area Utilised		115.8		73.6		118.3	
Total Area Installed							157.4

TABLE 7

Comparison between SOLUTION 2 and SOLUTION 3.

	Energy Cost ($/yr)	Capital Cost ($/yr)	Total Cost ($/yr)
SOLUTION 2	132,900	223,300	356,200
SOLUTION 3	144,100	182,900	327,000
SOLUTION 3 as a percentage of SOLUTION 2	108%	82%	92%

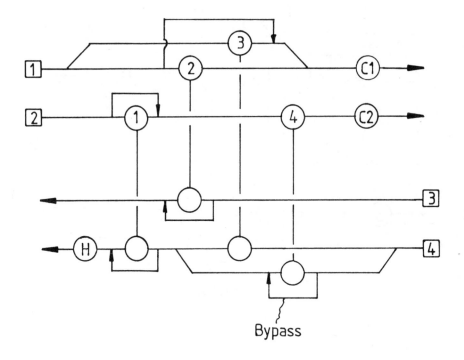

<u>FIGURE 1</u>

"SOLUTION 1", "the optimal configuration" as given by Floudas and
Grossmann (12). The flexibility required is met by the use of
bypasses (see Table 3). The minimum approach temperature (ΔT_{min}) is
kept at 10 deg C for all periods of operation and maximum energy
recovery is demanded. This effectively "freezes" the energy cost to
a pre-set value and the objective then is to minimise capital cost.
The capital cost for this solution is $269,400.

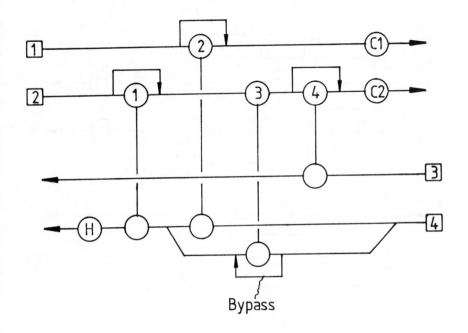

FIGURE 2

"SOLUTION 2" obtained using the Pinch Design Method. ΔT_{min} of 10 deg C is maintained for all periods of operation to enable a valid comparison with SOLUTION 1. The capital cost is reduced by 17%, see Table 5.

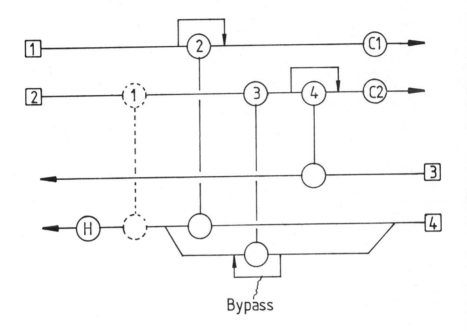

FIGURE 3

"SOLUTION 3" obtained using Pinch Technology and Sensitivity Tables.
Energy and Capital are allowed to "float" for minimum total annualised
cost across all periods of operation. Total cost is reduced by 8%
relative to SOLUTION 2, see Table 7.

HYDROCARBON PROCESSING CASE STUDY

A J O'Reilly*

Modern synthesis methods have been applied to an extant,
large scale hydrocarbon processing plant to achieve
maximum energy recovery at minimum capital investment.
The study has demonstrated how existing plant
constraints such as maximum safe column working pressure
and site utilities price structure have a significant
effect on determining retrofitting strategy and pay-out
time.

INTRODUCTION

This study has been undertaken to assess the scope for maximum energy
recovery with minimum capital investment, on an extant, large scale
hydrocarbon processing plant. The plant is of 1960's design, indicating that
considerable scope should exist for realising significant energy savings
using modern synthesis methods (1).

The process consists of feed purification prior to product formation,
followed by separation of LPG and final product purification. See figure 1
for the Base Case flowsheet.

Constraints such as existing plant layout, forced and forbidden stream
matches have by a process of elimination identified the stripper, light and
heavy HC separation columns upstream of reactor 2 for in-depth study
together with the LPG column. We observe that multiple utilities (fuel,
steam, hot oil) are required in this section of the plant. The quantities
of these utilities consumed here are 26.9, 93.3 and 44.1 GJ h⁻¹
respectively.

Selection of ΔT_{MIN}

Application of the procedure developed by Ahmad and Linnhoff (2) has shown
that a global ΔT_{MIN} of 10°C is a satisfactory value for optimal energy
recovery. See figure 2.

Analysis of Base Case Flowsheet

Figure 3 shows the composite T - H plots for the Base case flowsheet of
figure 1. Where necessary, stream data has been "safe side" linearised (1).

*Department of Chemical Engineering, Teesside Polytechnic, Borough Road,
Middlesbrough, Cleveland TS1 3BA.

The net hot utility is 127.7 GJ h^{-1}, a saving of 36.6 GJ h^{-1} over the original flowsheet, or about 25%. This is a substantial energy saving but a feasible network must be devised to determine how much of it can be realised.

We begin by considering the relatively significant quantity of steam that this section of plant consumes. See introduction. Since steam cost is greater than either hot oil or primary fuel cost (see appendix 1), then the appropriate strategy will most likely be to reduce steam consumption or exchange steam for one of the other sources of hot utility, or both.

We then observe in figure 3 that when the T-H intervals representing the three higher temperature reboiler streams are successively removed from the cold composite curve, the MER hot utility target is reduced in increments equal to these intervals. See dashed portion of figure 3 cold composite curve.

We now see from the modified cold composite curve that the appropriate strategy is heat exchange between reactor product and the light HC column reboiler, down to minimum driving force. Figure 4 shows the final "relaxed" version of the network for implementing this strategy. Energy relaxation and elimination of stream splits have, however, reduced the potential savings from 36.6 GJ h^{-1} to about 9.6 GJ h^{-1}. Although this result is disappointing, the scale of the plant is such that even this small residual saving yields £342,000 pa in reduced utilities costs, with a pay back of less than 12 months. See table 1 for details, noting that the savings accrue from a net reduction in STEAM cost, even though it is necessary to burn MORE fuel in the charge heater - and hence install more heater capacity.

The network of figure 4 has the advantage that it would cause minimum disturbance to existing plant-layout - essentially the addition of marginal extra charge heater capacity and the extra light HC column reboiler, whose duty is consistent with the ΔT limitation for nucleate boiling*. The existing charge preheater capacity is almost sufficient for the new configuration, requiring only the small additional increment in capital given in table 1. We note that the net utilities saving corresponds to the extraction of additional heat from the reactor product stream, lowering the temperature into the product cooler by about 18°C.

---

*If one assumes forced circulation for the light HC column reboiler, or a pump around with hot oil, then a separate calculation shows that by supplying the extra hot utility (29.59 GJ h^{-1}) as fuel to the charge heater, additional savings of £331000 pa are realised with a pay-out time of 15 months minimum. Whilst this may be a feasible development of the network in figure 4, the heavy HC column pressurisation scheme is considerably more attractive (see later).

Improved Flowsheet by Column Pressure Shift

Further inspection of figure 3 reveals that additional energy savings will
be possible if the heavy HC column can be pressurised so that its OH
condenser is able to supply reboiler duty for one of the other columns. The
ΔT range for the heavy HC OH condenser is extremely wide because this column
operates on subcooled reflux. This is necessary to prevent excessive loss
of reactor feed by venting to flare from the reflux accumulator. Any
strategy aimed at saving utility by a column pressure shift therefore
immediately implies a need for more sophisticated control hardware to
prevent additional reactor feed loss, though hopefully this constraint
should not be too onerous. However, mechanical design of the heavy HC
column must be such that it can satisfactorily operate at the higher
pressure envisaged (4 bar). Fortunately the plant management has been able
to give this assurance but this feature of the problem serves to illustrate
how influential existing plant constraints can be with respect to any
retrofitting strategy.

Figure 5 shows that a saving of 72 GJ h^{-1} or about 44%, is possible over the
Base case flowsheet, by means of the heavy HC column pressure shift. The
relaxed network (see figure 6) achieves recovery of about 56 GJ h^{-1}, or
about 78% of the maximum possible saving. As in the scheme of figure 4,
reactor product heat is utilised to supply reboiler duty, in this case for
both stripper and light HC column reboilers. The "network interactive"
effect of the heavy HC column pressure shift greatly enhances this
utilisation, increasing the distillate and hence reformer feed temperatures
considerably, see figures 6 and 7. See also figure 7 for the flowsheeted
version of the network. Obviously the strategy of figures 6 and 7 has
considerably greater capital implications than that of figure 4, although
the need for extra furnace capacity is avoided, but the magnitude of the
utilities savings is such that a very encouraging payout time is achieved
(about 2 months). See table 1, noting the advantageous "trade off" between
steam and hot oil.

A refinement of the above scheme entails further utilisation of reactor
product heat to preheat the heavy HC column feed, thus reducing the reboiler
duty. Table 1 shows the economic implications, indicating that this
additional strategy may just be feasible.

CONCLUSIONS

1.   Application of modern synthesis methods to this large scale plant has
      yielded significant energy savings with acceptably low pay out times,
      particularly so in the case of the column pressure shift strategy of
      figures 6, 7. Indeed, this study has provided further demonstration
      that intelligent use of systematic methods will guide the designer to
      the "best possible" flowsheet for energy recovery at minimum capital
      investment, according to the criteria for optimal ΔT_{MIN}. One should
      of course sound a note of caution in so far as all flowsheet synthesis
      strategies prompted by systematic methods are subject to "real life"
      constraints, such as column working pressure limitations. This is
      especially true for retro-fitting studies.

2.   The study has demonstrated how the price structure of site utilities
      may exert a profound influence on the strategy, savings and pay out
      times achieved for plant retrofits as suggested by systematic methods.

The drop in oil prices for example, over the past 12 months, although not proving excessively disadvantageous for the spectacular gain of the column pressure shift strategy, nevertheless almost doubles the pay back times of all schemes discussed in this study*.

*including the variant of figure 4 in which all hot utility is supplied as fuel to the charge heater.

REFERENCES

1. Process integration for the efficient use of energy, I.Chem.E User Guide, 1982.
2. Overall cost targets for heat exchanger networks. S Ahmad and B Linnhoff. First UK National Heat Transfer Conference, Leeds, 3 - 5 July, 1984.
3. Current costs of process equipment. Hall et al. Chemical Engineering, 5 April, 1982.

APPENDIX I

Capital and Utilities Cost Data Adjusted to Late 1986 Figures

Utilities	£/GJ
Steam	3.0
Fuel	1.6
Hot oil	1.85
Electricity	7.8

Capital (1,3)

Fired Heater:
(field erected)

$X \equiv$ fired duty, GJ h^{-1}

$C = 2.69*10^4 *X^{0.8}$ $\qquad X \leq 30$

$C = 2.30*10^4 *X^{0.85}$ $\qquad X \geq 30$

Heat Transfer PEC

a) Shell and Tube Equipment $\qquad A \equiv m^2$

$C = 3140* A^{0.219}$ $\qquad 5 < A < 30$

$C = 1180* A^{0.506}$ $\qquad 30 < A < 200$

$C = 333* A^{0.745}$ $\qquad 200 < A < 600$

b) Air Cooled Condensers $\qquad A \equiv m^2$ bare tube area

$\qquad\qquad\qquad\qquad\qquad\qquad N \equiv$ No bundles

$C = 1350* N * A^{0.872}$ $\qquad A \leq 10$

$C = 2760* N * A^{0.561}$ $\qquad A \geq 10$

c) Installation Factor = 3.5

TABLE 1

Cost Comparison: Base Case vs. Modified Flowsheets

(£/yr) Case

*Energy	BASE FIGURE 1	FIGURE 4 (vs.Base)	FIGURE 6 FIGURE 7 (vs.Base)	Heavy HC feed preheat (vs.FIGS 6,7)
Fuel	344000	457000	344000	344000
Hot Oil	652000	652500	1217000	1159000
Steam	2238000	**1797000	-	-
Electricity (fin fan Motors)	67000	53000	48000	48000
Total	3301500	2959000	1609000	1551000
Incremental saving		342000	1692000	58000

*Utilities affected

Capital PEC

Extra furnace capacity, extra (installed cost)		157000	existing	
Extra charge preheater capacity,		4 shells 6500 available	3 shells utilised	
reactor product exchangers:				
- light HC column reboiler		34300	13800	11000
- recycle gas preheat			18100	13900
- stripper reboiler			spare preheater shell utilised	
- heavy HC column feed				9100
Heavy HC column OH/light HC column reboiler			49000	
Total PEC (omitting furnace capacity)		40800	80900	34000
Total installed cost		300000	283200	119000
Pay back (months)		Approx 11	Approx 2	Approx 24

**See text and figure 4 for cost implications of replacing all steam to light HC column reboiler with fuel to fired heater.

ACKNOWLEDGEMENT

I wish to thank Bill Townsend, Don Yeaman and Mike Cooke, all of ICI plc, for their invaluable assistance throughout this study.

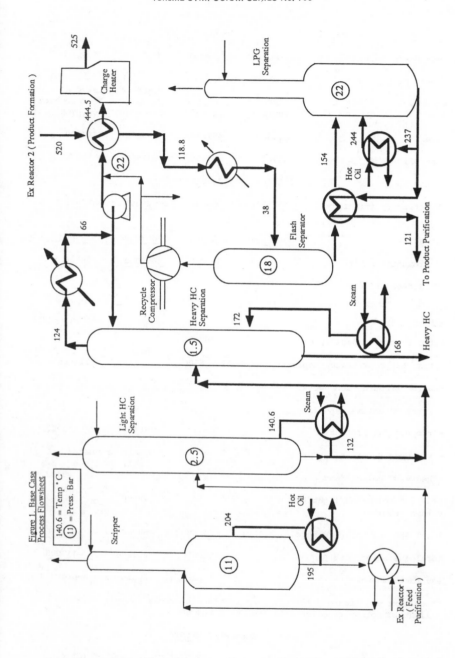

Figure 1. Base Case
Process Flowsheet

140.6 = Temp ° C
⑪ = Press. Bar

Figure 2 Selection of ΔT$_{MIN}$

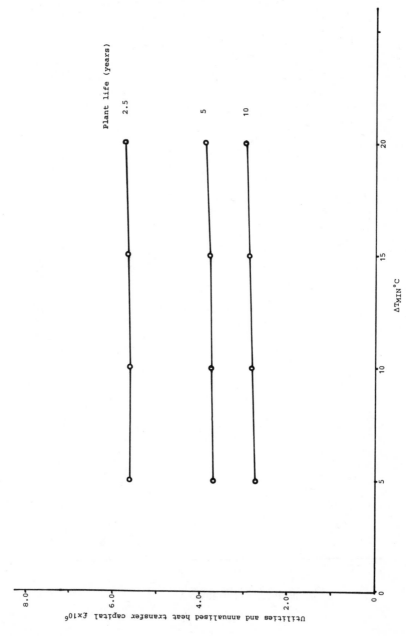

Plant life (years)

2.5

5

10

ΔT$_{MIN}$°C

Utilities and annualised heat transfer capital £x10^6

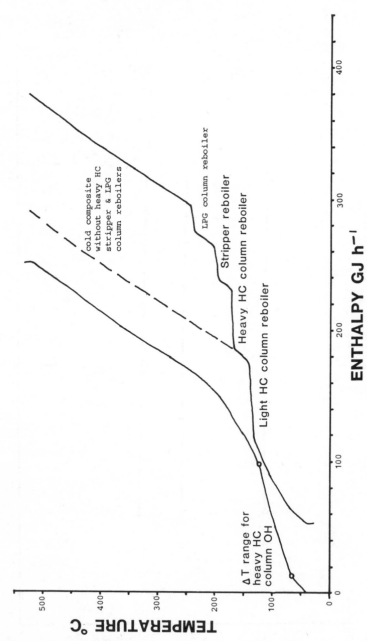

Figure 3 Individual T-H plots for "Base case" flowsheet.

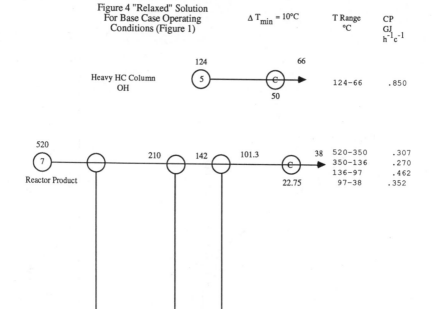

Figure 4 "Relaxed" Solution For Base Case Operating Conditions (Figure 1)

$\Delta T_{min} = 10°C$

T Range °C   CP GJ $h^{-1}c^{-1}$

Note :

i) LPG, stripper and heavy HC column reboilers are unaltered see Figure 1

ii) The bottoms-feed matches for the stripper and LPG columns are most appropriately left as individual subsets. See Figure 1

iii) If ALL Light HC column reboiler duty is supplied by reactor product, then additional energy savings of £331,000 p.a. are realised, with a minimum pay-back of 15 months for additional fired heater capacity.

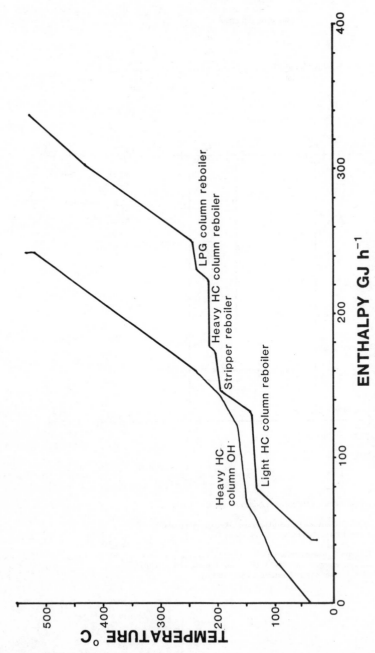

Figure 5 . Individual T–H plots for column pressure shift

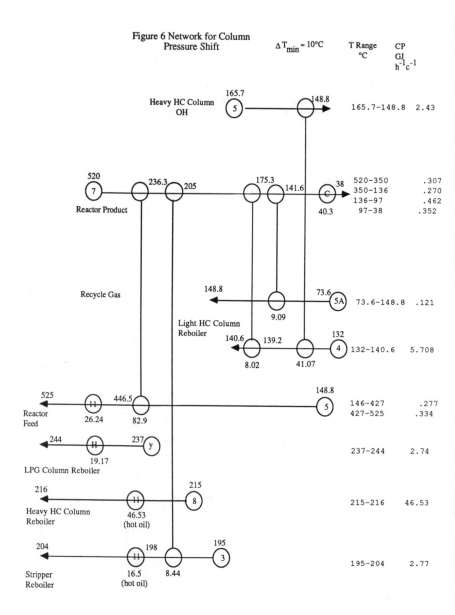

Figure 6 Network for Column Pressure Shift $\Delta T_{min} = 10°C$

T Range °C	CP GJ $h^{-1}c^{-1}$
165.7–148.8	2.43
520–350	.307
350–136	.270
136–97	.462
97–38	.352
73.6–148.8	.121
132–140.6	5.708
146–427	.277
427–525	.334
237–244	2.74
215–216	46.53
195–204	2.77

Heavy HC Column OH

Reactor Product

Recycle Gas

Light HC Column Reboiler

Reactor Feed

LPG Column Reboiler

Heavy HC Column Reboiler

Stripper Reboiler

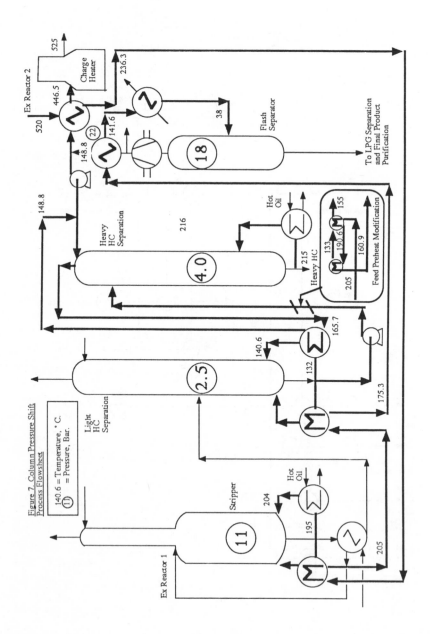

Figure 7 Column Pressure Shift Process Flowsheet

140.6 = Temperature, °C.
⑪ = Pressure, Bar.

ENERGY AND PROCESS INTEGRATION IN CONTINUOUS AND BATCH PROCESSES

I C Kemp and E K Macdonald*

Energy and Process Integration techniques have been developed from the original work on heat exchanger networks in continuous processes, to include the other plant sections – reaction, separation, heat and power systems – giving Total Process Plant Analysis. Extension to batch processes involves a third variable – time – in addition to temperature and heat flow. Time intervals analogous to temperature intervals are set up, and a "time pinch" can be identified. A time-dependent heat cascade is generated which gives comprehensive energy use and storage targets for single, repeated or parallel batch processes.

CONTINUOUS PROCESSES

Introduction

In the few years since it was introduced, Process Integration has become established as one of the most important design and analytical tools available to the process engineer. Although considerable experience is still required to apply them effectively and avoid the pitfalls, the concepts are elegant and simple. The major advance was that, for the first time, an engineer could find out exactly how much energy his plant should be using – a scientifically-based and rigorous, yet practical, target for his process, rather than a theoretical thermodynamic figure or a mere "experience value". He was also given the critical point for process design – the pinch – across which no heat should be transferred and from which design of a heat exchanger network should be started.

The heat flows in the plant – for hot streams giving up heat and cold streams accepting it – are represented by means of the hot and cold Composite Curves (Fig.1)[1]. The pinch is the point of closest temperature approach between the composite curves. The two curves can be combined into one – the Grand Composite Curve (G.C.C.) – by plotting the horizontal distance between the lines (with a temperature correction). This gives the net heat flux from hot sources (hot utilities and streams) to cold sinks (cold streams and utilities) throughout the process, and the pinch shows up clearly as the point where the graph touches the temperature axis, there being no net heat flow at the pinch by definition.

The graph can also be used to illustrate the fact that if any heat flow occurs across the pinch, the hot and cold utility use both increase by that amount; this is illustrated by the construction in Fig.2. The right-hand axis on the graph is for the conventional G.C.C. which corresponds to the composite curves

*Atkins-EPI (Energy and Process Integration Service), Process Engineering Research Centre, Building 488.T6, Harwell Laboratory, Didcot, Oxon, OX11 ORA.

in Fig.1: the flowsheet of the simple process from which these curves are derived is Fig.3. The construction shows the new (left-hand) axes if maximum heat recovery is not achieved, e.g. if the hot stream between 225°C and 150°C is not utilised for process heating, which leads to 0.5 GJ/hr of useful above-pinch waste heat being transferred across the pinch and thrown away in cooling water, and to an increase of 0.5 GJ/hr in hot and cold utility use, as shown. In other words, the construction shows clearly how far the process is off-target, which is not shown by the normal composite and grand composite curve plots.

Concepts and Process Changes

The most important achievement of Energy and Process Integration has not been to add to the already comprehensive variety of detailed design techniques for specific units and types of equipment, but to give a clear overview and show how the plant should be structured thermally.

There is a danger of forgetting this now that certain aspects of Process Integration techniques have been highly developed, with design methods and computer programs available for targeting, cost optimisation, network design, resilience and sensitivity analysis[2,3]. It is natural to extract the data for the present process, run the targeting calculations and carefully optimise a near-ideal network - only to miss an opportunity for a simple process change which could save far more energy at less cost. Typical examples are changes in distillation column pressures and temperatures, reflux ratios or sequencing; adding additional stages to a flash or evaporator system, or changing stage temperatures; eliminating unnecessary dilution or intermediate cooling of streams; and more efficient scheduling of batch processes.

In a typical process plant, cold feed streams are heated and pass into a chemical reactor; the resulting mixture of products and unreacted raw materials is divided up in a separation system, and the products are run down to storage. The plant can thus be divided into four sections; the reaction and separation systems, the heat exchanger network between products and feeds, and the residual heating and cooling systems.

These sections have a hierarchy of design[1,4] which is effectively illustrated by the "bag of onions" diagram, Figure 4. For each individual process plant, the basic sections are represented by an "onion"; the utility systems can be considered for the whole site. Process design tends to begin at the centre of the onion and work outwards. Many of the recently published developments in Process Integration techniques apply only to heat exchanger network analysis and design, a small part of the overall process. How do we extend the analysis to the other sections?

The key is to divide the Grand Composite Curve into several parts corresponding to the various systems. The reactor feed and product streams can be identified and plotted separately, and the same can be done for utility streams such as flue gas or boiler feedwater. All separation systems can be plotted as polygons. The remaining streams are known as the background process and their Grand Composite Curve can be plotted. To achieve maximum re-use of heat, the reaction and separation systems should be able to exchange heat with each other and with the background process.

In terms of the temperature-enthalpy graph, the different systems should fit above and below each other and the background process Grand Composite, not side by side. The detailed derivation of the analysis has been previously published[5,6].

186

These techniques give a rapid check on the positioning of the different systems. Since we are concerned with pre-optimisation there is usually no need for highly precise analysis; we are simply screening the possibilities to see which units are correctly placed in energy terms. Thus for distillation columns and evaporators, only the latent heat loads need usually be plotted, giving a simple rectangle; flash systems plot as triangles. If it is decided as a result to change the conditions of a separation system or other unit operation, the stream data set is changed and accurate revised targets are obtained. Most successful projects have been obtained on the three types of separation systems above, although similar methods can be applied to other separation processes[5] and to reactors[6].

BATCH PROCESSES

Introduction

Process Integration techniques were originally developed on continuous processes, for which they are ideally suited. However, the principles must also be adapted to cover batch processes to give a fully comprehensive analysis.

Batch processes present many potential difficulties for heat recovery; there are large intermittent flows so that large heat exchangers are required to recover moderate amounts of heat, and features such as direct steam injection, direct mixing and flash cooling are often appropriate. Often, too, they are on a smaller scale so that the monetary savings from reduced energy consumption are less - but on the other hand, this has meant that energy saving has been neglected on batch processes in the past so that very large percentage savings are often possible.

Published work on Process Integration has until now tended to cover only continuous processes, with batch processes mentioned only in passing, if at all. The behaviour of individual batch systems has also been considerably studied, but their interaction with other thermal systems has been neglected. Vaselanak et al.[7] studied heat integration between batch vessels, in particular the discharge and filling of vessels and whether flows were recirculated to the same tank or passed to a separate vessel. However, they assumed the heat was available at any time as desired, whereas in fact a major difficulty with batch processes is that the heat is only available intermittently so that precise scheduling may be required.

The simplest approach to the analysis of a batch system problem is to take the heating requirements of feed streams and the heat rejected from discharge streams, find the average enthalpy change per unit time and use that as a heat load in the analysis. Clearly, this "pseudo-continuous" approach is over-simplistic, as it makes no allowance for the different times at which heat is available or required, and the consequent sequencing requirements. It will therefore tend to produce over-optimistic targets.

For a rigorous analysis, we must add an additional variable - time - to those of heat flowrate and temperature which are being used already. Unfortunately it then becomes much harder to represent the concepts and the mathematics by simple graphs, as with the third variable present, a generalised analysis involves functions in three dimensions and the plots appear as surfaces rather than lines.

It is therefore easier to follow the explanation of the new EPI techniques if we start by holding one variable constant, and explore the relationship between the other two.

Temperature-Energy Graphs for Single Batches

Let us start by examining the heat flows in a process at specific times rather than simply taking the average over the entire period. At any given time, one can construct the equivalent of the composite and grand composite curves. We call these "snapshot" graphs. Either a plot for the preceding short time period (an "instantaneous" graph) or a plot for the whole period since zero time (a "cumulative" graph) can be drawn.

Take the simple process which was illustrated in Figures 1-3. This can easily be turned into an equivalent batch process by allocating appropriate time periods to each flow. This increases the instantaneous heat capacity flowrate (CP) but averaged over the whole batch period, the CP will be the same for the batch and continuous processes. We can therefore make a direct comparison between the rigorous batch analysis and the pseudo-continuous method. The resulting stream data is listed in Table 1, the overall batch period being taken as 1 hour.

At discrete time intervals, we can form the "snapshot" plots as shown in Fig. 5. The cumulative plots have been used as the instantaneous ones are trivial for this simple problem. Initially the cold stream 1 is filling the reactor and the hot stream 2 can supply only a little of the heat required. So halfway through the cycle we have a net heat deficit of 1.0 GJ, i.e. this amount of heat has had to be supplied up till then. After this, hot stream 3 comes into play and over the second half of the cycle releases 1.5 GJ. For a single stand-alone batch, we cannot make up for the heat we have already put in, and the energy released in the second part of the cycle cannot be re-used. We thus get the "batch composite" graphs shown in Figures 7(c) and 7(d).

Comparing Fig. 7(d), the batch composite curves after one complete cycle, with Figure 1, the composite curves for the continuous process and the pseudo-continuous analysis, we see that they are the same shape but the separation - and hence the energy target - is very different. The targets predicted by the pseudo-continuous analysis were 0.1 GJ/hr hot utility and 0.6 GJ/hr cold; the actual requirements for the single batch are 1.0 GJ/hr (ten times greater!) and 1.5 GJ/hr.

There is a strong conceptual similarity here between the thermodynamic "pinch" analysis and the time-dependent analysis. For just as in one case we have heat made available which cannot be used if it is below the pinch to supply heat to streams above the pinch - at higher temperature - so a net heat release after a time x cannot make up for a net heat demand which came before that time. Just as heat flow from a high to a low temperature is irreversible, so time is irreversible.

Exploiting this conceptual similarity, we can define a "time pinch" before which time net heat must be added to the process and after which net heat is released. In our example, this time pinch comes after 30 minutes. We can also produce a "time cascade" which is equivalent to the heat cascade and Problem Table in conventional Process Integration analysis. This is carried out for our example problem in Table 2. A simple analysis and calculation has enabled us to obtain the heat demands for a batch process.

The analysis is not only applicable to batch systems, though. All continuous processes require starting up, and some take hours or days (notably blast furnaces). The analysis above can be applied in just the same way by taking "snapshots" at different times during the startup process.

Repeated Batch Processes

On most plants, the discharge of one batch is followed directly by another being supplied. This gives the opportunity for using the heat released during the later stages of one batch to heat the next incoming feed. This will be feasible on a time basis although some energy storage will be required, e.g. by running the discharged product into a holding tank. Can we determine the feasibility of heat exchange in terms of temperature too?

We know that we can carry out a pseudo-continuous analysis by averaging the heat flows over the batch period. This yields the composite and grand composite curves which would occur for the equivalent continuous process – or for any process where there are no time restrictions on heat exchange. In other words, it gives the energy efficiency with maximum energy storage, so that heat from any part of the cycle can be carried over to any other part. So for repeated batches where there is also good energy storage, the pseudo-continuous analysis is a much better approximation to the target than it was for a single batch or with no energy storage allowed.

Also, several batch processes can be added together giving a graph of the sum of their heat contents. A continuous process is actually the limiting case of an infinite number of identical small batch processes with ideal heat exchange between all batches. So if energy storage opportunities are exploited to the full, the target for a continuous process can theoretically be achieved for multiply repeated batches. We now require a means of determining how much energy should be stored and at what times.

The Energy-Time Graph

Let us take a simple batch process as an example. A liquid mixture is passed from a storage tank at 0°C, via coils which heat it up to 40°, into a reactor vessel, and fills it over a period of 15 minutes. The full tank is then heated for 35 minutes; mixing and a mild reaction take place and the temperature rises to 60°C, the heating rate being gradually reduced to zero in the last 10 minutes as the reactants are exhausted. The product is then discharged at a constant rate over a 30-minute period. For simplicity of analysis, the heat of reaction, required temperature driving forces and heat losses are assumed to be negligible.

The course of the reaction could be represented on a three-dimensional surface with temperature, heat flow and time as axes. By taking sections of the surface, three two-dimensional graphs can be obtained. The graph of enthalpy against time is another useful tool in the analysis of batch systems.

Figure 6 shows the heat content of the system (the reactor and the feed and product storage tanks) over one cycle. The rate of heat rejection is less than the heat demand of the incoming feed, and the discharge will not be able to match the heat requirements of the feed if both begin simultaneously. However, if the discharge is begun earlier, the heat rejected initially can be stored for a short period and then used to heat the feed.

Application to Parallel Batches

Figure 7 shows two cycles superimposed; the sum of their heat contents is shown as a thick line and those for the individual cycles as a broken line. The peak value of the sum is 9 GJ, exactly the same as for a single cycle. The dips in the graph represent periods during which heat can be stored for future use in the process to give maximum energy efficiency. For the given sequence of operations, the maximum value of heat stored is 2 GJ, momentarily, after 30 minutes.

The sequence of operations was chosen to minimise heat storage, with the first cycle finishing 25 minutes after the second one starts. This gives a 30 minute delay between emptying the reactor completely and beginning to refill it, which reduces plant capacity. Obviously total throughput can be increased by reducing the turnround time between batches, but the energy efficiency then falls. For zero turnround time, with the operations in exact antiphase (40 minutes apart), the heat requirement over the two cycles rises to 13.5 GJ/hr and the heat storage requirement to achieve this is very much greater than in the first case, with a maximum value of 5 GJ/hr; zero turnround with the 25 minute phase difference makes the energy consumption even higher, peaking at 14.5 GJ after 40 minutes of cycle 1 and 15 minutes of cycle 2.

We can draw several conclusions from this example. First, the plot of total system heat content is a useful tool for showing us the cycle energy flows. Secondly, it gives an idea of the amount of energy storage which will be required to maximise energy efficiency, and the energy demand of a batch. Thirdly, accurate data on the exact timing of operations is needed for an analysis. Fourthly, phasing of parallel batch operations and turnround time are important factors.

Reduced turnround between operations leads to an increased overall production rate. In the example above this led to a rise in energy use, but it should not be assumed that there is a simple tradeoff. For some processes a reduced turnaround time leads to a fall in energy use because it improves the "fit" of the two processes' heat profiles; a project to achieve this will be doubly cost-effective.

Unfortunately, the analysis for this simple example is not adequate for a general system, as we also have the Second Law to contend with; the heat released from the batch process is not necessarily at a high enough temperature to fulfil the heating requirements of the next cycle (and this will be a particular problem with endothermic processes). For the simple case above, there was a single mixture of constant CP (heat capacity flowrate) so the energy content was always proportional to the temperature and quantity of the liquid in the reactor vessel and the use of the stored heat to warm the incoming feed and the reactor contents was always thermodynamically feasible (the reader may care to verify this by a simple calculation of the CP's). But for a generalised analysis, we need to explore the effect of temperature as a variable, by bringing in the time-dependent targeting calculations identified earlier.

General Time-Dependent Analysis

In Table 2 we obtained a table for the net heat released or required in each temperature interval. To allow for the effect of temperature, we need to break this down further in the general case so that the heat released from each temperature interval in each time interval is separately tabulated. This brings us back to a three-dimensional analysis. Temperature and time are independent variables, so we can form a grid of temperature and time intervals and find the heat flow between adjacent intervals. The result is a time-dependent heat cascade table.

The table gives a function f(T,t) which is the heat required by and released to the process at a temperature T and a time t. The function is divided up into discrete time and temperature intervals, and the simple Problem Table for continuous processes can be viewed as a special case where all time intervals are equal and coincident. The time intervals selected are those at which heat flows begin, end or change, and are precisely equivalent to the temperature intervals used in the calculation of the Problem Table, the Composite and Grand Composite Curves.

Let us again illustrate by means of an example. This is again based on a continuous process, and we will use the classic four-stream problem used as the core example in the I.Chem.E. Guide[1] (shown there as Table 2.1, page 17). Time periods for each flow have again been chosen, based on the likely form of the process, and this has yielded the batch process stream data shown in Table 3.

The first stage is to form heat cascades for each individual time interval; this will give us targets for a batch process where no heat storage is allowed (it is then immaterial whether the batch is single or repeated). The cascades can be formed in terms of heat flows in kW, and then multiplied by the time interval period to give the total heat loads in kWhr. Table 4 shows the enthalpy change in each individual interval. The maximum infeasibility in each heat cascade with zero hot utility supplied is found, and hot utility is added to produce the feasible cascades. The result is Table 5. The targets are 198 kWhr hot utility and 238 kWhr cold utility; the total heat exchange can be calculated to be 272 kWhr, so that over 60% of the cold stream heat requirements are being supplied by the hot streams. Each individual cascade has its own pinch or threshold temperature.

We now extend this to allow for the possibility of energy storage, so that heat can be time-cascaded as well as temperature-cascaded. Thus any heat in temperature interval T and time interval t can either be cascaded to the temperature interval below in the same time period, or to the next time interval at the same temperature. This gives a two-dimensional infeasible heat cascade table as in Table 6. The maximum infeasibility is 134 kWhr at 85°C and after 0.7 hours, and this is the overall pinch.

Because of the extra degree of freedom (heat can be temperature-cascaded or time-cascaded, but not both) it is not possible to form the feasible heat cascades by simply adding 134 kWhr to all the figures in the infeasible table. Moreover the largest infeasibility is not necessarily equal to the hot utility target, although it does turn out to be in this case. Since energy storage should preferably be kept to a minimum, the overall table can be formed by temperature-cascading heat in preference to time-cascading it whenever possible. The result is Table 7; the targets are 134 kWhr hot and 174 kWhr cold utility, and this implies that 64 kWhr heat storage is necessary, which turns out to be due to a transfer between the (0.3-0.5 hr) and (0.5-0.7 hr) time intervals at a range of temperatures between 140° and 25°C. So we have obtained energy use and heat storage targets for the single batch process.

The cascade table can be extended to further time intervals to obtain the results for the repeated batch process. However, the targets and pinch temperature are very easily found from the last column of the infeasible cascade, Table 6, which shows cumulative values after 1 hour. The largest infeasiblity is 20 kWhr at 85°C, so this gives the hot utility target and pinch temperature with maximum ideal energy storage. As expected, these are equal to the results for the equivalent continuous process. The energy storage can be deduced to be 178 kWhr.

Table 8 summarises the various results obtained for this process, covering the whole range from the situation where there is no heat exchange permitted at all to the best possible case where ideal heat storage within and between batches is possible. In this latter situation the pseudo-continuous analysis gives the correct targets, although it cannot distinguish between heat storage and heat exchange. The wide range of possible targets should be noted, showing how important are the effects of allowing energy storage and repeating batches.

Discussion of the Results

There are some significant differences between the results from the batch analysis and those for a continuous process. Most significant of all is that the pinch point is not absolute. Looking at Table 7, we find that some cold utility is being released before the time pinch and some above the temperature pinch; indeed, there are temperature intervals and time intervals in which both hot and cold utilities are used. In effect, the pinch is not a point, but a locus – the temperature pinch moves up in steps during the cycle.

The overall pinch does however have an important significance in the repeated batch process with ideal energy storage. No heat is cascaded across the temperature pinch in this case; conversely, no energy is stored across the time pinch.

There is one major difference between the effects of temperature and time. For heat transfer, some temperature driving force is required, and the rate of heat transfer is proportional to that driving force. Hence a minimum temperature difference, DT_{min}, is required in the analysis. This is not true of time – energy released at any time is available to carry out heating duties simultaneously – and so there is no need for a corresponding "minimum time difference" correction.

On the other hand, it is very easy to cascade heat between well-separated temperature intervals – driving forces are reduced and equipment is smaller, so it is a positive advantage – whereas to use heat in a much later time interval is inconvenient, as the energy will have to be stored somehow over the intervening time period.

There can be intermediate situations between ideal and zero heat storage where it is easy to store heat from one stream – e.g. in a holding tank – but other streams must be cooled and discharged immediately. A further rider is that the target obtained above allows any one stream to be heated at different times by different streams. It may be necessary to constrain some or all streams so that they only exchange against one other stream at all times, and this too must be allowed for in the Problem Table. Clearly there are considerably more factors and constraints to be taken into account in batch systems than in continuous systems, and the need for the process analyst to be skilled and experienced in using the techniques becomes even more important.

Having found the energy targets for batch processes with a specified scheduling (i.e. set times for each part of the operation), the final step is to identify scheduling changes. These are treated equivalently to process changes in conventional EPI analysis of continuous processes. By modifying the time-dependent stream data, a new target and revised composite and grand composite curves can be calculated.

CONCLUSIONS

The procedures described above represent a significant step forward in the use of Energy and Process Integration techniques for the analysis of batch processes. The techniques have been steadily developed from the initial work on heat exchanger networks in continuous processes. Using them, the experienced analyst can now obtain a clear overview of how all parts of a process interact in energy terms. Furthermore, analysis of the composite and grand composite curves points the way to beneficial process changes at an early stage.

The key to analysis of batch processes is the time-dependent heat cascade table, which is a development from the Problem Table for continuous processes, and gives four important results:

1. The heat supplied or rejected by a batch process at any given temperature and time.
2. The target energy consumption for the overall process.
3. The quantity and duration of the internal energy storage required to achieve the above target.
4. The temperature and time pinch location.

The results can be applied to single, repeated or parallel batch processes. Future work will expand on these concepts to develop systematic guidelines for batch plant design and retrofit.

The final conclusion to be drawn is that there are real, user-friendly methods for optimising process plant design at an early stage in the process design timetable. Engineers should take full advantage of these tools to shorten project lead times, reduce project costs, and build in optimum efficiency from the outset.

NOMENCLATURE

CP	Heat capacity flowrate of stream, MW/K or MJ/K/min
DT_{min}	Minimum allowable temperature difference, K
Q_H, Q_C	Hot and cold utility consumptions, MWhr or GJ
$Q_{H,MIN}, Q_{C,MIN}$	Minimum (target) hot/cold utility use, MWhr or GJ

REFERENCES

1.   A User Guide on Process Integration for the Efficient Use of Energy, 1982 (Institution of Chemical Engineers, Rugby).

2.   Tjoe T N and Linnhoff B, 1986.   Using Pinch Technology for Process Retrofit, Chem.Eng, April 28, 1986, 47-60.

3.   Kotjabasakis E and Linnhoff B, 1986.   Sensitivity Tables for the Design of Flexible Processes, CH.E.R&D, 64:3, 197-211 (May 1986).

4.   Douglas J M, 1985.   A Hierarchical Design Procedure for Process Synthesis, A.I.Ch.E.J, 31:3, 353-362.

5.   Kemp I C, 1986. Analysis of Separation Systems by Process Integration, J.Sep.Proc.Tech, Vol.7.

6.   Kemp I C, 1987. Total Process Plant Analysis by Process Integration. Paper presented at I.Chem.E. conference "Process Optimisation", Nottingham, April 1987 (I.Chem.E. Symposium Series No.100, 67-78).

7.   Vaselanak J A, Grossman I E and Westerberg A W, 1986. Heat Integration in Batch Processing. Ind.Eng.Chem.Process Des.Dev., 25, 357-366 (April 1986).

TABLE 1. STREAM DATA FOR PROBLEM SHOWN IN FIGURES 1 - 3

Batch time = 1 hour. Minimum temperature driving force, DT_{min} = 25°C.

No. of Stream	Hot/ Cold	Supply Temp.	Target Temp.	Averaged/Continuous		Instantaneous during batch		
				Heat Flow GJ/hr	CP MJ/minK	HT.Flow GJ/hr	Flow period min	CP MJ/minK
1	C	25	150	1.5	0.2	3	0 - 30	0.4
2	H	225	150	0.5	0.111	2	15 - 30	0.444
3	H	125	50	1.5	0.333	3	30 - 60	0.667

TABLE 2. TIME CASCADE FOR SIMPLE BATCH PROCESS EXAMPLE

Time-Period No. Times (min)	Heat Flows for period, GJ			Interval temp of flow	Cumulative utility use	
	Hot streams	Cold streams	Net		^{O}H	^{O}c
1 0 - 15	0.0	0.75(25-150)	-0.75	37.5-162.5	0.75	0.0
2 15 - 30	0.5(225-150)	0.75(25-150)	-0.25	37.5- 79.2	1.0	0.0
3 30 - 45	0.75(125-50)	0.0	+0.75	112.5- 37.5	1.0	0.75
4 45 - 60	0.75(125-50)	0.0	+0.75	112.5- 37.5	1.0	1.5

TABLE 3. STREAM DATA FOR EXAMPLE BASED ON IChemE GUIDE; DT_{min} = 10°C

Stream No./Type	Supply Temp. °C	Target Temp. °C	Interval Tmp Supply °C	Target °C	CP kW/K	Heat Flow kW	Start Time hr	Finish Time hr	Heat Load kWhr
1 C	20	135	25	140	-10	-1150	0.5	0.7	-230
2 H	170	60	165	55	4	440	0.25	1	330
3 C	80	140	85	145	-8	-480	0	0.5	-240
4 H	150	30	145	25	3	360	0.3	0.8	180

TABLE 4.   HEAT LOADS FOR EACH TEMPERATURE-TIME INTERVAL, kWhr

t	0	0.25	0.3	0.5	0.7	0.8	1.0
T	FROM HOT UTILITY						
165							
		0	4	16	16	8	16
145							
		-10	-1	-1	7	3.5	4
140							
		-110	-11	-11	-33	38.5	44
85							
		0	6	42	-18	21	24
55							
		0	0	18	-42	9	0
25							
					TO COLD UTILITY		

TABLE 5.   HEAT CASCADES FOR EACH TIME INTERVAL WITH NO HEAT STORAGE

t	0	0.25	0.3	0.5	0.7	0.8	1.0
T	FROM HOT UTILITY: 198 kWhr						
165	120	8	0		70	0	0
145	120	12	16		86	8	16
140	110	11	15		93	11.5	20
85	0	0	4		60	50	64
55	0	6	46		42	71	88
25	0	6	64		0	80	88
					TO COLD UTILITY: 238 kWhr		

TABLE 6.   TIME-DEPENDENT HEAT CASCADE TABLE - INFEASIBLE

t	0	0.25	0.3	0.5	0.7	0.8	1.0
T	FROM HOT UTILITY						
165	0	0	0	0	0	0	0
145	0	0	4	20	36	44	60
140	0	-10	-7	8	31	42.5	62.5
85	0	-120	-128	-124	-134	-84	-20
55	0	-120	-122	-76	-104	-33	55
25	0	-120	-122	-58	-128	-48	40
					TO COLD UTILITY		

TABLE 7. HEAT CASCADE WITH MINIMUM HOT UTILITY AND ENERGY STORAGE

t	0	0.25	0.3	0.5	0.7	0.8	1.0
T	FROM HOT UTILITY: 134 kWhr						
165	0	120	8	0	6	0	0
145		120	12	16	22	8	16
140		110	11	15	29	11.5	20
85		0	0	0	0	50	64
55		0	6	24	0	71	88
25		0	6	0	0	80	88
					TO COLD UTILITY: 174 kWhr		

TABLE 8. SUMMARY OF TARGETS FOR ENERGY USE AND HEAT STORAGE

Situation	Hot Utility	Cold Utility	Heat Exchange	Heat Storage
With utility heating/cooling only	470	510	0	0
Single/repeated batch, no storage	198	238	272	0
Single batch with energy storage	134	174	272	64
Repeated batch with energy storage	20	60	272	178
Psuedo-continuous analysis	20	60	(450)	

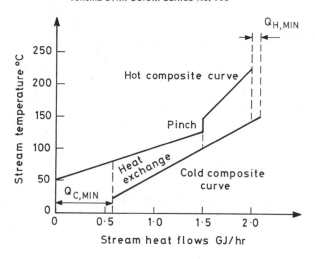

FIG.1. COMPOSITE CURVES

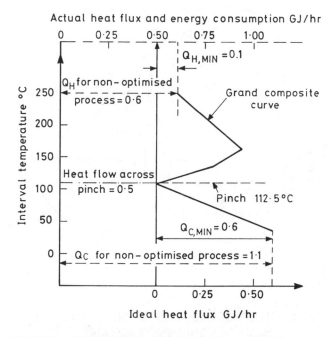

FIG.2. GRAND COMPOSITE CURVE (ΔT_{MIN} = 25°C) AND ITS RELATIONSHIP TO ACTUAL ENERGY CONSUMPTION

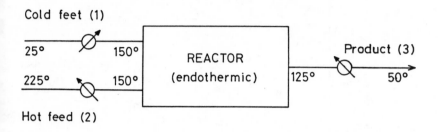

FIG. 3. PLANT FLOWSHEET FOR INTRODUCTORY EXAMPLE

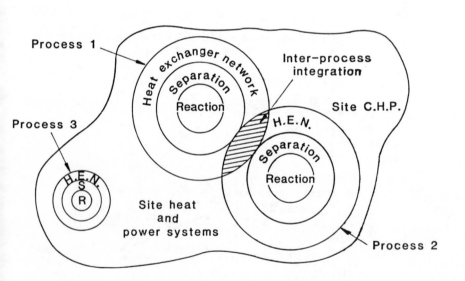

FIG. 4. INTERRELATIONSHIP OF SITE PROCESSES
(BAG-OF-ONIONS DIAGRAM)

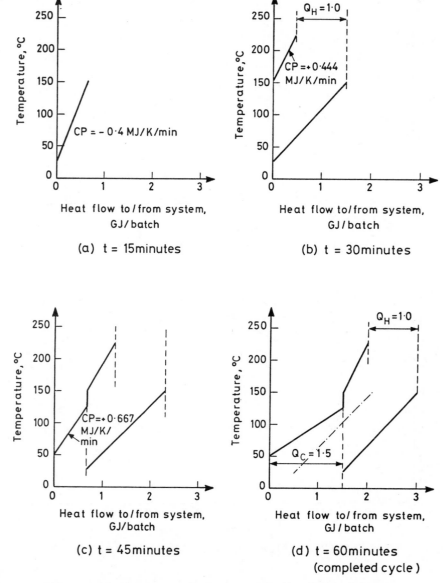

FIG.5. "SNAPSHOT" COMPOSITE CURVES FOR A BATCH
PROCESS AT DIFFERENT TIMES (ΔT_{MIN} = 25 °C)

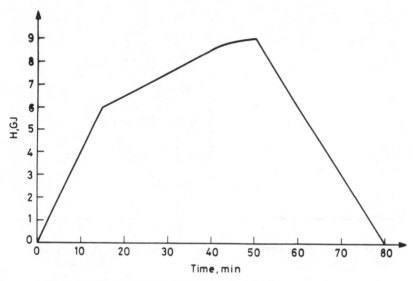

FIG. 6. SYSTEM ENERGY CONTENT PLOTTED AGAINST TIME

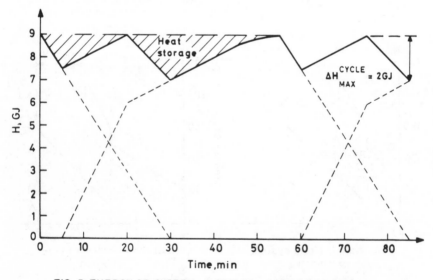

FIG. 7. ENERGY OF SYSTEM - 2 CYCLES 55 MINUTES APART

EFFICIENT USE OF ENERGY IN OIL REFINERY EXCHANGER NETWORKS SUBJECT TO FOULING

B D Crittenden*, S T Kolaczkowski* and R Varley**

Fouling of process plant leads to considerable financial penalties, particularly in terms of additional energy consumption and operability problems. MINERVA is a program which incorporates process-specific fouling models into a personal computer energy management and operability package, and is used to illustrate how oil refinery management can adopt operational strategies for minimising the impact of fouling on crude distillation units. The computer software is, however, readily adaptable to a much broader spectrum of process units in the oil, chemical and service industries.

INTRODUCTION

Nobody is certain what the financial penalties are when heat exchangers lose thermal efficiency as solid deposits build up and foul the transfer surfaces. In 1979 it was estimated that the annual cost to the UK industry was between £300M and £500M[1]. A more recent estimate[2] puts the annual cost of fouling and corrosion in the USA at between $3000M and $10,000M. For the oil industry, the Exxon Chemical Company estimated[3] in 1979 that the annual cost of fouling in a hypothetical but typical 100,000 barrels/day refinery was approximately $10M. Although a similar study of UK oil refineries has not been made, there is no reason to doubt that the Exxon estimates are broadly applicable here. Thus, with a total UK refinery capacity of around 1.8M barrels/day, the annual cost of fouling in UK refineries could well be as much as £180M. The Exxon estimates show that about half this cost is attributable to fouling in the crude distillation unit (CDU) in which all of the incoming crude oil is heated from ambient conditions to elevated temperatures in a network of shell and tube heat exchangers and furnaces. The financial burden arises from a combination of the following:

- the reduction in throughput as heat exchanger flow channels become blocked
- the additional energy which is required in the furnace in order to maintain the feed temperature to the crude oil distillation column
- the maintenance and cleaning costs of fouled exchangers
- the purchase of anti-foulant chemicals in an attempt to alleviate the severity of fouling

The Exxon estimates also show that the fraction of the total cost which can be

* School of Chemical Engineering, University of Bath, Bath BA2 7AY, UK
** Selectamaster Ltd, Bath, UK

attributed directly to additional energy consumption depends, *inter alia*, on the process unit in a refinery, but can vary from 20% for CDUs to 75% for vis-breakers.

Part of the flowsheet for a hypothetical, but typical, 100,000 barrels/day preheat exchanger train is shown in Figure 1. The preheat system as a whole is designed to optimise crude oil heating and product and pump-around cooling, but this is clearly hampered by fouling. Crystallisation, corrosion, chemical reaction and particulate fouling mechanisms are all believed to operate in the CDU exchangers. The problem is therefore complex, and any network optimisation simulation packages which are applied to exchangers that are likely to foul should contain scientific fouling models which can predict the dependence of fouling rates on time and on other key process parameters, such as temperature, flow rate, composition, equipment geometry *etc*. The development and testing of fundamental models of hydrocarbon fouling are discussed elsewhere[4-6].

In this paper it is demonstrated how a fouling model may be incorporated into a network simulation package to illustrate how the research effort in fouling may be applied by engineers and managers who are engaged in the design and operation of process plant.

EXCHANGER AND NETWORK SIMULATION

Many commercial computer packages are already available for assessing the steady-state design and performance not only of individual exchangers but also of networks of heat transfer equipment. Rigorous techniques are applied to evaluate physical properties, to compute heat transfer coefficients and to take into consideration mechanical design features such as tube diameters, pitch *etc*. However, less rigour is generally applied to the selection of fouling resistances. Fixed values are, at best, based on operating experience, but are more usually taken from TEMA standards[7]. Little scientific account is therefore taken at the design stage of the effects of key operating and geometric parameters on fouling resistances. Consequently, it is not unusual to find that heat exchangers are either grossly over- or under-surfaced. Existing commercial simulation packages do not contain predictive methods for re-evaluating fouling resistances with time. The laborious task of manually updating fouling resistances therefore takes little or no account of important changes in key operating variables which occur as fouling proceeds.

FOULING SIMULATION

The fouling simulation package MINERVA has been developed to complement the existing range of commercial packages through its ability to predict the dynamic behaviour not only of individual exchangers but also of existing and re-vamped series networks. MINERVA has been designed specifically to enable technical management to adopt operational and design strategies for minimising the impact of fouling on their plants.

With reference to Figure 2, the performance of a single exchanger with no phase changes is described by the following energy balances:

$$Q = M_1 Cp_1 \{(T_1)_o - (T_1)_i\} \tag{1}$$

$$Q = M_2 Cp_2 \{(T_2)_i - (T_2)_o\} \tag{2}$$

$$Q = U_o A_o F \Delta T_m \tag{3}$$

where

$$U_o = \left[\frac{1}{h_o} + R_{f_o} + R_w + \frac{A_o}{A_i} R_{f_i} + \frac{A_o}{A_i h_i} \right]^{-1} \tag{4}$$

Equations (1) - (3) are solved for each exchanger in the network to determine the outlet tube and shell side temperatures. Inside and outside film coefficients h_i and h_o, are determined from well established correlations[8]. Physical properties are described by equations and stored as subroutines. The fouling resistances R_{f_i} and R_{f_o}, dependent upon time are regularly updated by calculation from process-specific models which are called as subroutines by the main program. Since the rate of fouling is usually very low compared with the throughput of process fluid, quasi-steady-state conditions are assumed. That is, whilst the energy equations do not contain explicit time dependencies, they are updated periodically for changes which occur as a result of deposition.

The MINERVA suite of programs, Figure 3, is provided with the WIMPs (Windows, Icons, Mice, Pull-down menus) user interface which is designed to allow easy interaction for non-computer professionals. The suite has been written initially for a BBC personal computer with an Acorn 65C12 second processor. The most important module is the CALC program which embodies the fouling model and enables the user, *via* the WIMPs environment, to simulate fouling over a variable period from zero to 24 months, varying parameters such as flow rates, temperatures, physical properties, exchanger surface areas, configurations *etc*. Output from this program can be represented numerically and graphically as the fouling resistance for a specified time and as costs in cleaning, lost production, anti-foulant chemicals and increased furnace duty. Example output is shown in Figure 4. Additional features of the program are:

- the provision of flowchart and graph screen modes running independently of and concurrently with the fouling model
- a continuously scrollable flowchart representation of the exchanger network
- utilities to create special format extended graphics files and an edit program to enable the creation of flowcharts.

APPLICATION OF MINERVA

To illustrate the application of the fouling simulator for predictive purposes, a series of post-desalter heat exchangers is modelled for the typical but hypothetical crude CDU (Figure 1). Typical data for two of these exchangers, E5 and E8, are given in Figures 1 and 5 and in Table 1, and are used to determine the coefficients in a simplified version of the Crittenden and Kolaczkowski mass transfer and kinetics model of hydrocarbon fouling[4,5]. In this example model, the rate of deposition is assumed to be kinetically controlled, whilst deposit removal is assumed to occur *via* a fluid shear mechanism. In addition, fouling is assumed to occur on the tube-side (crude oil) only, *ie* $R_{f_o} = 0$. Hence,

$$\frac{dR_{f_i}}{dt} = A \exp(-E/RT_{f_i}) - \frac{C\,M_1^2\,R_{f_i}}{\rho\pi^2(a-x_f)^4} \tag{5}$$

A further simplification is that the effect of deposit build-up on fluid velocity and hence on shear removal is negligible. Thus x_f is set equal to zero in equation (5). In many cases, this is justified, since $x_f^4 \ll a^4$. Therefore,

$$\frac{dR_{f_i}}{dt} = A \exp(-E/RT_{f_i}) - \frac{C\,M_1^2\,R_{f_i}}{\rho\pi^2\,a^4} \tag{6}$$

Equation (6) can be integrated at constant M_1 to give:

$$R_{f_i}(t) = R_{f_i}(\infty)\{1-\exp(-\beta t)\} \tag{7}$$

where

$$R_{f_i}(\infty) = \frac{G}{u^2 \rho} \exp(-E/RT_{f_i}) \tag{8}$$

$$\beta = C u^2 \rho \tag{9}$$

$$G = A/C \tag{10}$$

An important feature of MINERVA is the requirement to evaluate fouling model coefficients from plant or experimental data. In this instance the historical plant data shown in Figure 5 are used to determine the coefficients A, C and E in equation (5). Application of equation (8) to the data of Figure 5 enables E and G to be determined (Table 1). From equations (6) and (10), when t = 0, *ie* when R_{f_i} = 0,

$$\left.\frac{dR_{f_i}}{dt}\right|_{t=0} = G C \exp(-E/RT_{f_i}) \tag{11}$$

Knowing G, the coefficient C (and hence A) is determined from the initial rate of fouling of one exchanger. In this particular example, the data given in Figure 5 for exchanger E8 are used, and the value of C is given in Table 1.

EXAMPLE SIMULATION RESULTS

The interactions between heat exchangers as they become fouled are complex. In order to demonstrate some of the more fundamental applications within the limited space of this paper, the following simplifications have been made in running MINERVA:

- changing pressure drop and its effect on reducing throughput is omitted
- shell-side inlet temperatures are held constant
- shell-side flow rates are maintained in direct proportions to the tube-side flow rate, *ie* if the crude oil flow rate decreases by 10%, then all shell-side flow rates decrease by 10%
- heat exchanger configuration correction factor (F in equation (3)) is set equal to unity for all exchangers
- physical properties for the gas-oil streams on the shell-side of exchangers E5, E7 and E8 are evaluated for a 30°API stream
- physical properties for the crude oil and atmospheric residue streams are evaluated for 40°API and 20°API streams, respectively
- all physical properties vary with temperature

Run 1 - Simulation of existing network

With a constant inlet temperature and flow rate to exchanger E5, the simulation results for the hypothetical but typical CDU (Figure 1) are shown in Figure 6. As expected from a model in which the deposition rate is kinetically controlled, the highest fouling resistances at any time are found in the hottest exchangers. The coolest exchangers, E5 and E6 appear to have reached asymptotic fouling conditions. The predictions in Figure 6 may be compared with the plant data in Figure 5. The irregular nature of the curves in Figure 5 reflects sudden plant upsets and short time-scale changes in process conditions, such as flow rate and crude type. Run 3 of this paper illustrates how MINERVA can handle the latter variations. MINERVA is capable of evaluating the annual costs of implementing a CDU shut-down programme for exchanger cleaning. Inputting a fuel oil cost of $100/Te, an exchanger train cleaning cost of $30,000 and a margin of $2/barrel to account for lost production over each 5 day shut-down, the results are shown in Figure 7. In this particular example, the debit associated with the lost production during shut-down outweighs the advantage to be gained in energy savings by cleaning within a 10 month period. In general, though, the conclusions drawn would depend very much on energy prices, production margins, the time-scale selected, and, in addition, whether or not additional penalties associated with throughput loss and pump-around limitations are included. For example, if pump capacity is insufficient to cope with additional pressure drops arising from fouling, then additional losses in throughput would be incurred, resulting in even greater financial losses.

Run 2 - Advantages of alternative exchangers

MINERVA is run assuming that management had the choice of using alternative exchangers in the position of E7. In this illustrative example, replacement exchangers with 4% smaller surface areas but with 35% higher tube-side velocities were considered, in order to reduce both the rate of fouling and the ultimate fouling resistance at E7. The comparative results, shown in Figure 8, indicate that energy savings can be made with slight reductions in exchanger surface area. Thus MINERVA can be used to review the relative merits, in terms of energy consumption and plant operability, of alternative exchangers and network configurations.

Run 3 - Effect of operating conditions

For the CDU network with the replacement E7 exchangers, MINERVA is run and the crude flow rate is changed after 2 months and again after 6 months. The resulting R_{f_i} values for exchangers E7 and E9 are shown in Figure 9. MINERVA can be used in this manner to study the effects of changes in operating parameters, such as the temperatures and flow rates of crude oil and pump-around streams, on energy requirements and operability. Changes in crude type can be studied as well, subject to the prior evaluation of fouling model coefficients. Use of MINERVA in this way assists the investigation of the causes of sudden high and low R_f values when plant operating conditions have been changed.

Run 4 - Cost effectiveness of anti-foulants

MINERVA may be used to assess the potential benefits of implementing an anti-fouling chemical programme. The anti-foulant vendor is asked to quantify how his product would reduce fouling rates. MINERVA prepares a base case, with no anti-foulant, from historial plant data. MINERVA is then run with the quantified fouling reduction parameter. Figure 10 illustrates such a comparison for exchanger E9 when a 50% reduction in the rate of fouling of all exchangers was specified by the vendor. For this illustrative example, anti-foulant costs were assumed to be $0.009 per barrel and totalled $273,000 over a 10 month period. The resulting energy savings are predicted to be $274,000 (with fuel oil at £100/Te) or $430,000 (with fuel oil at $160/Te). At the higher energy cost, substantial cost savings would be made. Even at the lower energy cost, the additional operability advantages would justify the application of the anti-foulant programme. MINERVA can also be used to monitor, on a regular basis, the performance of the anti-fouling treatment in the exchanger network.

CONCLUDING DISCUSSION

The concept of incorporating a fouling model into a network simulation package provides plant management with the ability to determine optimum exchanger cleaning cycles to evaluate commercial anti-fouling treatments, to evaluate network re-vamp strategies and to perform operability studies all with a greater degree of confidence. Although MINERVA has been written initially for operability studies in oil refinery crude distillation units, as a concept, it has a much broader application in all process industries that suffer from fouling. Its success as a fouling and operability simulator in any industry depends on the knowledge and ability required to write the process-specific fouling model subroutine. Thus MINERVA cannot be considered to be a panacea for fouling without the expenditure of much time and effort on the development of fundamental scientific models of fouling. Even when plant data are used to develop a fouling model, care must be taken when extrapolating to conditions outside of the range for which the model is developed. In such cases expert advice should be obtained. Currently, MINERVA is being used as a means to test fouling models for the crude distillation units of BP's oil refineries. This work forms part of a Science and Engineering Research Council/BP International Co-operative research project which is being carried out by the School of Chemical Engineering at Bath University.

REFERENCES

1   Thackery, P.A., 1979, Proc. Conf. 'Fouling - Science or Art?' *Inst Corr Sci and Tech/IChemE*, London, pp 1-9

2   Garrett-Price, B.A, Smith, S.A, Watts, R.L, Knudsen, J.G, Marner, W.J. and Suitor, J.A., 1985, 'Fouling of Heat Exchangers: Characteristics, Costs, Prevention, Control and Removal', *Noyes Publications*, Park Ridge, New Jersey, p 105

3   Van Nostrand, W.L, Leach, S.H. and Haluska, J.L, 1981, 'Fouling of Heat Transfer Equipment' edited by Somerscales,E.F.C. and Knudsen, J.G, *Hemisphere*, Washington, pp 619-643

4   Crittenden, B.D. and Kolaczkowski, S.T, 1979, Proc. Conf. 'Fouling - Science or Art?' *Inst Corr Sci and Tech/IChemE*, London, pp 169-187

5   Crittenden, B.D. and Kolaczkowski, S.T, 1979, 'Energy for Industry', edited by O'Callaghan, P.W, *Pergamon*, Oxford, pp 257-266

6   Crittenden, B.D, Kolaczkowski, S.T. and Hout, S.A, 1987, 'Modelling Hydrocarbon Fouling', *Chem Eng Res & Des*, 65, 171-179

7   Tubular Exchangers Manufacturers' Association, 1978, Standards of the Tubular Exchangers Manufacturers' Association, New York, pp 138-142

8   Kern, D.Q, 1950, 'Process Heat Transfer', *McGraw-Hill*, Tokyo

9   Ludwig, L.L, 1985 (January), *Hydrocarb. Process.* 64, 55-56

NOMENCLATURE

a	inside radius of tube	m
A	coefficient in equation (5)	$m^2 K\ w^{-1} month^{-1}$
A_i	inside heat transfer area	m^2
A_o	outside heat transfer area	m^2
C	coefficient in equation (5)	$m\ s^2 kg^{-1} month^{-1}$
Cp_1	cold stream specific heat	$kJ\ kg^{-1} K^{-1}$
Cp_2	hot stream specific heat	$kJ\ kg^{-1} K^{-1}$
E	coefficient in equation (5)	$kJ\ kmol^{-1}$
F	heat exchanger configuration correction factor	
G	coefficient defined by equation (10)	$kg\ m\ K\ w^{-1} s^{-2}$
h_i	inside film heat transfer coefficient	$kw\ m^{-2} K^{-1}$
h_o	outside film heat transfer coefficient	$kw\ m^{-2} K^{-1}$
M_1	cold stream mass flow rate	$kg\ s^{-1}$
M_2	hot stream mass flow rate	$kg\ s^{-1}$
Q	rate of heat transfer	kw
R	universal gas constant	$kJ\ kmol^{-1} K^{-1}$
R_{f_i}	inside fouling resistance	$(kw\ m^{-2} K^{-1})^{-1}$
R_{f_o}	outside fouling resistance	$(kw\ m^{-2} K^{-1})^{-1}$
R_w	wall resistance	$(kw\ m^{-2} K^{-1})^{-1}$
t	time	s, month
$(T_1)_i$	cold stream inlet temperature	K
$(T_1)_o$	cold stream outlet temperature	K
$(T_2)_i$	hot stream inlet temperature	K
$(T_2)_o$	hot stream outlet temperature	K
T_{f_i}	film temperature	K
u	velocity	$m\ s^{-1}$
U_o	overall heat transfer coefficient	$kw\ m^{-2} K^{-1}$
x_f	deposit thickness	m

Greek symbols

β	coefficient defined by equation (9)	month^{-1}
ΔT_m	log mean temperature difference	K
ρ	density	kg m^{-3}

Table 1   Coefficients and operation conditions

Operating parameters:

	E5	E8
Exchanger number	E5	E8
Tube side velocity (m s^{-1})	2.87	2.48
Tube side fluid density (kg m^{-3})	738.5	658.4
Clean tube wall temperature, T_{f_i} (K)	415	511
Asymptotic fouling resistance $R_f(\infty)$ (K m^{2}w^{-1})	0.00123	0.00704
Initial rate of fouling $(\frac{dR_f}{dt})_{t=0}$ (K m^{2}w^{-1}month^{-1})		0.001761

Coefficients:

$E = 36{,}936$ kJ kmol^{-1}

$G = 364{,}424$ kg m K w^{-1}s^{-2}

$C = 2.848\text{x}10^{-5}$ m s^{2}kg^{-1}month^{-1}

$A = 10.37$ m^{2}K w^{-1}month^{-1}

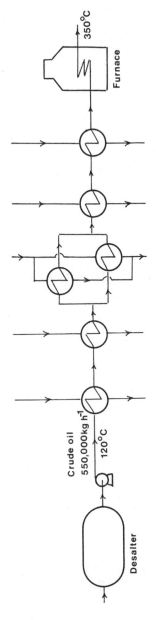

CODE		E5	E6	E7	E8	E9
tubes per bundle		1,548	1,350	1,350	2,010	2,010
tube i.d.	mm	15.39	15.39	15.39	15.39	15.39
tube o.d.	mm	19	19	19	19	19
tube pitch (square pitch)	mm	25.4	25.4	25.4	25.4	25.4
area	m²	564	492	492	733	733
number of tube passes		4	4	4	4	4
shell diameter	mm	1,219	1,143	1,143	1,372	1,372
shell-side fluid		LGO	Atmos residue	BPA	HGO	Atmos residue
shell-side inlet temperature	°C	270	270	280	310	350
shell-side flow rate	kg h⁻¹	130,000	300,000	175,000	260,000	300,000

Figure 1   Series of crude oil preheat exchangers in a hypothetical CDU

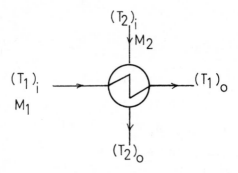

Figure 2   Arbitrary heat exchanger configuration

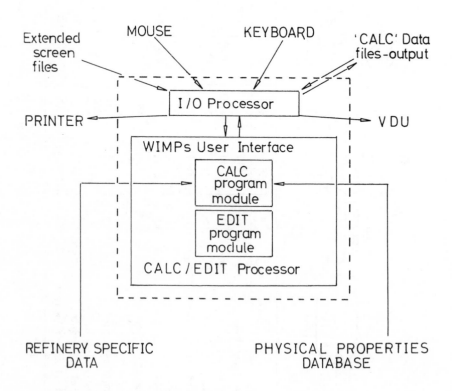

Figure 3   Architecture of MINERVA

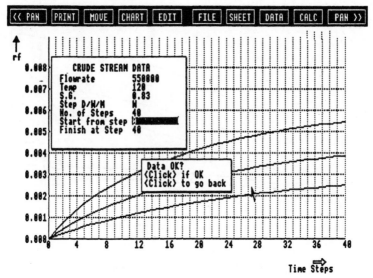

Figure 4   Typical graphical output from MINERVA

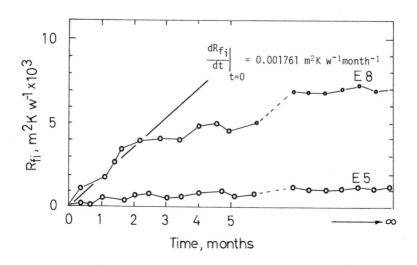

Figure 5   Historical fouling resistance data for
            exchangers E5 and E8 (M_1 = 152.8 kg s^{-1})

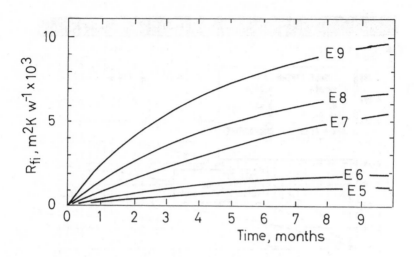

Figure 6   Fouling simulation of hypothetical CDU (Run 1)

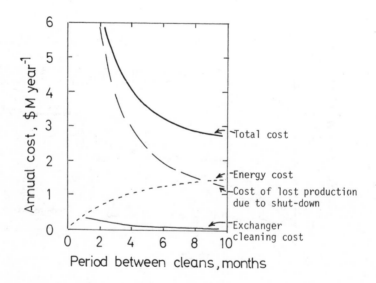

Figure 7   Annual costs as a function of exchanger
cleaning frequency (Run 1)

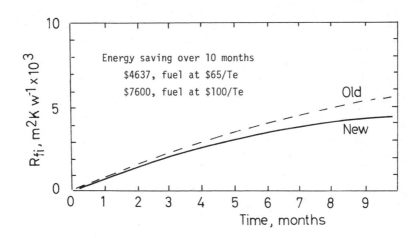

Figure 8   Effect of replacement exchanger (Run 2)

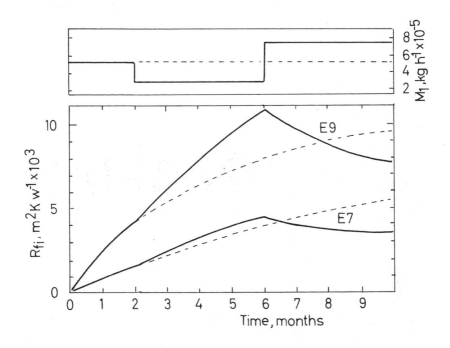

Figure 9   Effect of flow rate on fouling resistances (Run 3)

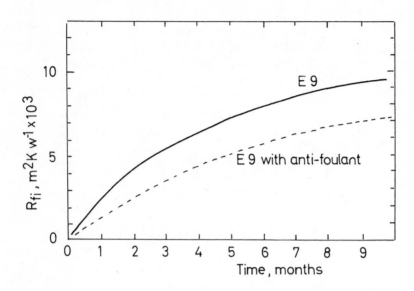

Figure 10   Application of anti-foulants (Run 4)

INTEGRATED FUEL CELLS SYSTEMS

D. J. Gunn* and E. Rhodes*

A brief review is given of the present state of development
of multi-megawatt phosphoric acid and molten carbonate fuel
cell systems. The phosphoric acid fuel cell operates at
$\sim 200^\circ$C, at which temperature phosphoric acid is extremely
corrosive. The molten carbonate fuel cell operates at 650°C
with an apparent limited flexibility for frequent startups and
shutdowns. The molten alkali acetate fuel cell, at present
in an early stage of development, is relatively non-corrosive,
with good flexibility in startup and shutdown at a temperature
of $\sim 300^\circ$C. The industrial potential of all fuel cells is
greatly dependent upon integrated developments in fuel
preparation. Compatible and integrated methods of fuel
preparation are outlined for each fuel cell system.

INTRODUCTION

Fuel cells (ref.1-5) are a means of converting chemical energy directly to
electrical energy, avoiding the energy deficiency limitations of the Carnot
cycle. Hydrogen is the ideal fuel, but hydrocarbons can also be used.
However, the direct use of hydrocarbon fuels is inefficient as carbon
deposition builds up relatively quickly in the porous gas diffusion
electrodes.

This difficulty may be overcome by integrating processes for the partial
oxidation or reforming of hydrocarbons in chemical reactors with the fuel cell,
a coupling that links established technologies for partial oxidation or
catalytic reforming with established designs of fuel cell. The possibilities
of energy integration are greatest with high temperature fuel cells when
combined with steam reforming since heat arising from the irreversibilities of
the operation of the fuel cell is then available at a useful temperature level
to meet some of the requirements of the endothermic reforming process. The
molten carbonate fuel cell (ref.6,7) operates at about 650°C, while primary/
secondary reformers operate in the range of 700 to 900°C. The carbonate fuel
cell has the advantage that carbon monoxide may be converted in the cell, and
the presence of significant concentrations of carbon dioxide in the fuel gases
represses loss of carbonate from the cell. Because of the undesirability of
frequent startups, it is envisaged that this cell will find most application
in central power stations.

The fuel cell may also incorporate a reforming catalyst integral with the cell,
but at present there are difficulties associated with sintering of the porous
nickel electrode in the reforming process.

*Department of Chemical Engineering,
 University College of Swansea, Singleton Park, Swansea, SA2 8PP.

Besides the high temperature molten salt fuel cell, phosphoric acid fuel cells (ref.8) using 40 molar phosphoric acid and operating at 190-220°C are also viable for multi-megawatt power stations.   The fuel is reformed hydrogen where sulphur compounds have been removed in a hydrodesulfurizer as the electrodes which consist of thin films of platinum deposited on porous carbon substrates are particularly sensitive to sulphur poisoning.   This cell is also susceptible to carbon monoxide poisoning.

The Bacon fuel cell (ref.1 and 8) was used successfully at kilowatt ratings in spacecraft applications, but required very low concentrations of sulphur in the fuel because of the deleterious effect on catalytic activity, and very low concentrations of carbon oxides because of reaction with the aqueous alkaline electrolyte.   Neither the carbonate nor the phosphoric acid fuel cells are as sensitive to the oxides of carbon and therefore there has been a substantial development of both of these cells within the past eight years, principally in the United States.

Two 4.5 MW phosphoric acid power plants have been built and operated, one in Tokyo and the second in New York, as part of an evaluation programme for commercial, on-site power plants by United Technologies Corporation.   A preliminary engineering design study for a 675 MW power plant has been carried out by General Electric based upon the molten carbonate fuel cell fuelled by gasification of coal.   These projects are the results of substantial development programmes involving cell testing for many thousands of hours and the expenditure of hundreds of millions of dollars.

The molten acetate fuel cell and solid oxide cells are less developed. Developed versions of the acetate fuel cell may be suitable and economic for dispersed power and combined heat and power systems.   Developed versions of the solid oxide cell may be suitable for central power stations.

Fuel cells suitable for combined heat and power systems

The basic principle of a fuel cell is that it enables the direct conversion of chemical energy to electrical energy without being subjected to the limitations of the Carnot cycle in conventional steam or gas turbine electrical power plant.   If a fuel, hydrogen, hydrocarbon or coal, is oxidised in oxygen or air, heat energy is evolved due to the rearrangement of the electrons in the outer shells of atoms as the new compounds are formed;   the electrons are now in a lower energy state than they were before.   This oxidation process can be carried out electrochemically if electrocatalysts in the form of electronically conducting cathodes and anodes are present in an electrolyte and the electric current which flows when electron transfer takes place during the chemical reaction is conducted through an external circuit to perform useful work.

In the electrochemical oxidation of hydrogen to water (Fig.1) using a suitable electrolyte for the transfer of the ionic charge in the cell the overall reaction is

$$H_2 + \tfrac{1}{2} O_2 \xrightarrow[\text{electrolyte}]{2e} H_2O$$

The standard Free Energy change ΔG_R^O of the reaction is related to the standard potential E of the cell by the Faraday Constant F (charge transferred per electron mole), such that

$$E = - \frac{\Delta G^O_R}{ZF}$$

where Z is the number of electrons transferred in the reaction.   In the fuel cell the reactions at the electrodes are:

At Fuel electrode - Anode - Standard Potential E^O_{Anode}

$$H_2 + (M) \rightarrow 2H^+ + 2e + (M)$$

M is electrocatalytically active metal such as Pt or Ni

At Oxygen electrode - Cathode - Standard Potential $E^O_{Cathode}$

$$\tfrac{1}{2} O_2 + H_2O + (M) \rightarrow 2OH^- - 2e + (M)$$

M is electrocatalytically active metal such as Ag or Ni

The standard cell potential $E^O = E^O_{cathode} - E^O_{Anode}$

    Under open circuit conditions (no current taken from the cell) and for standard conditions of 1 atm partial pressures of hydrogen and oxygen and unit activity of hydrogen ions in the electrolyte at 25°C the EMF E^O of the hydrogen fuel cell can be calculated as 1.23 volt from the standard Free Energy $\Delta G^O_{25^OC}$ of the chemical reaction.   Since $\Delta G = \Delta H - T\Delta S$ the EMF of the cell is temperature dependent and decreases to 1.12 V at 300°C and 1.01 V at 600°C. The actual EMF also depends on the partial pressures of reactants and products according to the Nernst Equation at temperature T

$$E_T = E^O_T - \frac{RT}{ZF} \ln\left[\frac{P_{H_2O}}{P_{H_2} P_{O_2}^{\tfrac{1}{2}}}\right]$$

    In reality the voltage of the cell is severely reduced by polarisation effects (η) when current is taken from the cell.   These effects are essentially due to the difficulty of electron transfer between metal electrode and ions (activation polarisation) and mass diffusion effects due to depletion of ions at the electrode surface (concentration polarisation).   As the current drain is increased internal cell resistance and resistances in the external connecting circuit become important.

$$V = \left(E^O_{Cath.} - \eta_{Cath.}\right) - \left(E^O_{Anode} - \eta_{Anode}\right) - iR_{cell} - iR_{circuit}$$

Polarisation and resistive effects lead not only to something approaching a 30% drop in cell potential at current densities of 100 ma/cm^2 but also to considerable heating of the fuel cell system.

Phosphoric acid fuel cell (PAFC)

These cells (Fig.2, a diagram of fuel cell construction ref.8) operate at 190°C in order to enhance the electrocatalytic activity of the platinum or platinum alloy catalyst, deposited at a rate of 0.25 mg/cm^2 on the porous graphite anode, and at double that loading of Pt-V alloy on the cathode.   At this temperature the 98 mol % phosphoric acid is extremely corrosive and is contained in a silicon carbide and Teflon matrix, about 0.02 cm thick, with very fine pores, in order to withstand pressure differentials of 6 kPa when filled with electrolyte.   Separator and current collector plates are of carbon.   Fuel and air supply manifolds supply the gases to the fuel gas and air passages adjacent to each of the anode and cathode plates in the cell stack respectively. There can be up to 500 cells in a stack and individual cell plates are about 0.3 to 0.4 m^2 in area.   Cell sealing is 'wet seal' using the phosphoric acid whilst the manifold seal is a Teflon caulking.

Most of the polarisation losses ($\sim$ 30%) occur at the cathode (Fig.3,ref.8). The cell performance can be improved by increasing the oxygen partial pressure and overall cell pressure to 8 bar; however, the temperature has then to be raised to 220°C in order to provide steam at a sufficient pressure for the fuel processor.   Fuels for the PAFC can be provided by steam reforming and catalytically converting CH_4, naphtha, methanol or coal gas to give a hydrogen rich fuel gas.   However, most of the CO must be removed as it is poisonous to the anode catalyst at about 1%, whilst desulphurisation down to 0.1 ppm is essential. (9,10)

There are a number of mechanisms which eventually cause the phosphoric acid fuel cells to fail in long-term use.   Prominent amongst these are the stability of the electrocatalyst and the resistance of the bipolar carbon plates to chemical attack by hot H_3PO_4, (11,12). A PAFC stack has been operated by United Technologies for 500 days with about 10% drop in voltage at a current density of 130 mA/cm^2 (13), but long term hot phosphoric acid corrosion is a major hazard.

Molten carbonate fuel cell (MCFC)

The molten carbonate fuel cell construction (Fig.4) is similar to that of the PAFC but the materials used are quite different and it operates at 650°C, somewhat above the eutectic temperature of the ternary Li, Na, K carbonate melt. Because of the higher operating temperature porous nickel containing 10% Cr is a satisfactory anode material, whilst porous nickel, whose surface is oxidised to nickel oxide and then doped with lithium oxide, is used for the cathode. The electrolyte can be the ternary alkali carbonate or, more often, a binary mixture containing 35 mol % Li_2CO_3 and 65 mol % K_2CO_3.   This mixture forms a paste at 650°C with Li aluminate and a "tile" when frozen. A crucial factor is the porosity of the paste, hence sub-micron lithium aluminate powder is required, as this determines the porosity of the electrolyte matrix tile.   The anode separator and collector plate is nickel but at 650°C this metal is oxidised at the cathode, hence corrosion resistant stainless steel is used.   The advantage of the MCFC over the PAFC is that CO from fuel gas reforming or coal gas does not spoil the system, but undergoes the water gas shift at 650°C, $CO + H_2O \rightarrow CO_2 + H_2$ providing further fuel.   Sulphur, however, is deleterious and must be removed from the fuel gases down to 1 ppm.   The carbonate electrolyte actually takes part in the electrode reactions

Anode	$CO_3^{2-} + H_2 \rightarrow CO_2 + H_2O + 2e$
Cathode	$\frac{1}{2} O_2 + CO_2 \rightarrow CO_3^{2-} - 2e$

It is, therefore, important that the oxidant gas should contain CO_2, usually in the ratio 2:1 of CO_2: O_2. An increase in pressure increases the current density as shown in Fig.5 (ref.8,fig.43.18). Optimisation of the system depends on the partial pressure of H_2 in the fuel gas, the extent of the fuel utilisation in the fuel gas (above 75% utilisation and current drain of 200 mA/cm^2, the voltage drop becomes unacceptable) and the correct CO_2 to O_2 ratio in the oxidant gases. Cell lifetimes of greater than 40,000 hr are predicted from existing data (J. R. Selman, ref.14) The usual cause of catastrophic failure before then is due to cracking of the electrolyte tile caused by thermal stresses and the consequent mixing of the fuel and oxidant gas streams. Other failures are due to electrolyte leakage and corrosion of component structures.

Molten acetate fuel cell (MAFC)

A system which is at present very much in the initial experimental stages is the use of the eutectic mixture 53.7 mol % CH_3 COOK and 46.3 mol % CH_3COO Na as the electrolyte (ref. 15-17). The eutectic mixture was formed into a paste with MgO powder and the cell was operated in the temperature range 270-320°C. The anode is porous nickel and the cathode was lithiated nickel oxide (Fig.6, ref.16). At these low temperatures unpromoted nickel or nickel oxide were not every effective electrocatalysts. However, corrosion and sealant difficulties were slight and there was no evidence of the electrolyte tile cracking on thermal cycling. With hydrogen fuel current densities of 8 mA/cm^2 at 0.5 volt were obtained (Fig.7, ref.17).

The molten acetate fuel cell could be the fuel cell system of the future. There are virtually no corrosion problems and the temperature of operation is low enough to avoid using energy over and above the resistive loss to keep the system molten. The molten salt system is neutral or slightly acidic and in the presence of water vapour is stable over days at temperatures below 300°C (18). The presence of CO_2 is not expected to be particularly deleterious as the free energy change involved in forming carbonates from aqueous acetate solutions is of the order of +140 kJ/gmol hence there is unlikely to be substitution in the melt. Another important fact is that the acetate eutectic mixture has only a slight positive change in volume on melting (or the order 1-2%)(ref.18), which means that the system can normally be repeatedly frozen and melted without cracking of the electrolyte tile enabling the system to be used as a reserve power supply.

Further investigations are proceeding at Swansea to investigate better catalytic materials and the effect of CO and CO_2 in the fuel supply.

High Temperature Solid Oxide Fuel Cells (HTSO) Fig.8 ref.8 operating at 1000°C have also been studied and are proving satisfactory for use with coal gas.

Integration of Fuel Cells and Fuel Processing Units

To obtain the maximum efficiency from a fuel cell electrical power system it is essential that the processing units to provide a hydrogen rich fuel be integrated with the fuel cells systems so that as much as possible of the heating value of the original fuel is recovered. Fig.9 shows a comparison of system efficiencies without waste heat recovery (ref.14). Each fuel cell requires chemical reactors and other processing units to prepare the fuel and to provide for efficient heat exchange and material transfer. Since sulphur and its compounds poison both nickel and platinum based catalysts desulphurisation is a common requirement for all types of cell, with removal of sulphur to 1 ppm necessary for nickel based catalysts, and removal down to 0.1 ppm recommended for platinum based catalysts. In a common arrangement hydrodesulphurisation is carried out by means of a cobalt/molybdenum/alumina catalyst followed by removal of hydrogen sulphide and organics by zinc oxide.

All types of cell will operate from coal-based or natural gas/naphtha based fuels but the types of fuel preparation depends upon the cell. Because of the more extensive plant and greater capital requirements coal-based systems have been proposed for central power stations of the order of 500 MW capacity. In this plant coal gasifiers are employed to manufacture enriched producer gas by means of a mixed air-stream oxidant stream which is then desulphurised. Natural gas, methane and naphtha fuels are steam-reformed to convert the hydrocarbons to a mixture of hydrogen and oxides of carbon. The subsequent processing of the fuel depends upon the type of cell.

Fuel preparation for the phosphoric acid cell

The PAFC cell converts hydrogen electrochemically, but it is poisoned by carbon monoxide. Since the degree of poisoning is a strong function of temperature, operation at a higher temperature is preferred because of reduced sensitivity to poisoning. The cell has primarily been considered for use as a local utility using methane or natural gas as feed, and a diagrammatic flow sheet for such a process is shown as Fig.10.

In the stage of primary reforming methane is reduced to about 1.5% by oxidation with steam, and a subsequent partial oxidation in air reduces methane further to about 0.2% with the exit gas leaving at 900°C. Modern reformers using nickel based catalysts can operate at pressures up to 40 bar, and in practice an operating pressure of 10 to 15 bar is sufficiently high to meet pressure losses in subsequent units and to provide fuel at the cell at a pressure sufficient for efficient operation.

The PAFC cell will operate in the temperature range to 250°C, but at lower temperatures the sensitivity to carbon monoxide poisoning is greater, while at higher temperatures the cell voltage falls and the corrosivity of the medium increases. The choice of temperature of operation appears to be greatly influenced by the operation of the water gas shift converters.

The exit gas from the reformer is cooled to about 400°C and enters the primary shift converter. Produce gases from this unit are cooled to about 200°C before entering the low temperature shift converter in which there is a small temperature rise with carbon monoxide reduced to about 1% in the exhaust gases. At this concentration reasonably long cell lives may be obtained at an operating temperature of 200-220°C, and this is the chosen temperature range since a lower temperature of operation would only be viable if the carbon monoxide concentration of the gas were reduced further.

In the 4.5 MW plant that recently completed trials in Tokyo, fuel gas passed directly from the low temperature shift converter into the cell, and water was condensed from the gas following reaction. Reforming of methane by steam produces water in excess of that required by the reformer.

$$CH_4 + 2H_2O \rightarrow CO_2 + 4H_2 \qquad \text{reforming and shift reactions}$$

$$4H_2 + 2O_2 \rightarrow 4H_2O \qquad \text{cell reaction}$$

and therefore water required for reforming was condensed, and excess water purged.

The condensed water was treated and entered a thermosiphon heat exchanger cooling the cell, with generated steam fed to the reformer. Exhaust gases containing unused hydrogen were burnt in the reformer to meet heat requirements for this reaction.

Current densities in the cells fall as the hydrogen concentration is reduced by reaction, and this is a serious problem since it is usually not economic to convert more than about 80% of hydrogen in the feed stream before removing product water. If carbon dioxide is removed from the gas stream hydrogen may be recycled as a cell fuel with a small purge stream to remove nitrogen that enters the system at the secondary reformer. However, the additional complexity of carbon dioxide removal and nitrogen purge may not be worthwhile since the re-use of exhaust hydrogen by burning is undoubtedly the simplest procedure.

Fuel preparation for the molten carbonate fuel cell

The MCFC has been proposed for use both with enriched producer gas and with reformed naphtha or methane feed streams. Since the cell operates at about $650^{\circ}C$ shift reaction beyond that already carried out in the primary and secondary reformers will reduce the carbon monoxide content and there may be a degree of electrochemical conversion of the carbon monoxide as well as the hydrogen.

In the 675 MW plant proposed by General Electric, coal in a coal water slurry is gasified in an oxygen blown converter to produce fuel gas at $1370^{\circ}C$ and 42 bar. The gas is cooled to $650^{\circ}C$ by passing through a series of regenerative heat exchangers condensing the water. The cooled gas passes through an ammonia scrubber, carbonyl sulphide converter and sulphur removal unit, and is reheated to $620^{\circ}C$ by the regenerative train. The pressure, now about 35 bar is reduced to 7 bar at $350^{\circ}C$ by a turbine, and is reheated before entering the cell stacks. The fuel exhaust is catalytically burnt and mixed with air to provide the correct CO_2/O_2 ratio for the fuel cell cathode, and this stream is exhausted through a gas turbine that produces 75 MW of electrical power. A further 150 MW is provided from the high pressure steam flows.

Fuel preparation for the molten acetate fuel cell

Although not yet commercially developed, difficulties with corrosion are much less than for either the PAFC or the MCFC. It is expected that the cell will be reasonably tolerant to carbon dioxide and therefore complete removal is not necessary, although downstream processing may require some removal. A schematic flow diagram is shown as Fig.11. Desulphurised gas leaving the secondary reformer at $900^{\circ}C$ is cooled to $400^{\circ}C$ where it enters the primary water gas shift converter, followed by secondary conversion to give a fuel gas containing hydrogen, carbon dioxide and a small amount of monoxide. If carbon dioxide removal is required the fuel gas is cooled to $30^{\circ}-40^{\circ}C$ where it enters a potassium carbonate scrubber; the emergent gas now contains about 1% carbon dioxide and 0.5% carbon monoxide. The fuel gas is heated to about $300^{\circ}C$ where it enters the fuel cells. When the fuel is reformed methane, the exhaust gases from the cell contain more water than required for reforming, but the reforming requirement may be met by recycling an appropriate fraction of the cell exhaust gas to the reformers, leaving water to be condensed from the remaining exhaust gas, with the residual hydrogen recycled as fuel.

Power and combined heat and power systems

Of the systems considered in this paper the PAFC is the most developed with operating hours in excess of 20,000 for individual plants, and efficiencies in

the field between 37 and 40%.   In addition, efficiency standards of the plant
are extremely good when used with steam-reformed fuels.   There is some
flexibility for combined heat and power at the design stage, and if the full
potential is realised, overall thermal efficiencies can approach 80%.   Further
development of the cell with respect to different operating processes, and
alternative methods of heat removal from the electrode stacks are likely to
improve efficiency and reliability further.

The extremely corrosive nature of concentrated phosphoric acid at
temperatures in excess of $200^{\circ}C$ is a disadvantage of the PAFC.   Reports on the
trials on the 4.5 MW units in Tokyo and in New York may not appear in the open
literature, but it is possible that carry-over of electrolyte from the cells
into piping and equipment may cause corrosion with consequent intermittent
operation of the unit.

Because of the operating temperature, and thermal stresses on the
electrolyte tile and electrode materials on starting up and shutting down, the
MCFC is more suited to steady base power loads rather than meeting fluctuating
demand.   However, estimated thermal efficiencies for the power plant are about
50%, while for combined heating and power schemes estimated efficiencies approach
90%.   Proven life test data for small single cells are now in excess of
10,000 hours with less than 10% fall in cell voltage.   Projected cell lives of
40,000 hours are anticipated, with cells operated at higher pressures to reduce
evaporative losses of carbonate.   It is anticipated that power stations based
upon the MCFC will come into service in the United States in the 1990's.

The acetate fuel cell has seen very little development as yet, but the
feasibility of the cell has been shown for small cells at low current densities.
There are indications that current densities may be significantly improved by
the use of more electrocatalytically active materials.

The environmental compatibility of the cell and reformed feeds should be
similar to the PAFC cell with less severe corrosion problems than either the
MCFC or the PAFC cells.   The removal of carbon dioxide from the fuel gases
depends upon the particular application of the cell.   If expectations of the
long-term tolerance of the cell to carbon dioxide are borne out, it may well be
feasible to operate a compact unit without CO_2 removal by removing water as
required from the cell exhaust gases to meet steam demands for reforming and
then burning residual gases to meet the energy requirements of the reformer.
Alternatively, carbon dioxide may be removed by absorption in potassium
carbonates and exhaust gases from the cell after removal of water may then be
recycled as fuel for the cells.

In operation of the cell solely as a power unit it is indicated that the
cell efficiency will be in the vicinity of 45%, but in combined heat and
power applications the net efficiency can be much higher when catalytic
oxidation of residual hydrogen in exhaust gases may prove attractive.

The development of power and combined schemes for the MAFC is dependent
upon the process strength of the cell itself.   Given this prerequisite there
are potential aptitudes of the MAFC for combined heat and power schemes, with
an operational flexibility that may be useful for some schemes.   The PAFC cell
appears to be a reliable dispersed power plant with capacities up to 10 MW.
For larger capacities and base loads the MCFC will probably be implemented
commercially in the 1990's with the solid oxide fuel cell a possibility in the
longer term.   To the United States and Japan the commitment to further research
and development is not in question, either now or in the foreseeable future.

International prospects for sales, licencing and construction of both power and combined heat and power schemes based upon fuel cells may improve considerably within the next few years, a wholly satisfactory reason for the present intensity of research and development.

References

1. A. B. Hart and G. J. Womack, "Fuel Cells", Chapman Hall, London 1967.

2. H. A. Liebhafsky and E. J. Cairns, "Fuel Cells and Fuel Batteries", McGraw-Hill, New York, 1968.

3. G. J. Young (ed.), "Fuel Cells, Vols. I & II", Reinhold, New York, 1960 and 1963.

4. B. S. Baker (ed.), "Hydrocarbon Fuel Cell Technology", Academic Press, New York, 1965.

5. J. O'M. Bockris and S. Srinavasan, "Fuel Cells - their Electrochemistry", McGraw-Hill, New York, 1969.

6. J. R. Selman and T. D. Claar (eds.), "Molten Carbonate Fuel Cell Technology", Electrochemical Society, New Jersey, 1982.

7. J. R. Selman and L. G. Macrinowski, "Fuel Cells", Ch.12, in "Molten Salt Technology", ed. D. G. Lovering,Plenium Press, New York, 1982.

8. P. Linden, Chap. 43, "Handbook of Batteries & Fuel Cells", McGraw-Hill, 1984.

9. M. Ratcliff, F. L. Pasey, D. K. Johnson and H. L. Chum, J.Electrochem.Soc., 132, 577 (1985).

10. "Renewable Fuels and Advanced Power Sources for Transportation Workshop", ed. H. L. Chum and S. Srinavasan, SERI/CP-234-1707. Solar Energy Research Institute, Golden, Colorado (1983).

11. J. T. Hoggins and M. L. Deviney, J.Electrochem.Soc., 131, 2610 (1984).

12. A. Kaufman, "Phosphoric Acid Fuel Cell Stack and System Development", Technical Progress Report Contract No. DE AC 03 78ET 15366, NASA Lewis Research Centre, 1979.

13. "Integral Cell Scale-up and Performance Verification", EPRI Project 842-4. Report no.EM1134, June 1979, Power Systems Division, United Technologies Corporation.

14. J. R. Selman, "The Molten Carbonate Fuel Cell", Preprint NATO Advanced Study Institute on Molten Salts, Camerino, Italy, 1986.

15. E. Rhodes, F. Dale and D. W. M. Williams, Proc. 4th International Symposium on Molten Salts, ed. M. Blander, Electrochemical Society, New Jersey, 1984 - 2, pp.676-693.

16. D. W. M. Williams, Ph.D. Thesis, University of Wales, 1973.

17. F. Dale, Ph.D. Thesis, University of Wales, 1976.

18. F. J. Hazlewood, E. Rhodes, A. R. Ubbelohde , Trans.Faraday Soc. 62, 3101, (1966).

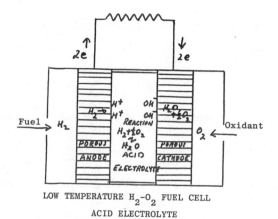

LOW TEMPERATURE H_2-O_2 FUEL CELL

ACID ELECTROLYTE

FIGURE 1.   FUEL CELL

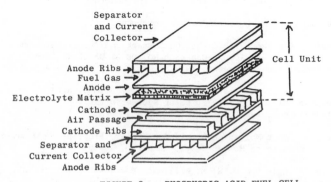

FIGURE 2.   PHOSPHORIC ACID FUEL CELL

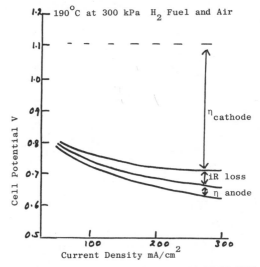

FIGURE 3   POLARISATION OF PHOSPHORIC ACID
             FUEL CELL

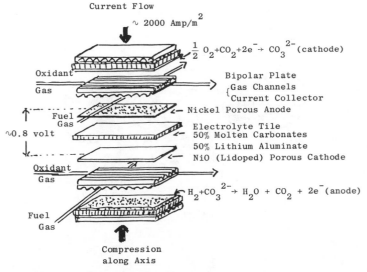

FIGURE 4.   MOLTEN CARBONATE FUEL CELL

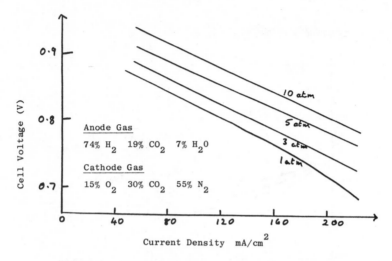

FIGURE 5   EFFECT OF PRESSURE ON MOLTEN CARBONATE FUEL CELL $650^{\circ}C$

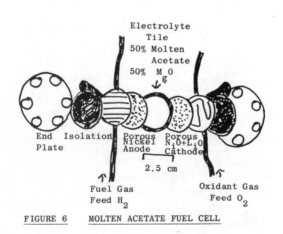

FIGURE 6   MOLTEN ACETATE FUEL CELL

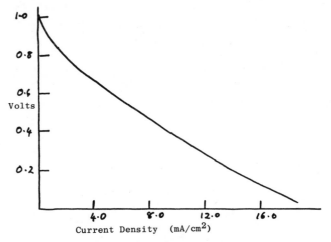

FIGURE 7   POLARISATION OF MOLTEN ACETATE FUEL CELL AT 305°C

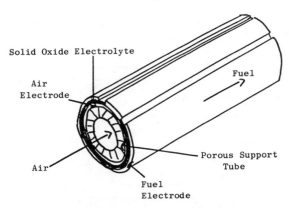

FIGURE 8   HIGH TEMPERATURE SOLID OXIDE FUEL CELL

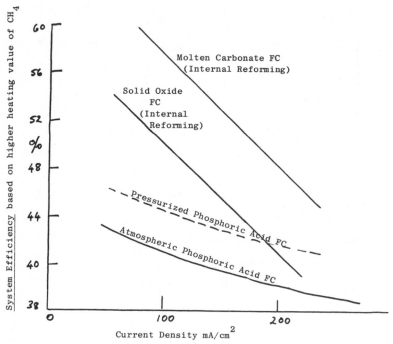

FIGURE 9   SYSTEM EFFICIENCIES

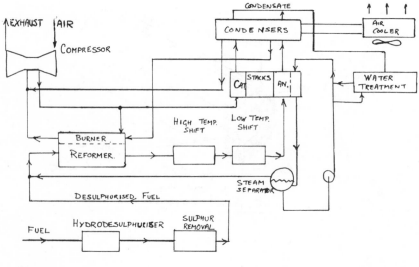

FIG. 10

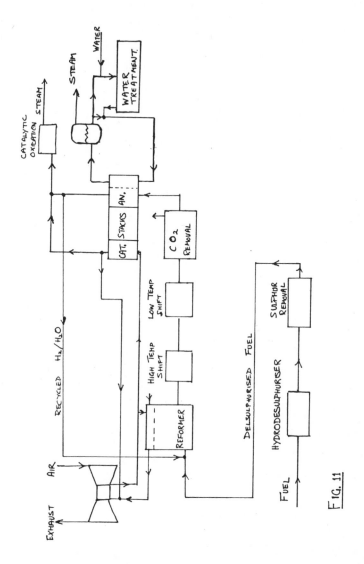

FIG. 11

WASTE HEAT RECOVERY USING MULTIPLE FLUIDIZED BEDS

T.F. Salam* and B.M. Gibbs*

Waste gases can be utilized for generating steam or preheating air. The recovered heat leads to an improvement in the overall efficiency of the process cycle. This paper describes the development of a heat recovery system using a circulating fluidized bed in which sand is circulated between two adjacent fluidized beds using jet pumps. The results showed that the fluidized bed system could be used as an efficient gas-to-gas heat exchanger with effectiveness approaching 80% of a perfect parallel flow heat exchanger. Heat recovery rates could be controlled by regulating the air flow to the jet pumps. Advantages include lack of moving parts, freedom from fouling and corrosion, and proven technology. A potential application is the recovery of waste heat from dust contaminated high temperature gases.

INTRODUCTION

Waste heat recovery, e.g. from flue gases, can significantly improve the overall efficiency of a process cycle, (reactors, furnaces, etc.), the recovered heat can be utilised in a number of applications, e.g. for raising steam, heating liquids or for preheating air. The heat recovery scheme which is eventually chosen is one which is easily incorporated into the existing plant operation and is generally justified in terms of payback periods. Currently, several heat recovery systems are commercially available and these include recuperators, regenerators, heat pipes, heat wheels and fluidized bed heat exchangers all of which have their own particular attributes.

Recuperators and regenerators are the two most commonly used air preheating systems, and can operate at temperatures in excess of 1000°C. A recent review, Reay (1), has shown that regenerators and recuperators, although capable of high temperature operations, have certain disadvantages regarding the susceptibility of the materials employed to degeneration, cracking, fouling and corrosion (of heat exchange surfaces). A major constraint in the design of heat exchangers is the low gas-to-surface heat transfer coefficient of between 30 to 100 W/m²K, which subsequently requires large heat exchange surface.

* Department of Fuel and Energy, University of Leeds, Leeds, LS2.9JT, U.K.

Fluidized beds have been extensively used in the petroleum, chemical and metal extraction industries at temperatures in excess of 1000°C and under harsh environments. Heat transfer coefficients as high as 700 W/m²K, have been reported. The subject of this paper is the development of a technique which utilizes parallel fluidized beds for recovering waste heat which might be available continuously or periodically for preheating combustion air or for the production of hot low dust content air for drying processes.

BACKGROUND TO FLUIDIZED BED HEAT EXCHANGERS

A fluidized bed is a direct contact gas-solid heat exchanger. Fluidizing gas is evenly supplied to the bed via a distributor in the form of flat plate or sparge pipe assembly. The velocity of the gas through the bed is such that the drag force of the particle is balanced by its weight, this balance exists over a range (due to the wide size distribution of the bed particles). A disengaging zone is provided to reduce elutriation of particles. Due to the highly mobile nature of the particles, a large surface area of the particles can be presented to the hot gas steam [3000-45000 m²/m³; Botterill (2)], resulting in very high heat transfer rates [6-23 W/m²K; Botterill (2)]. Details on fluidized bed technology can be found in many textbooks; Botterill (2) and Kunii and Levenspiel (3).

As a heat exchanger, the fluidized bed has a number of potential advantages. These are:-

(i)   Isothermal bed conditions - bed temperature is generally uniform, therefore there is negligible temperature gradient.
(ii)   Non-Fouling - the gentle scouring action of the particles keeps the exchange surfaces (e.g. tube bundles) clean.
(iii)   Pollution control potential - the fluidized bed acts as a natural particle filter. This could be an important feature because, e.g. an uncontrolled glass furnace emits approximately 0.75 gm/kg of particulates [Doyle and Donaldson (4)].
(iv)   Thermal capacity - particles normally used (e.g. sand and alumina) have very large volumetric heat capacities and there is potential for storing heat.

A fluid bed heat exchanger, however does have some potential disadvantages. The pressure drop through the unit will be higher (due to pressure drop across the bed and distributor) than other types of heat transfer equipment. Gas residence times are not uniform (i.e. for gaseous pollution control). In thermodynamic terms, the total amount of heat that can be transferred is limited to that of a perfect parallel flow heat exchanger. The latter can be overcome by combining the attributers of a fluid bed heat exchanger with those of a conventional heat exchanger.

The parallel fluidized bed heat exchanger is illustrated in Fig.1 which shows a plan view of the heat exchanger. Jet pumps are used to circulate bed solids (usually sand) between two interconnected fluidized beds. This enables the fluidized beds to operate at a flowrate most suited to the process requirements and the heat exchange rate to be governed by the rate of solid circulation induced by the jet pumps. In an industrial unit, hot flue gas (which would replace the in-bed electric heater) would be used to preheat Bed 1 (which is fluidized). The hot sand particles from Bed 1 are then circulated between Beds 1 and 2 by the jet pumps located in the dividing wall. Cold fluidizing air entering Bed 2 is subsequently heated. Some interchange of the fluidizing gas will take place, but previous experiments, Salam (4), have shown this entrainment to be small and can be almost eliminated by careful design. Thermodynamically, the system can be treated as a rotary regenerative heat exchanger where the circulating sand replaces the rotating heat wheel.

EXPERIMENTAL RIG AND PROCEDURES

The experimental unit (constructed from stainless steel) comprised of two adjacent fluidized beds, separated by an insulated partition wall. Each fluidized bed compartment of cross-sectional dimension 30 x 15 cm contains a bed of sand of mean diameter 262 μm, at a depth of 30 cm. The height of the unit was 90 cm. Fluidizing air is introduced into each bed via a tuyere type distributor plate (Fig. 2). Bed solids are circulated between the two beds by two jet pumps located in the dividing wall. Each pump circulates solids in opposite directions as shown schematically in Fig. 2. The vertical distance of the jet pump could be varied (heights of 10 or 20 cm.) according to the experimental conditions under investigation. The jet pumps were 5 cm apart. Slide valves were located at the entrance of the jet pumps. Vertical baffles could be placed between the jet pumps to increase the residence time of the solids in each bed to prevent their return before transferring the stored heat. An electric heating element was located in one of the fluidized beds in order to simulate a fluidized bed that was being heated by a hot gas stream.

Fluidizing air was supplied from a compressor and metered by means of rotameters. Compressed air was also used as the working fluid of the jet pumps and each air flow was metered by a rotameter and pressure gauge respectively. The power supplied to the electric heating element could be controlled by a variable transformer.

The fluidizing air supply (to the wind-boxes), bed and hot gas temperatures were monitored by thermocouples, the output of which was connected to a data logging system, and subsequently displayed on a mo nitor.

In order to determine the heat exchange rate, the two beds were fluidised at the appropriate fluidizing velocity (2.5-5.0 U/Umf). Bed 1 containing the in-bed heater was preheated to a temperature of about 105°C. When both Bed 1 and Bed 2 had reached steady state conditions, the jet pumps were turned on, (jet pump air flowrates between 300-900 cm³/s were investigated), the slide valves removed and solids circulated between beds. The temperature change in each bed was recorded by the data logger.

The ratio of jet pump flowrate to fluidized bed flowrate covered a range between 1.8 and 12.7%. The object of the experimental work was to determine the flexibility and effectiveness of using parallel fluidized beds as gas-to-gas heat exchangers.

RESULTS AND DISCUSSION

When the jet pumps were turned on and the solid circulation initiated between the two beds, the temperature of Bed 1 decreased
whilst the temperature of Bed 2 increased. This is shown in
Figs. 3 a and b, which are plots of the average temperature (from
four thermocouple readings) of Beds 1 and 2 versus time for different jet pump flowrate. The difference between the final exit
temperatures of Beds 1 and 2 depends on the circulation rate that
can be achieved and it is the solid circulation rate that principally governs the response and the effectiveness of the heat exchanger.

The solid circulation rates are evaluated from measurements of
bed temperature and fluidizing velocity and can be calculated
using the time dependent heat balance equations for Beds 1 and 2.
These can be written as:

$$M_1 C_p \, dT_1/dt = Q_1 + V_1 C_g \, T_1 - \dot{m} \, C_p (T_1 - T_2) + Q_e \qquad (1)$$

$$M_2 C_p \, dT_2/dt \quad Q_2 + V_2 C_g \, T_2 - \dot{m} \, C_p (T_1 - T_2) \qquad (2)$$

Under steady state conditions, $dT_2/dt = 0$, and Equation 2 can be
written as:

$$\dot{m} = V_2 C_g (T_2 - T_{in})/C_p (T_1 - T_2) \qquad (3)$$

The solid circulation rate '$\dot{m}$', was found to be dependent on
principally the jet pump flowrate and fluidizing velocity. Full
details of solid circulation are given elsewhere, Salam (5).

In order to evaluate the performance of the parallel fluidized
bed heat exchanger, its effectiveness was compared with that of
a perfect parallel flow heat exchanger. The heat exchanger
effectiveness is defined as the ratio of the actual rate of heat
transfer in a given heat exchanger to the maximum possible rate
of heat transfer. The heat exchanger effectiveness of the fluidized bed (ε_{tb}), can be written as:

$$\varepsilon_{tb} = \frac{C_h \, (T_1' - T_1)}{C_{min} (T_1' - T_2')}$$

where C_{min} is the smaller of the $\dot{m}_1 C_g'$ and $\dot{m}_2 C_g'$ magnitudes and
T_1' is the temperature of Bed 1 before heat circulation is in-
duced.

For a perfect, parallel flow single stage heat exchanger (ε_{pf}),
the effectiveness is approximately 50% with equal hot and cold
fluid thermal capacities and approximtely 67% when the thermal
capacity of the hot fluid is half that of the cold fluid.

Subsequently the heat exchanger effectiveness defined by

Equation 5 was plotted against jet pump flowrate.

$$E = \frac{\varepsilon_{fb}}{\varepsilon_{pf}}$$

Figs. 4 a and b show the dependence of E on the various operating parameters, e.g. fluidizing velocity of Bed 1/Bed 2, jet pump flowrate and position and the position of vertical baffles. It should be noted that any parameter that has an adverse effect on the solid circulation rate will lead to a reduction in the value of E.

It can be seen that when the jet pump flowrate was increased, the value of E rises sharply, this increase being greater for B.G.1 where the jet pump is further away (20 cm) from the distributor plate. Also, when the jet pump is closer (10 cm) to the distributor plate (B.G.2), the value of E reaches a limiting value much earlier. This is attributed to hydrodynamic effects of the distributor near the jet pump entrance which leads to a reduction in the circulation of solids between the parallel fluidized beds.

By comparing Figs. 4 a and b, it can be seen that when a vertical baffle is placed between the jet pumps, the effectiveness of the heat exchanger is increased and did not reach a limiting value over the range of experimental conditions investigated. The reason for this is the increase in solid residence time of the hot particle by the prevention of solid recirculation before attaining steady state temperatures.

The effect of fluidizing velocity on heat exchanger effectiveness is small, although there is an increase in system response due to the increase in solid circulation rates with increasing fluidizing velocity.

The result of reducing the flowrate or effectively the heat capacity of one of the fluids by half can also be seen from Figs. 4 a and b. The effect of halving one of the flows is to effectively increase the effectiveness from approximately 39% to 52%, however when this is compared with the effectiveness of a perfect parallel flow heat exchanger, it can be seen that the parallel fluidized bed heat exchanger effectiveness is actually approaching 80% of the maximum effectiveness possible. However, from the plots it can be seen that higher effectiveness can be attained by using higher jet pump flowrates.

CONCLUSIONS

The fluidized bed gas-to-gas heat exchanger was operated successfully and at an effectiveness approaching 80% of a perfect parallel flow heat exchanger. Heat recovery rates could be easily controlled by means of the solid circulation rate. The proposed heat exchanger has no moving parts which can fail, it is free from fouling and sand or alumina which is used as the heat transfer medium, is inert and will not degenerate in the presence of corrosive gases. Its potential application will be for recovering heat from hot dirty flue gases, which could lead to fouling, corrosion and thermal degradation. Processes such as glass furnaces, coal combustors, etc., could benefit from this type of preliminary heat recovery technique.

NOMENCLATURE

C_p	=	Specific heat of sand (kJ/kg K).
C_g	=	Specific heat of air (kJ/m³ K).
$C_g{}'$	=	Specific heat of air (kJ/kg K).
E	=	Ratio of heat exchanger effectiveness (-).
M_1, M_2	=	Mass of sand in Beds 1 and 2 respectively (kg).
$\dot{m}$	=	Solid circulation rate (kg/s).
$\dot{m}_1, \dot{m}_2$	=	Mass flowrate of air into Beds 1 and 2 (kg/s).
Q_1, Q_2	=	Air heat input into Beds 1 and 2 respectively (kW).
Q_e	=	Electric heat input into Bed 1 (kW).
Q_{jp}	=	Air flowrate into jet pump (cm³/s).
R	=	Ratio of Bed 1/Bed 2 fluidizing velocity (-).
T_1, T_2	=	Temperature of Beds 1 and 2 (°C).
T_{in}	=	Inlet air temperature of Bed 1 (°C).
T_1', T_2'	=	Initial temperatures of Bed 1 and 2 (°C).
t	=	Time (sec.).
U	=	Superficial fluidizing velocity (m/s).
U_{mf}	=	Minimum fluidizing velocity (m/s)
V_1, V_2	=	Volumetric air flowrate into Beds 1 and 2 (m³/s).
V_{jp}	=	Volumetric air flowrate into jet pump (m³/s).
ε_{fb}	=	Effectiveness of fluidized bed heat exchanger (-).
ε_{pf}	=	Effectiveness of perfect parallel flow heat exchanger.

REFERENCES

1. Reay, D.A., A Review of Gas-to-Gas Heat Recovery Systems. J. of Heat Recovery Systems, Vol. 1, p.3-41, 1980.
2. Botterill, J.S.M., Fluid Bed Heat Transfer, Academic Press, 1975.
3. Kunii, D. and Levenspiel, D., Fluidization Engineering, Krugir Pub. Co., Inc., 1969.
4. Doyle, E.E. and Donaldson, L.W., Batch Preheating via a Fluidized Bed Offers Improved Melter Operation, Glass Industry, p.18, July, 1985.
5. Salam, T.F., Ph.D. Thesis, University of Leeds, 1986.

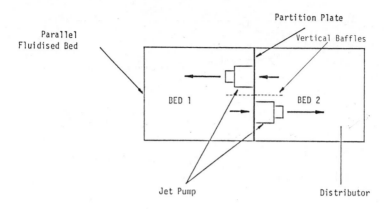

Fig. 1:   Plan of heat exchanger.

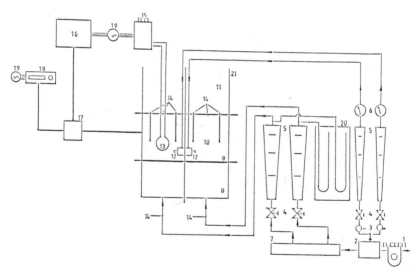

1. Oil and Water Filter; 2. Air Flow Splitter; 3. Quick Shut Valves;
4. Needle Valves; 5. Rotameters; 6. Bourdan Gauges; 7. Large Air Flow
Splitter; 8. Windbox; 9. Bubble Cap Distributer Plate; 10. Fluidized
Bed Zone; 11. Disengaging Zone; 12. Jet Pumps; 13. Electric Heater;
14. Chromel Alumel Thermocouples; 15. Variac; 16. Date Logging System;
17. Thermocouple Wire Splitter; 18. Digitron Digital Thermometer;
19. A.C. Supply; 20. Water Manometers; 21. Stainless Steel Fluidized
Bed.

Fig. 2:   Schematic diagram of experimental rig and ancillaries.

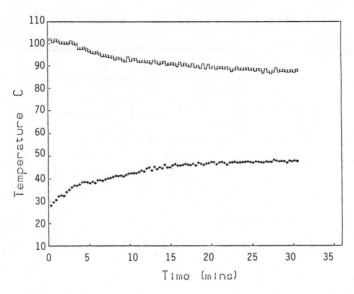

Fig. 3a: Average bed temperature versus time (Q_{jp} = 300 cm^{3}s^{-1})

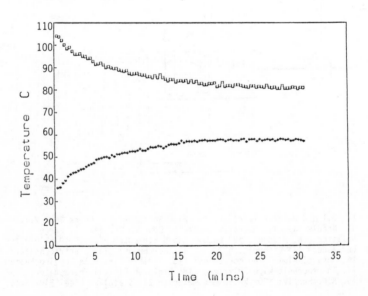

Fig. 3b: Average bed temperature versus time (Q_{jp} = 900 cm^{3}s^{-1})

(a)

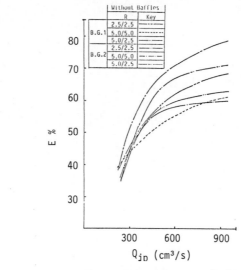

Air flowrate through jet pump (cm³/s)

(b)

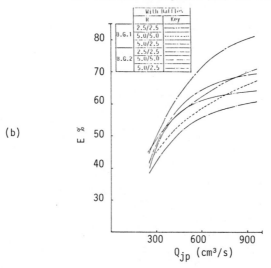

Air flowrate through jet pump (cm³?s)

Fig. 4 (a and b): Plot of heat exchanger effectiveness (as a percentage) versus jet pump flowrate.
(a) Jet pump at two positions (20 and 10 cm from distributor) without baffles.
(b) Jet pump at two positions (20 and 10 cm from distributor) with baffles.

THE PRESCRIPTION FOR SPEEDY WASTE HEAT RECOVERY

G Srinivasan, BSc, CEng, FIChemE, MInstE, ACGI
WS Atkins Management Consultants, Epsom, Surrey

SYNOPSIS

Heat recovery is a low priority for industry. Hot furnace gas are still passed into the atmosphere while many companies pour hot effluents down the drain. This paper discusses the designing, specifying and installing of heat recovery schemes, and draws on experience from the Management of the DTI's Heat Recovery Scheme. Equipment selection must be made with due regard for potential corrosion, erosion and fouling of heat exchange surfaces. Reference to some Case Studies from the DTI Scheme highlights some of the other factors which affect the technical feasibility and cost effectiveness of projects.

WASTE HEAT RECOVERY - THE PROBLEMS FACED BY INDUSTRY

Generally industry accepts that cutting energy costs through waste heat recovery makes good economic sense. However there continues to be a marked reluctance for companies to actually invest in heat recovery equipment, and often when the money is spent the anticipated savings are not achieved. Allowing feasibility studies to gather dust on an Energy Manager's bookshelf does nothing to promote the economic health of a company. The cure to the problem of cutting energy cost is the adoption of a positive approach. However in the competition for scarce resources, capital projects which are soley concerned with energy savings are less likely to be sanctioned than more glamorous ones which are associated with production improvements.

In fairness, many companies have appointed committed Energy Managers who have developed energy conservation programmes, the first phase of which normally consists of a detailed energy audit. Government funding is available for this work if it is carried out by independent Consultants.

They will produce a report analysing the factory operating conditions and, most importantly, identifying various possible energy saving measures. This is normally arranged as a 'hit-list' categorised according to cost effectiveness and the level of actual investment which is needed. Companies have been generally willing to implement the low cost measures immediately since these normally have a rapid payback period and there is little or no technical or finanical risk. Examples of such measures included improved control systems, better standards of maintenance, improved insulation etc.

The next phase of a Company's programme calls for a significant level of capital investment and it is at this stage that the best laid plans are often set aside. In most cases the initial heat recovery ideas are obvious, and the project economics can be quickly established through a few simple calculations. All too often however the good ideas are not pursued any further. What is the reason for this reticence, and what is the cure?

THE NEED FOR PROPER EVALUATION

One of the main difficulties faced by engineers in industry is that capital sanction for heat recovery projects is not readily forthcoming unless a convincing case can be presented. The initial enthusiasm for a project needs to be supported by a careful study of the various options, and a consideration of the overall system to ensure that what is being proposed will work in practice. Reasonably accurate costings need to be obtained, but nearly always these are underestimated. Another problem is that engineers, and this includes Consultants, tend to produce feasibility reports which contain too much irrelevant technical data, do not address key problem areas (such as the possible fouling of heat exchanges surfaces) and do not provide sufficiently clear recommendations to allow an investment decision to be made with confidence. The ivory tower approach will always be viewed with suspicion and this is why a systematic and logical design approach is essential if the correct heat recovery decisions are to be made.

The alternative approach of delegating such studies to equipment suppliers should not be regarded as a cure to the problem. The apparently free design advice which is provided by manufacturers is naturally biased around their standard kits and is generally based on inadequate data and insufficient effort.

It is also unfortunate that over the last few years there have been some instances of heat recovery projects which have failed, mostly through poor engineering, or incorrectly selected materials of construction. This limited equipment life stems from cutting corners during the design stages. It is accepted that such occurrences are the exception rather than the rule, but even so, heat recovery, perhaps unfairly, does not enjoy a good reputation.

THE DTI HEAT RECOVERY SCHEME

Some four years ago the Department of Trade and Industry in recognising the foregoing problems faced by industry launched the Industrial Heat Recovery Consultancy Scheme (IHRCS). This offered high energy users in industry the benefit of funded consultancy studies carried out by independent experts. The DTI entrusted the technical and financial management of the IHRCS to a firm of energy consultants and the author has acted as Scheme Manager. This has allowed first hand experience to be derived of industry's present uptake of heat recovery projects and the way in which they are designed and specified. The following sections discuss, in more detail, the various steps which are necessary in carrying out a proper design of a heat recovery system. Reference is then made to some Case Studies from the IHRCS to show how this prescription for speedy waste heat recovery, i.e., proper attention to design detail, ensures the eventual health of the installed plant.

STAGES OF DESIGN FOR HEAT RECOVERY SYSTEMS

(i) Quantification of Heat Source

This can generally be determined by direct measurement of the flow and temperature of the gas or liquid stream being discharged. It is also important to verify the temperature profile of the waste fluid. A constant temperature level is likely to be a better proposition for heat recovery than a heat source at a varying and inconsistent temperature.

Other operating conditions should be noted. Any dust or acidic fumes entrained in an exhaust stream will affect the final equipment selection while the moisture content of a gas stream can significantly improve the heat recovery potential if there is latent heat recovery through condensation. The recovery of as little as 10% of the latent heat from a drier exit gas stream can render an unattractive project into one which is clearly cost effective.

The collection of data is particularly important when retrofitting equipment for, often, the flows of exhaust gas or hot effluent are quite different from design values. It is also whilst carrying out on-site trials that any scope for improving plant performance by altering process conditions becomes apparent. Clearly there is little point in recovering waste heat which should not be present in the first place.

(ii) Selection and Measurement of Heat Sink

The quantification of the heat requirements of the area in which it is proposed to re-use the waste heat involve similar measurements to those described in (i) above. A common problem with heat recovery systems is a mis-match between the heat source and sink, not only in terms of heat load, but in terms of cycle times. An example of the former problem arose during one of the IHRCS funded Design Studies. It involved the condensation of a steam/petrol vapour mixture from a solvent recovery plant using a water cooled shell and tube heat exchanger. The amount of the recoverable heat greatly exceeded (i.e. by a factor of five times) all of the available heat sinks. Even so, the proposed project was shown to be cost effective. The mis-match on batch processes can be overcome through the use of buffer storage tanks.

(iii)   **Consideration of Various Options**

On a factory site in which production involves several processing steps there may be a series of heat recovery options to be considered before a preferred scheme can be finally selected. Process integration techniques can be used to identify the most effective heat exchange system from a thermodynamic point of view and provides a good basis for developing subsequent designs. Apart from matching the heat sources and sinks, there are other quite fundamental choices facing the designer, e.g. there is the decision to be made between a heat pump system and a passive heat recovery device.   It is generally difficult to justify a heat pump on an industrial application, because of the requirement for a rapid payback - usually 2 years or less which in the author's opinion is an unreasonably strict requirement bearing in mind that energy savings, once made, are available for the future life of the plant.   However in the case of swimming pools the dilemma becomes more interesting.   In terms of simple payback the favoured arrangement is a straightforward heat exchange system to preheat incoming air with the warm exhaust.   (i.e. 1 year vs 4 years.)   However, almost double the savings can be achieved with a heat pump and if a discounted cash flow analysis is made over a 10 year period then the heat pump solution is much more attractive.

(iv)   **Selection of Heat Recovery Technique**

Except in those rare cases where the exhaust heat can be re-used directly, such as the preheating of metal scrap with flue gases, a heat exchanger is needed to recover the heat.   There are numerous devices available for this purpose ranging from thermal wheels to heat pipes and from recuperators to run-around systems. The various types of static heat exchanger include plate, spiral and shell and tube type units.

There are various pros and cons associated with each specific equipment type and a discussion of these can be found in any standard text on heat recovery.   The suitability of a particular type of heat exchanger must be carefully considered, but often, the personal preference of the designers, based on successful past experience, may be an overriding factor, and justifiably so.

(v)   **Development of Heat Recovery System**

Although the heat exchanger is the core item of a heat recovery system its cost is generally only a fraction of that of the total project, particularly if it is a turnkey one.

A complete installation may require new or uprated fans, pumps, tanks, interconnecting ducting and pipng, instrumentation and a system of controls involving modulating valves and electrical interlocks.   This stage of design requires the preparation of detailed engineering flow diagrams to define the way in which the plant is intended to work.

(vi)   **Detailed Engineering Design**

At this stage the practical engineering problems need to be resolved. The layout of equipment, the prevention of fouling and blocking and the selection of the correct materials of construction are typical design considerations.

It can be seen from the foregoing discussion that short cuts should not be taken when installing heat recovery equipment, a maxim which applies when embarking on any engineering project requiring capital expenditure. In the past, for some reason, companies have assumed that normal project disciplines need not be applied to heat recovery plant.

CASE STUDY 1 INADEQUATE PLANT DATA

It has already been emphasised that the development of meaningful designs of heat recovery plant depends on knowing, with reasonable certainty, the operating conditions on which the designs can be based. Often the obtaining of plant data is easier said than done. Some key process parameters cannot be measured directly, and often, in such cases the current plant operation does not correspond to that defined in the original specifications.

An example of that situation arose on a study of a small SO_3 production unit on which it was proposed to recover heat from the sulphur combustion and the SO_2 to SO_3 conversion reactions. On a normal scale sulphuric acid plant these particular duties represent essential gas cooling duties to ensure 95%+ efficiency, and so heat recovery (usually in the form of h.p. steam) is not only a bonus but is an established process feature.

On the plant in question the necessary cooling was achieved by allowing the process gas to pass through sections of unlagged pipe exposed to the ambient air. However, there was only a minimum of installed instrumentation and the information provided by the Client relating to the sulphur burning rate, the SO_3 production rate and the air flow rate was inconsistent. Details of these could not be safely or readily obtained through on-site measurement. Therefore the plant operating conditions, and in particular, the air flow rate had to be inferred from the temperature conditions and the construction of the 'equilibrium-operating' diagram as shown in Figure 1. This assumes a reasonable approach to equilibrium in each of the three reaction stages and an overall plant efficiency of about 90%.

The concentration of SO_2 in the gas stream leaving the sulphur furnace significantly affects performance and is typically in the range 6 to 10% by volume. As well as direct gas sampling, the temperature of the exhaust gas from the sulphur furnace normally gives an accurate indication of gas strength, but a suspect thermocouple reading rendered this cross-check invalid.

The diagram which was felt to be the 'best-fit' of the operating conditions allowed the plant air flow rate to be deduced and hence the potential value of the recovered heat. The calculated conditions gave a higher than expected air flow rate with a correspondingly low concentration of SO_2 in the feed gas. It was therefore necessary, in specifying the duties for the heat exchanger to check if and how performance would be affected in the event that the assumed conditions were not actually correct.

CASE STUDY 2 - HEAT PUMP CYCLES

Apart from giving longer than normally accepted payback periods, heat pump projects require a careful study of all the possible arrangements to ensure that the finally accepted scheme is properly optimised thermodynamically and moreover is a practical proposition. These points are borne out by the following Case Study.

The project is concerned with the recovery of 2½ MW of heat in the overhead vapour stream from an alcohol still. As a first step, a small amount of the heat is being used to preheat the inlet feed stream to the distillation column.

In further developing the ideas for heat recovery it was recognised that it was best to re-use the heat within the distillation process itself. The source and sink are well matched in size and exactly matched in terms of time availability and requirement. However, the temperature levels are such that some heat pump arrangement is required; the temperature of the overhead vapour is 78°C while the boiling temperature at the bottom of the still is 104°C.

The following four options were considered:

Option A Open Cycle Heat Pump

This arrangement which is shown diagrammatically in Figure 2.A consists of using the overhead water/alcohol mixture itself as the working fluid. This is compressed and passed to the reboiler in which the necessary heat input to the column is provided. However, the handling of hot, inflammable gas poses a safety risk, while there is also the need to ensure that there is no contamination of the product by compressor lube oil.

Option B - Closed Cycle Heat Pump

This is shown in Figure 2.B. The latent heat of the condensing water/alcohol mixture is absorbed by the boiling working fluid which is then passed to the reboiler. The normally used halocarbons tend to break down under 100°C+ temperature conditions and will give a poor performance due to operation near the critical point. Special refrigerant R114 could be used but there is insufficient long term operating data available to recommend its use for continuously operating process plant.

Alternatively, water could be used as the working fluid but this requires sub- atmosphere evaporation, a high pressure ratio and a de-superheater.

Option C Open Cycle Heat Pump Using Thermocompressor

This is shown in Figure 2.C. The thermocompressor is a device which uses a flow of high pressure steam to compress a flow of low pressure steam to give a combined flow at an intermediate pressure. In this case the low pressure steam is that which is flashing off the heated cooling water leaving the overheads condenser. A practical problem with this arrangement is erosion of the thermocompressor venturi.

Option D Open Cycle Heat Pump Using Mechanical Compressor

This is similar to option C but uses a mechanical compressor and in this case is the preferred option. A special type of heat exchanger is used for condensing the overheads while some boiling of the cooling water occurs because of the sub-atmospheric operation. The centrifugal compressor is a high cost high efficiency unit while the Rootes type is a low cost low efficiency machine which is best suited for low pressure ratio duties.

The project economics depend on the ratio of electricity to fuel costs, and with the present low oil prices, the estimated payback period is over 5 years. However the operating company concerned is seriously considering the implementation of this project in the event of fuel prices returning to their former high levels.

CASE STUDY 3 - FRUGALITY FROM FUMES

On a new plant it is much easier to justify arrangements for heat recovery even though these will not be justifiable in a retrofit situation. An example of such a case is shown in Figure 3. The heart of the process is a series of adhesive drying ovens. The exhaust air from these is laden with solvent and plasticiser fumes. These have to be recovered or disposed of. With the calorific value of the exhausted air being sufficient to satisfy the heat requirements of all the new requirements, the route chosen for treating the exhaust was a thermal oxidiser. The exit temperature of the combustion products was 760°C.

Consideration of the overall process, with heat recovery in mind, led to various changes being made to the initially proposed designs of oven and oxidiser. The air extraction rate in the oven was reduced in order to minimise the level of excess air for the combustion process in the thermal oxidiser. However the extent to which the air rate could be reduced was limited for safety reasons since it was necessary to maintain fume levels below 25% of the Lomer Explosive Limit. The speed of the exhaust fan is modulated according to fume level. Extraction rates during start-up are 20% of normal running to further reduce fuel use in the oxidiser.

On the oxidiser, the burner turn-down ratio was substantially increased to reduce fuel costs since the minimum requirement was for a pilot ignition source only. The additional savings from this measure will be several tens of thousands of pounds per annum.

The finally selected heat recovery system consists of a cascade of three units in series, with the lowest temperature being the last in the train. The exhaust gas/thermal oil heater precedes the air preheater to enable maximum heat addition to the thermal fluid. The selection of sutably sized heat exchangers has meant that all the energy input for the overheating is provided by the oxidiser exhaust gas stream.

The following control features are included:

High solvent by-pass to control the combustion temperature in the oxidiser, a high temperature by-pass around the first heat exchanger to protect the thermal fluid from decomposition and a second by-pass on the economiser to prevent condensation in the fans and flue.

CONCLUSIONS

Experience has shown that, with heat recovery projects, as in any management situation, the need for a positive approach and good communication between the participating parties is of paramount importance. There are very few problems which defy solution but often unresolved fouling and corrosion problems are used as an excuse for dropping a project in which there is only a half-hearted interest.

It is also important to take a long term view that energy savings, once made, are available throughout the future life of a plant. The application of a rigid payback criterion of 2 years is short-sighted and could mean that many good projects are not implemented. Those companies which adopt a more relaxed attitude to capital investment reap the benefits in the long run and also gain a useful competitive edge over their rivals.

REGENERATORS - "APPROACHING THE ULTIMATE IN HEAT RECOVERY DEVICES"

N.H. Malik*, M.S. Saimbi* and B.S. Sidhu*

British Gas plc has developed, in conjunction with a manufacturer, a compact regenerative burner system suitable for use in a wide variety of high temperature process heating applications. Around 350 regenerative burners have now been installed, the system proving to be particularly valuable in applications where the flue gases are contaminated. Fuel savings of up to 65% have been achieved usually allowing the capital cost of installation to be recovered in less than two years. This paper discusses the initial development work and evaluation of a production prototype. A comparison is drawn between laboratory performance measurements and the predictions of a mathematical model. The first installation of the system is described.

INTRODUCTION

Rising fuel prices and increased competition in the marketplace have, in recent years, prompted many industrial users to examine the ways in which their heating requirements can be provided cost effectively. Development of more energy-efficient combustion equipment has therefore attained considerable momentum, particularly in the area of industrial high temperature process heating. Traditional burners and furnaces usually deliver only a small proportion of the available energy in the fuel as heat to the stock, the majority of the energy being lost through the combustion products.

Efficiency can be improved by recovering heat from the combustion products and using it either to preheat the furnace load or to preheat incoming combustion air. The second option is by far the most widely applicable in retrofit situations. The potential fuel savings, as a function of air preheat temperature, are shown in figure 1. The metallic recuperator, operating in the lower shaded area of figure 1, is currently the most commonly used air preheat device, continuously transferring heat between flue products and incoming air across a metallic interface. The recuperative burner[1,2] is an example of such a device, which combines the functions of a burner, recuperator and flue in a single compact unit. Around 2500 recuperative burners have now been sold by licensed manufacturers.

The potential fuel savings due to the preheating of air are shown in Figure 1. Currently available metallic recuperators operating at high temperatures have a performance which falls into the lower shaded area with the maximum air preheat being limited by the service temperature of available materials[2]. This results in fuel savings of up to 35% but still leaves a

* British Gas plc, Midlands Research Station, Wharf Lane,
  Solihull, West Midlands, B91 2JW.

substantial amount of the heat content of the combustion products unrecovered. Ceramic materials are able to withstand high temperatures but they are difficult to form into the complex configurations required for a compact recuperator and are subject to failure at quite low levels of tensile stress. Their high temperature properties are therefore best used in simple shapes such as the packing of a regenerator. Regenerators work on the principle of short term heat storage and therefore operate in a totally different way to recuperators. In a regenerator, a matrix of ceramic material is used as the heat store and is alternately heated by the flue products and cooled by the combustion air.

Two types of regenerator have traditionally been used. The rotary regenerator[3] consists of a cylindrical arrangement of ceramic material which rotates about an axis parallel to the gas flows. Heat from the combustion product stream is taken up by the matrix and transferred to the incoming air as the cylinder rotates. In this arrangement, it is difficult to provide satisfactory sealing between the gas streams and the cylinders are also liable to mechanical failure induced by thermal stresses. Static bed regenerators are widely used in the glass manufacturing industry and consist of very large and costly arrangements of brick chequerwork which can withstand temperatures of up to 1600°C. These regenerators have an acceptable heat exchange performance as a result of the large size of packing involved, but are unsuitable for the majority of high temperature plant as they cannot be retrofitted.

Through consideration of currently available methods of heat recovery an opportunity for the development of a compact, high-effectiveness switched static bed regenerator system which would be suitable for use on both existing and new furnaces was identified, and a programme of work was initiated in order to carry out studies into the thermal design of such a system.

DESIGN CONSIDERATIONS

The performance of regenerators is governed by a number of parameters including bed geometry, overall heat transfer coefficients, storage characteristics of the bed, the flowrates of the gases and the cycle time[4,5]. During each cycle the bed and fluid temperatures are continually changing and detailed analysis and optimisation of such a system can only realistically be carried out with the aid of a computer based mathematical model. However, performance can usefully be characterised through the use of two dimensionless groups.

Dimensionless length $\quad \lambda = \dfrac{hAL}{G\,C_g}$

Dimensionless period $\quad \pi = \dfrac{hAP}{M_s C_s}$

The relationships between these parameters and regenerator effectiveness are available in graphical form[6] for balanced, symmetrical regenerators and indicate that for maximum effectiveness, the dimensionless length should be as high as possible and the dimensionless period small. These requirements can be met by using a bed packing which has a high heat transfer surface area together with high density and specific heat to provide substantial heat storage. Thin-section packing is desirable as it ensures that conduction effects are minimised and the maximum heat storage capacity is utilised. This is difficult with thicker sections due to the low thermal conductivity of ceramics.

Values of the two dimensionless parameters are limited by practical considerations. The achievement of high dimensionless length requires high heat transfer coefficients which result in a high pressure drop through the bed and are of the order of 2.5 kPa. This is a major departure from conventional recuperator practice where flue side pressure loss is minimised to allow the system to operate under natural draught or the suction created by a low pressure eductor and thus avoid the use of costly hot gas fans with high maintenance requirements. But this is not possible in a compact regenerator where flue gases and air flow through the same bed and pressure losses are therefore similar. The level of pressure loss in a compact regenerator need not, however, pose difficulties because if the regenerator effectiveness is high, the outlet flue gas temperature will be sufficiently low to allow extraction through conventional cold air fans.

In order to achieve a high effectiveness, the value of the non-dimensional period should be small. This can best be achieved by reducing the reversal period, but is limited by the need to ensure satisfactory lives for control components such as valves relays and switches. Furthermore after each reversal, the flow patterns of combustion products in the furnace can take some time to establish and failure to maintain these patterns for a substantial proportion of each cycle could result in the level of temperature uniformity being reduced and the product quality being compromised.

The performance of a recuperative or regenerative heat recovery system is conventionally expressed in terms of effectiveness, a measure of the amount of heat recovered as a proportion of the total recoverable heat content of the flue gases. For simplicity, effectiveness can be evaluated using an expression containing flue gas and air temperatures alone, avoiding the difficulty of measuring flow rates.

$$\text{Heat recovered} \quad q_a = (\dot{m}C)_a (T_{ao} - T_{ai}) \tag{1}$$

$$= (\dot{m}C)_f (T_{fi} - T_{fo}) \tag{2}$$

For a heat recovery system in which the outgoing combustion products are used to heat the incoming air for combustion, the heat available in the combustion products is greater than can be absorbed by the air. This is due to the fact that the product of mass flow and the specific heat for the combustion products stream is higher than that for the air. Since the maximum temperature to which the air can be heated is equal to the flue gas temperature at inlet to the heat exchanger, the maximum recoverable amount of heat is given by:

$$q_{max} = (\dot{m}C)_a (T_{fi} - T_{ai}) \tag{3}$$

Defining the effectiveness, η, as $\dfrac{q}{q_{max}}$ and substituting

$$= \frac{(\dot{m}C)_a (T_{ao} - T_{ai})}{(\dot{m}C)_{min} (T_{fi} - T_{ai}} \tag{4}$$

Therefore $\qquad \eta = \dfrac{T_{ao} - T_{ai}}{T_{fi} - T_{ai}}$ $\qquad\qquad$ (5)

This expression can be readily evaluated and is particularly useful for quoting performance. For a typical gas fired appliance with good air/gas ratio control, an ideal heat recovery device, having an effectiveness of 1, would recover just over 80% of the waste heat.

REGENERATOR MATHEMATICAL MODEL

As described in the previous section, regenerator design requires a knowledge of the heat transfer characteristics of the bed and its pressure drop, the former usually being the controlling factor. The normal procedure is therefore to size the bed for the required thermal performance. Pressure drop is then evaluated using Ergun type correlations[7], obtained under laboratory conditions, to ensure that it does not exceed the capacity of commonly available suction fans. If this condition is not met, bed size and packing geometry are varied and the process repeated.

A mathematical model, to describe the behaviour of compact regenerators, was developed and it includes conservation of energy equations for the two fluid streams and the bed material. Their initial and boundary conditions and the pressure drop equation for the bed. The basic assumptions made are:

1) Constant physical properties for the bed material and fluids.
2) No conduction effects in the solid or gas phases
3) Plug flow
4) Constant heat transfer coefficient along the length of the bed.
5) Adiabatic regenerator walls

PROTOTYPE UNIT

Thermal design of a prototype system was finalised using the above procedures. A schematic diagram of the system, built for evaluation under laboratory conditions, is shown in figure 2, and the heat storage bed is shown in detail in figure 3. Zirconia cylinders 20mm long and 20mm in diameter with an average wall thickness of 4mm were used as the packing material, arranged in staggered rows to ensure that there was no clear gas flow path through the packing. The bed containers were vacuum formed ceramic tubes insulated with fibre blanket and the bed material was supported by perforated stainless steel plates, made possible by the low gas temperature at this point. The advantages perceived for the chosen packing material were acceptable cost, adequate strength at high temperature, good oxidation resistance and low pressure drop. The packing was designed for the following duty:

1) Burner heat input 150 kW
2) Maximum flue gas temperature 1400°C
3) Reversal period 150s
4) Effectiveness in excess of 70%

The use of high air preheat temperatures suggested that extremely arduous conditions would prevail in the region of the burner, an air preheat of 1000°C raising adiabatic flame temperature by 350°C. It was therefore necessary to design an appropriate burner[9] which could be fabricated entirely in high temperature refractory materials. This burner has a short flame and high combustion product exit velocity, both beneficial features in establishing the highly turbulent, recirculating flow within a furnace which promotes good temperature uniformity. Such burner characteristics allow the required flow patterns to be established using a minimum number of burners, thus reducing capital costs.

The burner produced differs from conventional practice in the respect that it does not have a central nozzle for flame stabilisation, this function being performed by an expansion step. Such an arrangement results in an attractively low burner pressure drop. A further requirement of the burner was to effectively insulate incoming fuel gas from the high temperatures within the burner to prevent cracking and carbon deposition.

The prototype system was installed on a laboratory furnace and subjected to an extensive test program; an important part of which was to assess the long-term effects of cyclic thermal stress on the various components. The thermal performance of the system is illustrated by typical temperature profiles for the system shown in figure 4. Using mean values of preheat, the effectiveness of the prototype was calculated to be 76% under the specified conditions, rising to 84% when the reversal period was reduced to 60s. This work was crucial in proving the concept of small scale, compact regenerative burners having a high thermal effectiveness.

FURTHER DEVELOPMENT

A vital part of the development of any new technology is the mechanism chosen for its transfer into industry. Collaboration between British Gas and manufacturers has proven to be a successful way of accomplishing this transfer and manufacturers often now become involved as soon as a development has reached the "proof of concept" stage. This was the case with the regenerative burner system, further development taking place in conjunction with Hotwork Development Ltd. There are a number of advantages in such arrangements: the manufacturers input ensures that development proceeds along lines appropriate to successful commercial exploitation, suitable production engineering methods can be evolved at an early stage and the new product can be given the company identity of a manufacturer, who is usually well known in the field.

The major change in design which was made as a result of the production engineering exercise is the replacement of zirconia cylinders with alumina balls of 15mm diameter. This new packing avoids the laborious stacking procedure necessary with the original packing as the balls can simply be poured into the regenerator. With appropriate provision, the bed material can also be removed easily for cleaning, allowing the system to be used on applications where the flue gases are contaminated. This is a valuable attribute of the regenerator as the use of metallic recuperators on such processes is virtually precluded by their susceptibility to corrosion and blockage.

When the new design had been proven in the laboratory, it was considered appropriate to undertake a field trial to demonstrate the performance of the regenerative burner system under industrial conditions. The application chosen was a soda lime glass pot furnace used in the production of lamp shades. The furnace is charged once per day whilst at a temperature of 1150°C. The temperature is then raised to 1400°C for melting and refining before being returned to 1150°C for the glass to be gathered and worked.

Monitoring of regenerator performance showed that air preheats up-to 1300°C could be attained with the furnace at 1400°C. Prior to conversion, a refractory recuperator provided a maximum air preheat of around 700°C. With the regenerators an additional fuel saving of 20% was obtained, and the system worked reliably and satisfactorily under arduous conditions. The trials also confirmed the need for regular bed cleaning at intervals of around four weeks, but this could be completed in one hour and without taking the furnace out of service. A visual indication of the need for bed cleaning is provided by a system which monitors bed pressure drop.

The work which has taken place since this trial has been mainly concerned with development of a range of burner sizes (currently 300 kW to 4MW) and with increasing the range of application of the system.

DETAILED THERMAL PERFORMANCE

During the course of evaluation of the regenerative system, detailed measurements of the thermal performance have been carried out in order to validate the mathematical models used for the design of regenerators and also to confirm the air preheat temperatures achieved in practice. The tests were carried out on a standard pair of production regenerative burners, with the size of the bed reduced by 30%. This was necessary in order to obtain a defined geometry which could be easily modelled for validation purposes.

The design of the test facility has required a considerable amount of sophistication. This is primarily due to the fact that the most parameters are changing quite rapidly within each cycle. The collection of the data by manual recording methods is therefore impractical and computer based logging methods, complete with software to perform the required calculations on line, have had to be used. For instance, these calculations include the determination of the thermal effectiveness using air and combustion products temperatures, which have been averaged from values taken at intervals of 5 seconds. These values are examined to establish the time when cyclic steady state conditions had been achieved, so that detailed measurements to define the performance can be taken.

The most important performance parameter for a regenerative burner is thermal effectiveness, values of which are defined in this paper according to equation (5). The measured effectiveness is plotted against reversal period for the prototype system (zirconia rings) and the new system (alumina balls) in figures 5 and 6 respectively. Also shown on these figures are the predictions of the mathematical model, which show good agreement with the measurements. The main point which emerges from the comparison is that with the original packing, reversal period has a marked effect on thermal performance whereas with the new packing the effect is small. This is true for both measured and predicted performance and can be explained by further consideration of the fundamental parameters involved in regenerator design.

The main factor influencing thermal effectiveness is the dimensionless length, assuming that dimensionless period is small, say less than 5. In this case, for a balanced, symmetrical regenerator, effectiveness is related to dimensionless length through the approximate relationship[10].

$$\frac{\lambda}{\lambda + 2} \tag{6}$$

Table 1 gives comparative figures of dimensionless length and period for both types of packing. The values of dimensionless length are similar but dimensionless period is significantly lower for the alumina balls as a result of the higher density and lower voidage which is obtained with the spherical packing. The higher values of dimensionless period for zirconia cylinders produce greater dependence of effectiveness on reversal period.

Figure 6 also shows the expected increase in effectiveness as heat input rate is decreased. The regenerator is, of course, sized for a required effectiveness at maximum input, but reducing the input under turndown conditions creates, effectively, an over-capacity of the regenerator bed and an increase in dimensionless length with consequent increase in effectiveness.

CONCLUSIONS

About 350 regenerative burners have now been installed in a variety of applications, the system proving to be particularly valuable where the flue gases are contaminated. Fuel savings of up to 65% have been achieved on these installations, usually allowing the capital cost of conversion to be recovered in less than two years. The data acquired during development and subsequent commercial exploitation of the regenerative burner system suggests that careful design and selection of operating conditions will yield a thermal effectiveness of 90%. When this is coupled with other advantages, it leads to the conclusion that the regenerative burner is a major advance in heat recovery equipment, and is certainly very close to the ultimate.

Acknowledgements

This paper is published with the permission of British Gas Plc. The authors wish to express their thanks to Professor P.J. Heggs for the provision of mathematical modelling information. Thanks are also given to their colleagues at the Midlands Research Station who have contributed to the developments described and also to Dr. R. Pugh and others who have assisted in the preparation of this paper.

References

1) DAVIES, R.M., MASTERS, J. and WEBB, R.J.
The use of modelling techniques in the development and application of recuperative burners
J. Inst. Gas E. 9(6) 1979

2) MASTERS, J. and WEBB, R.J.
The development of a recuperative burner for gas fired furnaces
Proc. R. Soc. London. A393 19-49 (1984)

3) TABOREK, J., HEWITT, G.F. and AFGAN, N.
Heat exchangers, theory and practice
McGraw Hill 1983

4) KULAKOWSKI, B.T. and SCHMIDT, F.W.
Explicit design of balanced regenerators
Heat Transfer Eng. Vol. 3, No. 3-4 1982

5) HEGGS, P.J.
Transfer processes used in packings in regenerators.
Ph.D. Thesis, Univ. of Leeds. 1967

6) SCHMIDT, F.W. and WILLMOTT, A.J.
Thermal energy storage and regeneration
McGraw Hill 1981

7) ERGUN, S.
Fluid Flow through packed columns
Chem Eng. Proc. Vol. 18, 1952

8) HANDLEY, D. and HEGGS, P.J.
Momentum and heat transfer mechanisms in regular shaped packings. Trans.
Inst. Chem. Eng. Vol. 46, 1968

9) WARD, T. and WEBB, R.J.
Regenerative burners for use in high temperature furnaces
Inst. Gas Eng. 51st Annual meeting, 1985. Communication 1273

10) HEGGS, P.J. and HOLLINS, S.J.
Development of an apparatus to investigate the thermal characteristics of
regenerators
First U.K. National Conference on Heat Transfer, Univ. of Leeds, 1984

Nomenclature

A	Packing surface area/volume	m^2/m^3
G	Gas mass velocity	Kg/m^2s
h	Gas to packing heat transfer coefficient	$W/m^2\ K$
C	Specific heat	$J/kg\ K$
$\dot{m}$	Gas flow rate	kg/s
M	Packing bulk density	kg/m^3
L	Packed bed length	m
P	Period between changeover	s
q	Heat recovered	J/S
T	Absolute temperature	K
λ	Dimensionless bed length	–
π	Dimensionless period	–
η	Thermal effectiveness	–
S	Solid	

Subscripts

a	Air
f	Flue gas
i	inlet
o	outlet
g	gas

Table 1 : Comparative values of thermal effectiveness (η), dimensionless length (λ) and period (π) for the original proto-type and the new systems

Change-over period (s)	Original prototype system 180 kW			New System 75 kW			195 kW			315 kW		
	Thermal effectiveness	λ	π	Thermal effectiveness	λ	π	Thermal effectiveness	λ	π	Thermal effectiveness	λ	π
30	0.88	7	1.6	0.88	8.6	0.1	0.82	5.7	0.1	0.74	4.5	0.2
60	0.87	7	3.3	0.88	8.6	0.2	0.82	5.7	0.3	0.74	4.5	0.4
120	0.83	7	6.6	0.88	8.6	0.5	0.82	5.7	0.7	0.74	4.5	0.8
240	0.67	7	13.3	0.88	8.6	1.0	0.81	5.7	1.5	0.73	4.5	1.6
360	-	-	-	0.88	8.6	1.5	0.80	5.7	2.3	0.72	4.5	2.5
480	-	-	-	0.88	8.6	2.0	0.79	5.7	3.0	0.70	4.5	3.3
600	-	-	-	0.87	8.6	2.5	0.77	5.7	3.8	0.67	4.5	4.2

257

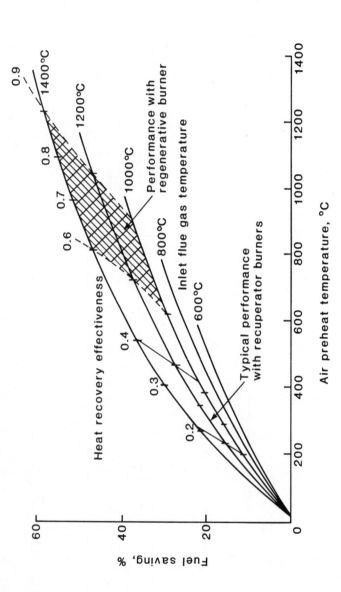

FIG 1   THE EFFECT OF HEAT RECOVERY EFFECTIVENESS, FURNACE TEMPERATURE AND AIR PREHEAT ON FUEL SAVINGS

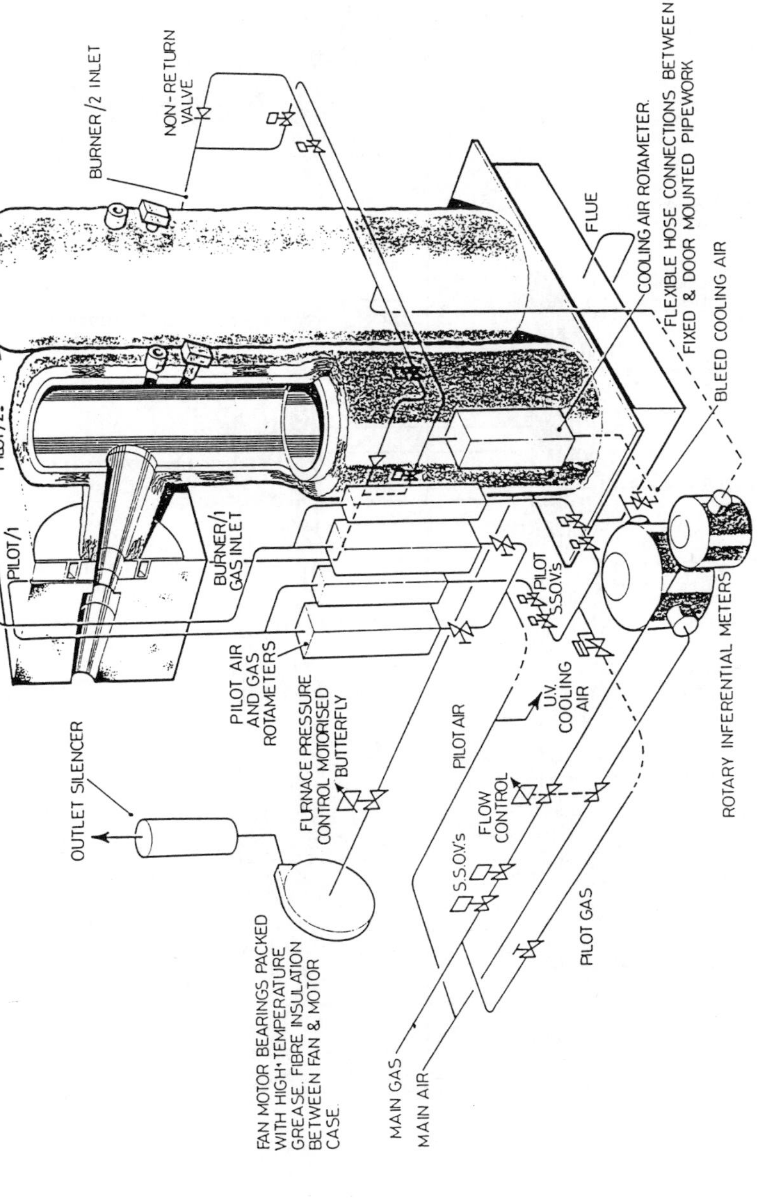

FIG. 2 A SCHEMATIC DIAGRAM OF THE PROTOTYPE REGENERATIVE BURNER SYSTEM

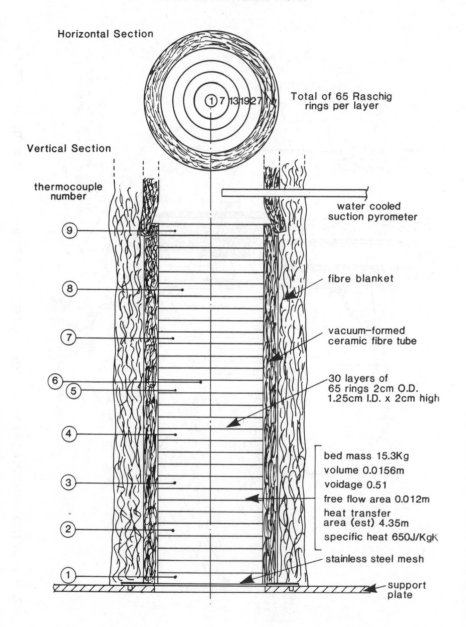

Horizontal Section

Total of 65 Raschig
rings per layer

Vertical Section

thermocouple
number

water cooled
suction pyrometer

fibre blanket

vacuum-formed
ceramic fibre tube

30 layers of
65 rings 2cm O.D.
1.25cm I.D. x 2cm high

bed mass 15.3Kg
volume 0.0156m
voidage 0.51
free flow area 0.012m
heat transfer
area (est) 4.35m
specific heat 650J/KgK

stainless steel mesh

support
plate

FIG 3 CROSS-SECTION OF HEAT STORE APPROX. FULL SIZE

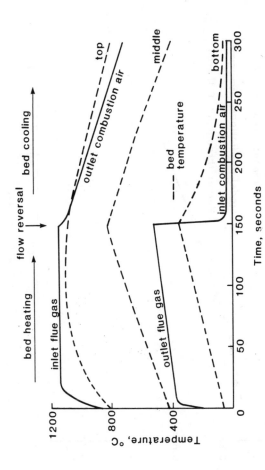

FIG 4 TYPICAL TIME TEMPERATURE HISTORIES FOR PROTOTYPE SYSTEM

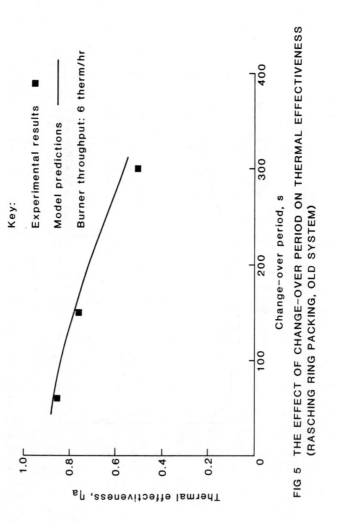

FIG 5 THE EFFECT OF CHANGE-OVER PERIOD ON THERMAL EFFECTIVENESS
(RASCHING RING PACKING, OLD SYSTEM)

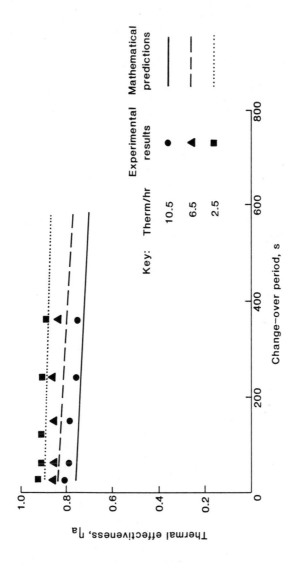

FIG 6 THE EFFECT OF CHANGE-OVER PERIOD ON THERMAL EFFECTIVENESS
(NEW SYSTEM)

THE COMBUSTION OF WASTE IN FLUIDISED BED BOILERS
TWO CASE HISTORIES

1) F. ELLIS AND C. ARMITAGE   2) J. O'NEIL AND D. HOBSON
3) R. SHIPLEY AND D. GOODWIN

Two fluidised bed boilers supplied by Foster
Wheeler to Industrial Users have successfully
burned waste fuels. This paper describes the
waste fuels and the development of the combustion systems by the boiler owners, Woolcombers
Ltd., Bradford and Robinsons of Chesterfield.
The two wastes have different properties and the
problems encountered in handling and burning
them are reported. The first stage of development has been completed and savings due to
reductions in primary fuel and disposal costs
have been identified.

INTRODUCTION

Foster Wheeler Power Products Ltd. is a major contractor of
boiler plant to the industrial and utility markets. The
company supplies equipment to burn the complete range of
liquid, gaseous and solid fuels. This paper describes the
design and operation of two smaller industrial atmospheric
fluidised bed boilers in which systems have been developed
for the combustion of in-house wastes.

The boilers which were developed in collaboration with the
N.C.B. (British Coal) are of the watertube design producing
steam for space heating and process use.

Rising fuel prices have led to the increasing attraction of
waste fuels, and several methods are available for the
combustion of such materials. Both Woolcombers Ltd. and
Robinson and Son's favoured fluidised bed technology as it
was suited to burn coal products. Fuel costs at the time
of project assessment justified the installation of the
boiler plant, burning coal alone. Further savings have been
demonstrated from the combustion of the waste material.

1) FOSTER WHEELER POWER PRODUCTS LTD., JOSEPH'S WELL,
   LEEDS, LS3 1AB
2) WOOLCOMBERS LTD., FAIRWEATHER GREEN MILLS, BRADFORD,
   BD8 0NY.
3) ROBINSON & SONS, WHEATBRIDGE, CHESTERFIELD, S40 2AD.

The waste combustion systems have been developed by the respective companies. Both preferred to install a system and then develop it from "hands on" experience gained by actual operation of the equipment in a commercial environment. The alternative was to undertake a programme of design and pilot trials before exposing the development in the operating boiler plant.

Woolcombers Ltd., part of the Illingworth Morris Group, Bradford burn a highly fouling, high water content woolgrease. The heavy fouling problem has been overcome and the low calorific value effluent burnt, providing an alternative disposal method. The combustion system is in its final stage of development.

Robinson and Son's Chesterfield burn a bulky cellulosic material. Handling of the waste is labour intensive but it burns easily and cleanly within the boiler.

Both projects were sponsored by the Energy Efficiency Office of the Department of Energy through the Energy Efficiency Demonstration Scheme.

These brief case histories highlight the flexibility of fluidised bed technology to burn waste fuels, with diverse properties, substituting a portion of the primary fuel input.

BOILER DESIGN

The main operating parameters of the two boilers are given in table 1. The boilers are from a standard range of designs for coal firing. No special features were necessary in either unit to enable waste to be burned.

The design of the 10.2 MW unit at Robinson and Son's is described in figure 1.

The boiler was constructed in three modules, comprising a generating bank with single drum and two fluidised bed modules. The design of the 7.1 MW unit at Woolcombers is similar to that of the 10.2 MW unit, however only a single fluidised bed module is required.

The fluidised bed design, licenced from British Coal is the atmospheric, shallow bed technology. The beds are nominally sized 2.0m x 2.5m. The distributor is constructed from watercooled surface, which forms an integral part of the boiler circulating system. Standpipes on the fins between the tubes, forming the floor, distribute the fluidising air. The bed material consists of sized silica sand, at a static bed height of nominally 200 - 250mm. An inclined tube bank is positioned within the furnace, such that it is almost fully immersed in the fluidised bed at the maximum fluidising velocity. As the velocity is lowered, the amount of bed expansion reduces, and the tube bank is progressively uncovered. The bank is selected such that, at maximum output the bed temperature is 950 deg C. and the excess air level is 30%. A minimum turndown of 2:1 is possible per bed and thus the larger boiler is capable of a 4:1 turndown by slumping one bed. Coal is delivered from bunkers by gravity through the furnace sidewalls. The coal presently burned at both sites is a washed singles.

The generating bank is of a straight tube construction between upper and lower headers. Horizontal baffles form multiple passes through the bank before the gas exits to an economiser and subsequent gas clean up equipment.

The boilers are operated under a balanced draught, resulting in a slightly negative furnace pressure. Start up is achieved by preheating the air at the inlet to the plenum using natural gas or diesel oil burners. Coal feed to the bed is gradually increased until normal bed operating temperatures are reached, and the burners are switched off. Fully automatic start up and load following is provided by the use of a solid state programmable logic controller. Load following is based on a step change philosophy in which the coal to air ratio remains fixed.

The Woolcombers 7.1MW unit was commissioned in 1982 and has since operated for over 22,000 hours.

Commissioning of the Robinson and Sons unit was completed in January 1985 and it has since operated for over 9000 hours.

CASE HISTORY 1
WOOLCOMBERS LTD.

Woolcombers Ltd. is a wool processing company in Bradford, West Yorkshire. Energy consumption accounts for approximately 10% of the company's production costs and in 1982 Woolcombers installed a fluidised bed boiler as a means of reducing these costs. The fluidised bed boiler was selected because of its ability to burn both coal and a waste effluent produced in the wool scouring process. The company is subject to trade effluent charges for the disposal of the waste, a woolgrease with a positive calorific value.

Woolcombers produce a wool top which is the material from which Spinners produce yarn. In the production of top the raw wool is scoured in five bowl sets using hot water as the cleaning medium. The effluent from each of the stages is principly water contaminated with grease, sand and vegetable matter. The concentration of contaminants reduces in the effluent from each of the five stages.

Woolcombers originally discharged all of the effluent to drain, a service for which they are charged. They now concentrate the liquor from the first bowl. Figure 2 is a diagrammatic representation of the process. The effluent is initially settled and centrifuged to remove sand and lanolin, leaving a solution containing up to 94% water. This solution is stored and pumped on a continuous basis through a shell and tube heat exchanger before entering a two stage flash tank. The system is energy efficient utilising steam from the boiler, and returning the condensate to the hotwell. Low pressure steam produced from the flash tanks is used for space heating. A back pressure of 3 bar g is maintained on the system.

Problems have been encountered with the heat exchanger. Initially, the liquor passed through the shellside and the steam through the tubes. A substantial short fall in performance occurred due to fouling of the outside of the tube surface. The exchanger was redesigned and a new tube bundle installed with the liquor passing through the tubes and the steam through the shell. The increased effluent velocity now maintains the surface in a clean condition.

Rapid erosion of the heat exchanger and associated pipework occurs due to the abrasive action of sand in the effluent. Although settled and centrifuged, fine particles remain in suspension and frequent maintenance is required to keep the plant operational. Long term maintenance requirements are still being evaluated.

The resultant liquor has been concentrated to around 64% water. This reduction in volume allows it to be disposed of at a reduced cost, or alternatively to be burned in the boiler.

An analysis of the woolgrease is given in table 2. It was anticipated that, on combustion, fouling of heating surfaces would be a problem, due to the high proportion of the alkali metals, particularly sodium and potassium. The net calorific value is low, due to the large percentage of ash and also, particularly, water. However trials performed at the outset of the project confirmed that the waste could be successfully burned.

Woolcombers Ltd. have developed a combustion system, principly by carrying out full scale trials in the operating plant. Initially the liquor was pumped directly, without atomisation, through a single nozzle located at a position just above the active area of the fluidised bed. (see figure 3, position A). Some of the material burned with a flame that carried into the furnace exit screen tubes and after 2-3 hours severe fouling had occurred. The remaining liquor poured into the bed and within 24 hours the bed had defluidised due to clinkering from local overheating and/or fouling. At this time, there was no bed management system operational to allow on-line cleaning/regrading of the bed material.

Foster Wheeler Power Products carried out a series of
deposition tests, firing coal and woolgrease with and without
additives in their test facility at Hartlepool. The tests
identified that combustion was improved by introducing the
woolgrease in a fine spray and with air injection. There was
indication that the use of additives had some beneficial
effect on the deposits, producing a lighter formation.

Woolcombers Ltd. designed a nozzle and introduced it into the
freeboard of the boiler furnace. The nozzle incorporated
primary and secondary air. Higher air pressures were found
to result in better atomisation and combustion, however they
also resulted in longer flame lengths. Trials were performed
with the primary and secondary air until optimum settings
were found suitable for the boiler width. Additives were
also mixed with the woolgrease and although fouling of the
furnace exit screen tubes still occurred, the deposits were
soft, friable and easily removed. A steam lance, introduced
through the furnace sidewall, removed most of the deposits
from the screen tubes.

Finally the burner was repositioned in the expanded bed zone.
(see figure 3, position B) Two nozzles were installed in the
front corners of the furnace, directed diagonally to maximise
flame length. Trials were performed to optimise air
injection and nozzle diameter. A minimum hole diameter of
4.7mm was established to avoid blockage.

The arrangement was most successful and 225 litres per hour
of woolgrease has now been burned on a regular basis for more
than 4000 hours. Up to 450 litres per hour has been burnt
over short periods of time without any detrimental effect to
the boiler. A bed management system is now operational and
fouling and defluidisation of the bed rarely occurs.

The woolgrease rate is controlled manually and separately
from the boiler automatic control. The system is operated by
the regular boiler operator. Proposed future development
will integrate the waste firing into the boiler control
system allowing unsupervised operation.

Furnace exit gas temperatures increase by up to 30 deg C.
Boiler exit gas temperature increased by up to 20 deg C.
within the first few hours of firing woolgrease. The
temperature settles and only returns to it's normal level
after approximately 24 hours of operation without woolgrease
firing. The increase in temperature is attributed to fouling
of the boiler surfaces from the woolgrease. This fouling has
been noted during boiler inspections. The deposits are very
light and it is considered that they are gradually removed
during operation without woolgrease, allowing the gas exit
temperatures to revert to their original, lower values.

Trials have been performed to prove the effectiveness of additives and to assess boiler performance. Primary fuel savings have been shown regularly when burning woolgrease. Table 3 illustrates a typical operating period, during which a primary fuel saving of 9.3% was recorded.

The cost of replacing the original evaporator unit and the, as yet, unevaluated maintenance and operating costs do not allow the meaningful calculation of a payback period for the project. However, if the system can be developed to dispose of the total liquor produced, the savings from disposal costs and the reduction in primary fuel will be in excess of £100K/p.a. The resultant payback period for the heat exchanger and associated equipment would then be less than 12 months. A detailed economic assessment is soon to be published as part of the Energy Efficiency Demonstration Scheme.

Current Status

The two nozzle, air atomised system is installed in the front corners of the furnace, located just above bed height.

As much as 20,000 litre of Woolgrease have been successfully burned in a week at a maximum continuous rate of 225 1/h. Problems are still encountered with fouling of tube surfaces and very occasionally with solidification of the bed. The evaporator unit now requires repairing every three/four months and complete replacement will soon be necessary.

CASE HISTORY 2
ROBINSON AND SONS

Robinson and Son's, one of Britains largest privately owned manufacturing companies, mainly produces cotton wool and pulp cellulose products such as nappies.

Originally, steam was raised by gas fired shell boilers, however as with Woolcombers Ltd., energy costs instigated a change in fuel to coal. The ability to burn the large quantities of waste, produced in-house, favoured a fluidised bed. The steam produced is used in the manufacturing process and for space heating. The boiler is designed for the future addition of a turbine generator.

The waste produced from the manufacturing process is a mixture of cardboard, paper, cotton wool and a small quantity of plastic.

An analysis of the constituents is given in table 4.

In any one load, the proportion of constituents varies greatly and hence so does the bulk density.

The amount of plastic was initially of concern due to the evolution of chlorine and its potential corrosive effects. It was anticipated that the furnace flue gasses would contain in excess of 800 ppm of HCL. Under these conditions the furnace tube surfaces would be most at risk. It was concluded, though, that the low metal temperatures (220 deg C.) would ensure that the problem was not serious. Tube surfaces have been monitored regularly during the operation of the boiler and no evidence of hydrogen chloride corrosion has been found.

Probably, the major problem in the utilisation of the waste is that on a volume basis the material has a relatively low heat content in comparison to other fuels (typically 2,000 MJ/m3, compared with 24,000 MJ/m3 for coal).

Four methods of handling the waste were considered:

   o   Compressing to form a hard lump
   o   Shredding to give a light fluffy material
   o   Shredding and conditioning to form a compact
        material
   o   Shredding, conditioning and mixing with bed material
        to form a compact dense material.

The simple shredding of the material was selected as the proposed handling method. The shredding process is simple and the resultant material is easily transportable by pneumatic methods.

Combustion tests were carried out in a small scale test facility. The material was fed over a fluidised bed. Whilst some of the materials dropped into the bed, the majority burned rapidly in suspension.

Robinson and Son's installed the handling system as shown diagrammatically in figure 4. Waste material is deposited in large plastic bags in each of the various factories. Lorries collect the bags and deliver them to the boiler house. The bags are manhandled from the lorries to the conveyor serving the shredder. The shredder is not sized to accept complete bags of waste, and the operator is required to split them as they are deposited on the conveyor. Manning levels have not been increased for the waste firing facility. All of the waste handling is carried out by the boiler operators in conjunction with their other duties.

After shredding, the material is blown to a large drum which acts as a holding vessel. It is agitated by a rotating blade to prevent any compacting and ensure a homogenous supply. Screw feeders meter the material from the drum into another pneumatic system, which transports it directly into the boiler furnace. Presently only one furnace is being utilised, the waste being admitted through ports in the two front corners, at a position just above the bed level. (figure 5) It is intended to extend the system to incorporate waste firing in the second furnace, the necessary ports and screw feeders were installed with the original equipment.

Up to 40 tonnes per week of waste is produced within Robinson and Son's. Unless burned in the boiler, this material is dumped with an inherent disposal cost. Presently approximately 10 tonne per week of material is burned and has been for over 4,000 hours. The burning of more is limited by the storage and handling problems due to the bulky nature of the material.

In order to limit excess air levels the final transport air was taken from the F.D. fan. It was found that this limited the fluidising air to the bed, affected bed performance and resulted in high bed temperatures. Subsequently individual fans were installed for the transport lines. Control of the total air is by manual adjustment and care must be taken to retain design excess air levels whilst maintaining acceptable conditions within the bed.

Due to the nature of the fine shredded material, it has been found to "pack" in some area's of the transport system. In this condition it forms a solid mass which is very difficult to remove and continues to accumulate. Minor modifications, particularly an increase in transport air resolved this problem.

Performance tests have been carried out on the plant when burning waste. The results are given in table 5. A reduction of 6.6% in coal usage has been identified when burning waste at 130 kg/h equivalent to 6.7% of total heat input.

The savings made in both waste disposal and coal usage have substantiated the capital expenditure in the plant. An estimated economic analysis of the waste handling system is given in table 6 and shows a payback period of 14.8 months.

Current Status

The system operates satisfactorily and up to 10 tonne per week of waste can be burned on a continuous basis. The actual throughput is limited by the availability of an operator to supply waste to the handling system. It is proposed to increase the automation and hence capacity of the waste firing facility and to integrate it fully into the boiler microprocessor control.

CONCLUSIONS

The first stage of development has been completed. Waste handling and firing systems have been developed for the combustion of two waste fuels with diverse properties. Fluidised bed boilers have been shown that they are capable of successfully burning waste fuels.

Savings due to a reduction in primary fuel and disposal costs have been identified, indicating an acceptable payback period for the capital invested. However, a full economic analysis cannot yet be compiled, as the developments are not completed and long term effects cannot yet be assessed.

Before further investment can be made, work is required to consolidate the experience gained and reach a full understanding of the combustion and handling criterion. Periods of monitoring and/or trials may be required to assist in this assessment. Further sponsorship may also be necessary to promote the progress, and release information of use in other applications.

Future developments will address maintenance problems and the integration of the combustion systems into the boiler control, thus allowing unsupervised operation. The main development will be to increase combustion rates and thus cost savings.

ACKNOWLEDGEMENT

This paper is published with the permission of Foster Wheeler Power Products Ltd., Woolcombers Ltd., and Robinson and Sons Ltd., but the views expressed are those of the authors and not necessarily those of the respective companies.

TABLE 1
NOMINAL BOILER OPERATING CONDITIONS

		WOOLCOMBERS	ROBINSON & SON'S
Boiler Duty	(MW)	7.1	10.2
Boiler Output	(Tonne/h)	11.3	13.6
Steam Outlet Pressure	(barg)	12.4	13.8
Steam Condition		Saturated	Saturated
Gross Efficiency on GCV (coal only)	(%)	83.3	81.5
Feedwater Temperature	(deg.C)	77	65
Economiser Exit Gas Temperature	(deg.C)	170	170
Design Fluidising Velocity	(m/s)	2.4	2.2
Excess Air	(%)	30	30
Bed Temperature Operating Range	(deg.C)	800-950	800-950
Static Bed Height	(mm)	250	210

FIGURE 1
MODULAR WATER TUBE BOILER

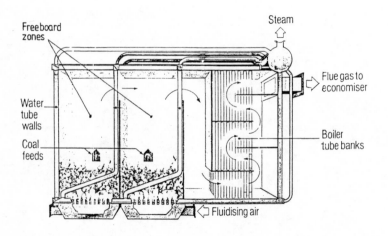

FIGURE 2
DIAGRAM OF WOOLGREASE CONCENTRATION
SYSTEM AT WOOLCOMBERS LTD.

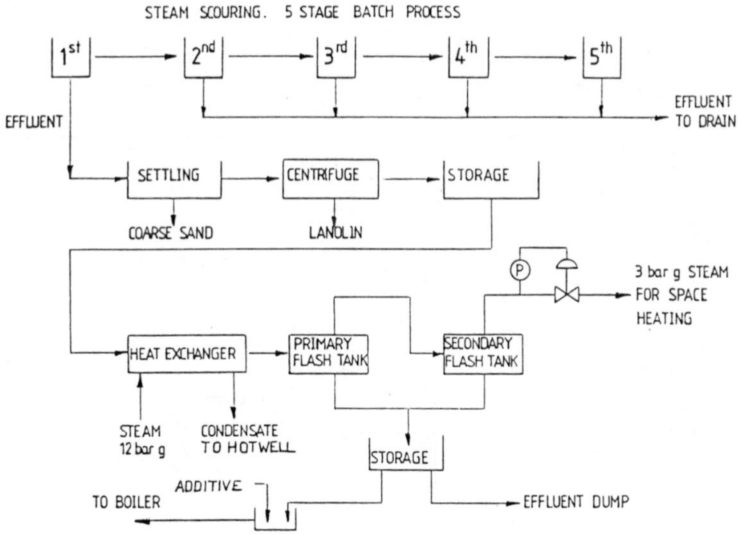

STEAM SCOURING. 5 STAGE BATCH PROCESS

TABLE 2
ANALYSIS OF WOOLGREASE EFFLUENT

DRY ANALYSIS:

ASH	%	31
ORGANIC MATTER	%	69

ELEMENTAL ANALYSIS:

	mg/l
SILICON	3,300
ALUMINIUM	2,450
IRON	1,490
SODIUM	3,700
CALCIUM	1,600
MAGNESIUM	680
PHOSPHOROUS	20
POTASSIUM	52,050

GROSS CALORIFIC VALUE   20,930 kj/kg DRY

TYPICAL WATER CONTENT   64% (as fired)

FIGURE 3
LOCATION OF WOOLGREASE INJECTION POINTS
WOOLCOMBERS F.B. BOILER

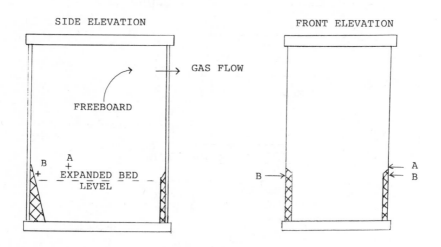

SIDE ELEVATION

FRONT ELEVATION

A   INITIAL, SINGLE NOZZLE INJECTION POINT
B   RELOCATION ABOVE EXPANDED BED LEVEL

TABLE 3
TYPICAL OPERATING DATA
WOOLCOMBERS F.B. BOILER

PERIOD	STEAM RAISED T	COAL BURNED T	WOOLGREASE BURNED l	STEAM COAL RATIO	TYPICAL F.E.T. DEG.C	TYPICAL STACK TEMP. DEG.C
A	1,318	135	-	9.76	720	168
B	782.8	73.4	21,140	10.66	750	179

PRIMARY FUEL SAVING FROM WOOLGREASE = 9.3%

TABLE 4
ANALYSIS OF WASTE MATERIAL
ROBINSON AND SON'S

Constituent Analysis

	wt % (dry)
Cardboard	9
Plastic	20
Cotton Wool	71
Moisture Content	4 (as received)

Calorific Values

	kj/kg (dry)
Cardboard	15,550
Plastic	34,200
Rag Waste	15,700

Size Distribution

Cardboard	up to 50mm Square
Plastic	up to 30mm Square
Cotton Wool	up to 100mm Cube

FIGURE 4
DIAGRAM OF WASTE HANDLING SYSTEM AT
ROBINSONS & SON'S LTD

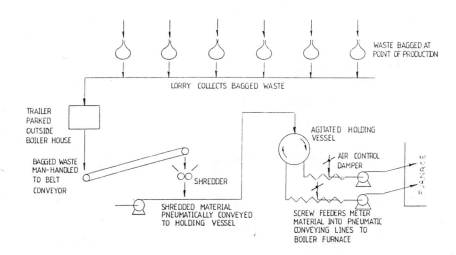

277

FIGURE 5
LOCATION OF WASTE FIRING PORTS
ROBINSON & SON'S F.B. BOILER

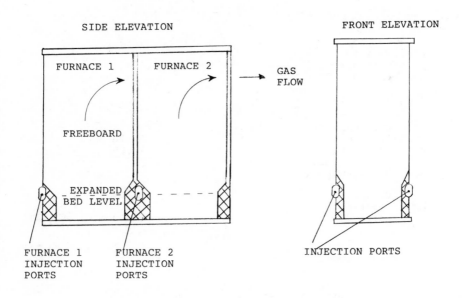

TABLE 5
TYPICAL OPERATING DATA
ROBINSON & SON'S F.B. BOILER

PERIOD	HEAT INPUT MW	GROSS EFFICIENCY %	COAL BURNED T	WASTE BURNED T	TYPICAL F.E.T. DEG.C.	TYPICAL STACK TEMP. DEG.C.
A	10.57	80.8	6.38	–	660	170
B	10.60	80.5	5.96	0.67	680	170

PRIMARY FUEL SAVING FROM WASTE = 6.6%

TABLE 6
ESTIMATED PAYBACK ANALYSIS
WASTE HANDLING SYSTEM

ROBINSON & SON'S F.B. BOILER

CAPITAL AND INSTALLATION COST	£31,000
SAVINGS:	
WASTE DISPOSAL	£13,000
PRIMARY FUEL	£22,700
COSTS:	
POWER	£10,600
NET ANNUAL SAVING	£25,100
PAYBACK 14.8 MONTHS	

BASIS:

1) 4000 Waste burning hours per annum
2) 6% reduction on primary fuel
3) System maintenance not included
4) System manned by existing boiler operators

PERFORMANCE OF A 600 KW PUMP-AROUND HEAT RECOVERY
SYSTEM ON HUMID PNEUMATIC CONVEYOR EXHAUSTS

Dr. Ian Eastham* Ph.D., B.Tech., M.I.Chem.E., C.Eng.

A pump-around heat recovery plant is described which
was commissioned in November 1986. It is performing as
predicted and likely to pay back in just over one year.
Special engineering features are described to overcome
the problem of dust build up on the wet surface of a
conventional heat exchanger. The features include a
control system which ensures condensation is only
allowed to occur at a high rate, whatever the heat output, to flush the surface clean. The operating data
from the first full winter will be presented.

2.   INTRODUCTION

Many industrial processes use air drawn from the workspace for conveying,
cooling, dilution, drying, milling and stripping for example. Much is
discharged into the atmosphere from the roof. Precleaning before
discharge is normal to remove powders and dusts, sometimes to remove
gaseous or solvent pickup. Contamination often precludes consideration
as a waste heat source.

The heat it contains originates from the workspace in winter, motor
power from fans and mills or process stream heat when drying and
cooling.

Where the nature of the process precludes taking cold outside air
directly, e.g. multiple suction points which draw air from the workspace, the make-up to the building has to be heated to a comfortable
working temperature, whether the make up leaks in through doors and
windows or is ducted.

A simple heat recovery system which can accommodate remote exhaust
and inlet positions and with the engineering complexity to overcome
contamination can payback in one year.

* EnviroSystems Limited, Lancaster

3.   SPECIFICATION

The specification for this foodstuffs mill, which was being revamped was
to supply 33m³/s air to make up the volume exhausted by the conveying
system.   It was proposed to heat it to 14°C minimum, when ambient air
temperature was at -1°C.   The running cost of achieving this with gas
fired boilers would be approximately £40K in the average year.

Any heat recovery system to save this primary energy would need to be
fitted to air exhausted from 4 bag filters, consequently a quantity of
fine dust was usually present and any heat recovery exchangers had
to accommodate the possibility of a bag bursting.

4.   THEORY

Simple heat exchange between a circulating liquid and warm process
exhaust will recover heat into the liquid.   By pumping that liquid to
a second heat exchanger in the building air supply the circulating
liquid is cooled thus warming the supply air.   The cooled liquid picks
up more heat on being returned to the exhaust and a continuous process
results.

The size of heat exchange equipment required depends on the temperature
difference between the exhaust (the heat source) and the air supply
(the heat sink).   Experience in non-process buildings indicates that
temperature differences less than 15°C make such a system uneconomic
and very susceptible to fouling of the heat transfer surface.

Selection of the circulation rate of liquid heat transfer medium requires
iterative calculation but the heat transfer performance of the range of
sizes of air to liquid heat exchangers is not normally accurately
available for computation.   Hence the theoretical ideal is not readily
found.

Data from the operating plant is being gathered to show the effect of
changing circulation rate.   The analysis of such data will show the
optimum for the heat exchanger installed here and permit further
computation on future designs.

5.   DESIGN

Half of the air consumed by the milling system was rejected on the
roof via 4 exhaust vents at 33°C, the rest being rejected via 8 other
rooftop vents at temperatures near 20°C.   Since the 4 warm vents
contained the majority of the heat derived from the milling equipment
power, and the higher exhaust temperature reduced the capital cost
of the heat transfer equipment, it was decided to heat the incoming air
from the heat recovered from only half of the flow, which was the
warmer.

The production process caused the evaporation of water into the conveying air such that the chosen 4 exhausts had a dewpoint of approximately 23°C so 60% of the heat to be recovered at full load had to be from the condensation of water.

The key to a successful operation would be the design of a heat recovery heat exchanger which could accommodate the slight dust burden simultaneously with condensation.

The heat output of the system is driven only by the temperature difference between the ambient air and the workspace. Therefore long periods of operation at part capacity were possible. This was seen to create the risk of operating on the edge of the condensation regime when some of the heat exchanger surface was only misting over and not fully wetted. The control system was specified to ensure that those of the 4 heat recovery units on the roof which were in service were operating either dry or flooded with condensate. The critical feature of the control specification is shown in Figure 5.1.

No reliable strategy could be formulated to survive the dust burden from a filter bag bursting, so the heat exchangers were hinged for quick release.

A diagram of the scheme is shown in Figure 5.2.

6. CONTROL

The control system is based on a standard Building Energy Management microprocessor controlled unit manufactured by JEL Limited.

The function of the system is sevenfold :

1. To maintain the air supply in the process space at 14°C.

2. To provide automatic start up sequence for the air make up and heat recovery system.

3. To control condensation on the heat recovery units.

4. To minimise pump operating hours to save electricity running costs.

5. To measure operating parameters, mostly temperatures.

6. To store operating data.

7. To manage standby and breakdown.

6.1. Supply air temperature control

Two PID control loops in the JEL unit provide direct digital control of the flow rate of water/ethylene glycol mixture to the twin, parallel inlet air heat exchangers.

6.2. Automatic start-up

The two 30 kw air supply fans are started Star-Delta, automatically in sequence against a closed discharge damper. When the pressure rise across each reaches the running condition the damper opens and the second fan sequence begins.

If ambient air temperature is below 14°C the lead circulator pump is started. Then the required number of roof top heat recovery units are switched into service (ON/OFF only).

6.3. Condensation Control

Since heat demand is related only to ambient temperature, the sensible or condensation heat recovery mode can be controlled on that temperature alone. The condensation control sequence in Figure 5.1. has the effect on each heat exchanger shown in Table 6.1.

The control sequence ensures that each heat exchanger is recovering only sensible heat or has at least 61 litres/hour of condensate flushing the surface.

6.4. Pump operating control

Since no heat is required when ambient air exceeds 14°C, the lead circulator is switched off. When ambient temperature falls below -1°C the standby circulator is commissioned to run in parallel to give +10% over design capacity from the heating system by improving temperature driving forces. Commissioning the second pump at only peak demand saves up to £1,000 of pump electrical usage a year.

6.5. Temperature Monitoring and Recording

The table shows the range of temperature sensors and their function :

No. of Probes	Purpose	Logging allocation*
2	Control of Air Temperature	2
1	Control of Roof Units	1
8	Monitoring Water/Glycol	4
8	Monitoring Discharge Ducts	4
2	Monitoring Building Space	2

* To always contains previous week's data, 13 data points can be accommodated in memory, logging each hour.

6.6. Standby and Breakdown Management

Differential pressure cells detect fan and pump failure. On fan failure its discharge damper closes. On pump failure the standby is automatically started.

7.   PERFORMANCE

Since the new air supply system brought air specifically to the process
area for the first time some weeks elapsed before the full effect of this
facility was appreciated.

It was realised that the process space optimum temperature was 14°C.
This allowed best control of processing parameters.

During initial operation (i.e. when process space was not controlled
down to 14°C), it was observed that the PID control of duct temperature
did not balance the available heating fluid between the two air supply
intake systems at all thermal conditions. Consequently, some air outlets
did not supply air at the same temperatures due to lamination in the fan
supply plenum. This condition is characterised in Figure 7.2.

The heating capacity of the scheme was found to vary approximately
proportionally to the temperature difference between outside air
temperature (the heat sink) and the process exhausts (the heat source)
see Figure 7.1. Data from more than one season will be needed to
improve the correlation.

After one season the combined effect of the managed air supply (ducted)
and the heat recovery system has benefitted two process conditions,
final product moisture control and yield.

8.   ECONOMICS

The capital cost of the Heat Recovery System described here is illustrated
in Figure 8.1. and Table 8.1. The simple payback shows less than
two years for plant over 17m³/sec of make up air.

The justification for supplying make-up air in the first instance is
considered separate but involves factors such as control of air quality
(filtration), absolute pressure in the workspace to keep doors closed
or safe to open and draughts from all angles. Payback times also
including the capital cost of the equipment to supply the air, being two
fans, inlet filter and main ductwork are less than 3.0 years 20m³/sec
capacity.

The capital cost for the Heat Recovery System alone includes all features
described in the paper, except fans, filters and supply air ductwork.

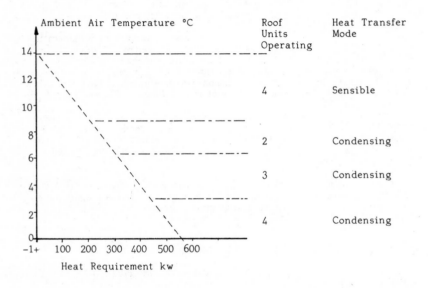

Figure 5.1.   Heat Requirement v Ambient Air Temperature

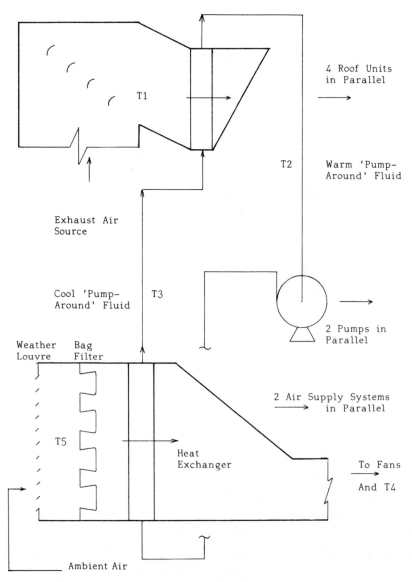

Figure 5.2.   The Flow Scheme

Table 6.1.

Ambient Temperature	Exchanger Capacity Utilised				Heat Quality
	A	B	C	D	
14 – 9°C	0–25%	0–25%	0–25%	0–25%	Sensible
9–6.5°C	40%	40%	OFF	OFF	Sensible
	66–100%	66–100%	OFF	OFF	Condensation
6.5–3°C	40%	40%	40%	OFF	Sensible
	66–100%	66–100%	66–100%	OFF	Condensation
3°C →	40%	40%	40%	40%	Sensible
	75–100%	75–100%	75–100%	75–100%	Condensation

Note :   The dust build up risk is perceived to be between 35% of load per exchanger to 60% load. After which 61 litres/hour of condensate flows from each unit keeping the surface flushed.

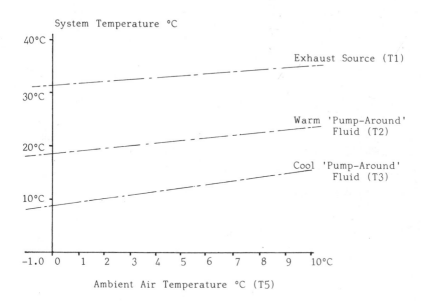

System Temperature °C

Exhaust Source (T1)

Warm 'Pump–Around' Fluid (T2)

Cool 'Pump–Around' Fluid (T3)

Ambient Air Temperature °C (T5)

T1 to T5 refer to locations on the scheme diagram Figure 5.2. Process floor condition maintained at 14°C.

Figure 7.1.   Thermal Performance of System

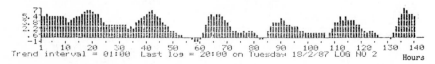

Ambient Air (T5) - SINK

Exhaust Air (T1) - SOURCE

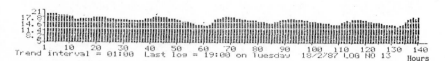

Process Space (T4)

Figure 7.2   140 Hour Logs off JEL PC 8000 Controller. Thirteen such Logs record
Space, Liquid and Air Temperatures each Hour.

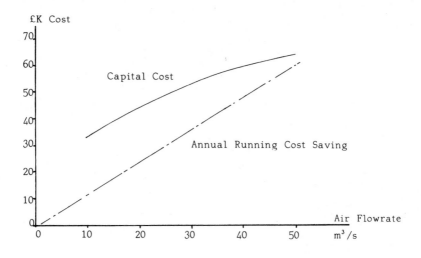

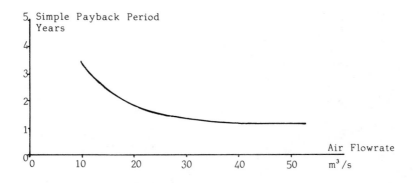

Figure 8.1.   Economics of Pump–Around Heat
Recovery From Process Exhausts

TABLE 8.1. ECONOMICS

Air Flow m³/s	Capital Costs (£K) Heat Recovery System	Annual Heating Cost* (£K)	Simple Payback (Yrs)
50	63	61	1.00
33	56	40	1.4
16	42	20	2.1
8	34	10	3.4

*   Gas Boilers, Steam Heaters

DECREASING FUEL CONSUMPTION IN BOILERS BY WASTE
HEAT RECOVERY AND REDUCED STANDBY LOSSES

B.M. Gibbs*

The objective of this study was to assess the performance
of an economiser which was designed to give enhanced fuel
savings by utilising a built-in flue damper to reduce
standby heat losses in addition to conventional heat recovery. Standby savings, predicted by a cooling model, were
found to be dependent on the boiler operating time and burner firing cycles. Heat recovery from the boiler flue gases
was dependent on exhaust gas temperature and excess air
level. Overall, the fuel savings that could be achieved in
practice were found to be in the range 6-16% depending on
the operating conditions prevailing.

INTRODUCTION

A substantial proportion of the available energy supplied to a
boiler is contained in the flue gases that are exhausted to the
atmosphere through the stack. By fitting an economiser,
designed to recover additional heat from the boiler exhaust
gases, boiler efficiency can be increased and fuel consumption
reduced. A conventional economiser is therefore a heat
exchanger in which the heat removed from the exhaust gases is
transferred to the boiler feedwater. The extent to which the
exhaust gas energy can be recovered is governed by the dew-point
of the combustion products in order to avoid severe corrosion of
the economiser or stack. This limitation restricts the typical
heat recovery, as equivalent fuel savings, to about 3-6%. It
has therefore become necessary to develop alternative
economisers that give enhanced fuel savings and that are
therefore considerably more economically viable.

This paper is concerned with an investigation into the
fuel savings that can be achieved in a fuel-fired package boiler
by the use of an economiser which incorporates a flue damper,
linked via a controller, to the burner. The fuel savings are
accomplished by two means.

1. By direction heat recovery from the flue gases leaving the
boiler and

* Department of Fuel and Energy, University of Leeds,
Leeds, LS2 9JT.

2.   By reduction in boiler draught and cooling losses by closing the flue damper which is activated on burner shut down. These losses occur during normal burner cycling and standby periods.

ECONOMISER CONSTRUCTION AND OPERATION

Economiser construction

The economiser, Fig. 1, consists of a cylindrical shell containing a bundle of small diameter flue gas tubes which surrounds a larger central flue gas tube. The tubes are secured by two end plates. Water flows through the annulus formed by the central flue tube and outer cylindrical shell casing.

A damper is fitted into the central flue tube which when shut, diverts the hot combustion products through the tube bundle resulting in heat exchange between the flue gases and water flowing through the shell. A second damper is fitted into the exhaust gas stream which is closed when the burner is not firing thereby reducing the cooling and draught losses from the boiler. Both dampers are motor operated and linked to the burner by a controller.

Operating sequence of the dampers

The operating sequence of the dampers in a firing cycle, or after the boiler is switched on, is as follows, and is shown in Fig. 2.

1.   The flue gas damper (Fig. 1, item 8) and the by-pass damper (item 7) are set to the open position and the boiler and flue ways allowed to purge for a pre-set period of time.

2.   The burner is ignited (with the dampers open) enabling the combustion products to have unrestricted flow through the central flue tube. Once the flue gas temperature reaches a pre-determined level, the by-pass damper (7) is closed diverting the combustion products through the tube bundle in the shell. Thus water passing through the economiser will now be heated by the combustion products normally exhausted to waste from the boiler.

3.   On burner shut down, during a firing cycle or when the boiler is turned off, the flue gas damper (8) and by-pass damper are closed after a pre-set delay. During the time that the damper remains closed, boiler cooling losses are diminished resulting in a reduction in standby heat losses and a consequential fuel saving.

FUEL SAVINGS DUE TO FLUE GAS DAMPER

The fuel savings that can be achieved by the flue damper are deduced from an analysis of the heat losses (Fig. 2) that occur, with and without the flue damper, after the burner is shut off during normal burner cycling and standby periods. The cooling characteristics of a typical boiler were obtained from tests on a number of boilers. In each test the boiler was allowed to reach steady state conditions before the burner was shut off. Fig. 2 shows the typical cooling behaviour that occurs with and without the flue damper during a burner firing cycle.

Boiler cooling losses on burner shut down - no flue damper

It was established from the cooling tests that the transient heat losses could best be approximated by an exponential equation of the form

$$q_1 e^{-t/\tau 1} \tag{1}$$

where q_1 represents the initial heat lost, above the steady state level q_{s1} which remains after the transient has ceased (see Fig. 2); and τ_1 is the boiler time constant. At time t after the burner has shut off the heat loss, $q(t)$, will thus be given by

$$q(t) = q_1 e^{-t/\tau 1} + q_{s1} \tag{2}$$

Hence the total heat lost, Q_{1F} that occurs during a burner cycle, ignoring purging, is found by integrating equation (2) over the shut off period, which yields

$$Q_{1F} = q_1 \tau_1 (1-e^{-t/\tau 1}) + q_{s1} t \tag{3}$$

For a standby period of duration t_s that is greater than $10\tau_1$, the total heat lost, Q_{1s}, is simply the steady state cooling losses or

$$Q_{1s} = q_{s1} t_s \tag{4}$$

Boiler cooling losses with flue damper

The flue damper is closed some t_d seconds after the burner is shut off. As a result of the flue damper, the boiler cooling losses are reduced as indicated by Fig. 2. Again for exponential cooling behaviour, the heat lost at time t after the burner is shut off is given by;

$$t < t_d \quad q(t) = q_1 e^{-t/\tau 1} + q_{s1} \text{ (before damper activated)}$$

$$t > td \quad q(t) = q_2 e^{-t/\tau 2} + q_{s2} \text{ (damper activated)}$$

where q_2 represents the initial heat loss when the damper is shut and τ_2 the time constant with the damper installed.

q_{s2} is the steady state heat losses to the boiler flue gases when the flue damper is installed.

The total heat lost, Q_{2F}, during a firing cycle is again found by integrating over the total burner shut off period (ignoring purging), which gives for $\tau_2 \gg t$.

$$Q_{2F} = q_1\tau_1(1-e^{-td/\tau 1}) + q_{s1}t_d + q_{s2}(t-t_d) \tag{5}$$

During a long standby period of t_s the heat loss, Q_{2S}, with the flue damper installed will be given by

$$Q_{2S} = q_{s2}\, t_s \qquad (6)$$

Total heat saved, Q_T, due to flue damper

This is best considered as the sum of two components;

Q_F, the heat savings achieved during normal boiler operation, with regular burner cycling on and off, and

Q_{SB}, the heat saving obtained during boiler standby periods.

Heat saving during burner cycling, Q_F

The heat saved for a burner cycle is simply the difference between the cooling losses with and without the flue damper during one burner cycle, or

$$(Q_{1F} - Q_{2F})$$

Thus for a boiler operating for N h/day with a burner cycling n times/h the total daily heat saving, Q_F, is given by

$$Q_F = Nn\,(Q_{1F} - Q_{2F}) \qquad (7)$$

Heat saving during boiler standby periods, Q_{SB}

When the boiler is on standby for T h/day, the total heat savings, Q_{SB}, will be

$$Q_{SB} = (Q_{1S} - Q_{2S})$$

$$= (q_{s1} - q_{s2})\,(3600\ T) \qquad (8)$$

Total daily heat saving due to flue damper, Q_T

The total heat saved/day is simply the sum of the savings obtained during boiler cycling and standby periods,

$$Q_T = (Q_F + Q_{SB}) \qquad (9)$$

Fuel savings due to damper

If the boiler has an efficiency of η_1 and burns a fuel with a calorific value, CV, then the fuel saved, m_s, resulting from the installation of the flue damper is given by

$$m_s = \frac{Q_T}{\eta_1 CV}$$

If the original fuel consumption is m_1, then the fractional fuel saving, F_d, is therefore given by

$$F_d = \frac{m_s}{m_1} = \frac{Q_T}{m_1 \eta_1 \, CV} \qquad (10)$$

Predicted fuel savings using the boiler cooling model

The next step is to use the model to predict the fuel savings that could be achieved by a flue damper activated on burner shut down. Table 1 shows that range of values, covering typical boiler operation, that were used to obtain predicted fuel savings. The boiler time constant was estimated from transient measurements on a gas fired boiler.

Table 1: Values used for computations

Boiler operating efficiency	$\eta = 0.8$
Boiler time constant	$\tau_1 = 60s$
Initial cooling losses	$(q_1 + q_{s1}) = 5.5\%$ (Q)
Steady state losses – no damper	$q_{s1} = 1.5\%$
Steady state losses – with damper	$q_{s2} = 0.5\%$
Burner purging time/cycle	$t_p = 60s$
Damper delay time	$t_d = 30s$
Boiler load	max.

All calculations were carried out by means of a computer programme. The computed fuel savings as a function of dimensionless firing time (equivalent to the fraction of a cycle that the burner fires) are shown in Figure 3 and 4.

The influence of boiler operating time on the fuel saving is shown in Fig. 3 for a burner firing 15 times an hour. The fuel savings can be substantial. The longer the boiler stands-by, the larger are the fuel savings. Thus a boiler operating for 12 hours, firing typically for about one-third of a cycle has a fuel saving of 5.5% compared to 1.7% for 24 hours operation.

Decreasing the number of firing cycles per hour (n) increases the fuel saving, Figure 4, but only by about 5-10%.

FUEL SAVINGS BY DIRECT HEAT RECOVERY

Measurement of Direct Heat Recovery

The direct heat recovered in the economiser heat exchanger was measured over a wide range of boiler exhaust conditions. In these tests, the economiser was connected to a hot gas generator designed to simulate boiler exhaust gas conditions and which provided a metered supply of hot air in the temperature range of $150{-}400^\circ C$. The water-side of the economiser was supplied with hot water in the temperature range of $35{-}80^\circ C$. The flowrate of water was measured by a rotameter. The economiser inlet and outlet gas temperatures were measured by suction pyrometers, and the inlet and outlet water temperatures measured by direct insertion mercury in glass thermometers.

During the tests, the bypass damper was closed and the exhaust damper opened so that the hot air passed through the heat exchanger tubes located in the shell. The heat recovered in the economiser was measured over a wide range of hot gas flowrates and temperatures in order to simulate the exhaust conditions of a boiler in the size range 100-200 kW, which was the boiler size for which the economiser was originally designed.

Fuel Savings

Having determined the heat that can be recovered by the passage of hot gas through the economiser, the next step is to utilise these data in order to predict the fuel savings that would be achieved by direct heat recovery when the economiser is fitted onto a boiler. In order to calculate the fuel savings the following procedure is used:

(1) For a given fuel, operating excess air level (obtained from the CO_2 content of the flue gas) and the boiler exhaust temperature, calculate the boiler efficiency η_1, (see ref.1).

(2) Estimate the fuel flowrate, m_1, required to provide the heat load Q for a boiler of efficiency η_1.

(3) Calculate the flue gas flowrate from the fuel flowrate and operating excess air level.

(4) Obtain the heat recovered in the economiser, q, for the calculated flue gas flow and boiler exhaust temperature, from the experimental data.

(5) Use the heat recovered in the economiser, q, to determine the new boiler efficiency, η_2, from the relationship.

$$\eta_2 = (1 + \frac{q}{Q})\ \eta_1$$

(6) Calculate the new fuel flowrate, m_2, resulting from the improved boiler efficiency from

$$m_2 = m_1\ \eta_1/\eta_2 \qquad (11)$$

Hence the fuel saved, F_s, is given by

$$F_s = 1 - \eta_1/\eta_2 \qquad (12)$$

Predicted fuel savings and boiler efficiencies

Figs. 5 and 6 show the improved boiler efficiencies and fuel savings that would be achieved by direct heat recovery, if the economiser was fitted onto a boiler with a heat output of

100-200 kW, firing on fuel oil at an excess air level of between 20-40%.

OVERALL FUEL SAVINGS

This section shows how the total fuel savings achieved by the economiser, involving direct heat recovery and standby savings, can be calculated.

When only the economiser heat transfer surfaces are utilised, the boiler efficiency is increased from η_1 to η_2 and the new fuel flowrate m_2 is obtained from equation (11), i.e.

$$m_2 = m_1 \eta_1 / \eta_2$$

The flue damper however results in the fuel consumption being further reduced from m_2 to m_t. Applying equation (10) enables m_t to be calculated from the fuel saving F_d, due to the flue damper, which yields

$$m_t = m_2 (1-F_d)$$

Hence if we combine the above equations m_t can be expressed in terms of the original fuel consumption m_1 which gives

$$m_t = m_1 \cdot \eta_1 / \eta_2 \cdot (1-F_d)$$

and the total fuel savings F_t is therefore given by

$$F_t = 1-(1-F_d)\eta_1/\eta_2 \qquad (13)$$

Estimating the total fuel savings

To simplify the calculation procedure it is possible to utilise graphical plots of boiler efficiency (Fig. 5) for a given fuel and computed standby savings (Fig. 3 and 4) to obtain an estimate of the likely overall fuel savings. All that are required is a knowledge of the boiler operating data (fuel type, exhaust temperature, excess air level) operating time, and burner cycle history. The following example illustrates the method of calculation for the fuel savings.

A 150 kW oil fired boiler operates with 20% excess air and the exhaust gas temperature is 300°C. The burner is known to cycle on average 5 times per hour and fire for approximately one third of the cycle. Estimate the fuel savings if the boiler operates 12 hours per day and an economiser with a flue damper was fitted onto the boiler.

(1) The boiler efficiency with and without the economiser is estimated from Fig. 5 (which applies only for an oil fired boiler).

For an exhaust temperature of 300°C
$\qquad \eta_1 = 0.81$ without economiser.
$\qquad \eta_2 = 0.86$ with economiser.

(2)   Determine the fuel savings, F_d, resulting from the flue damper using Fig. 4. (Note this graph applies for a boiler of efficiency = 0.80).

$$F_d = 0.059 \quad \text{at } \eta_1 = 0.8.$$
$$F_d = 0.059 \times 0.80/0.81 = 0.0583 \text{ at } \eta_1 = 0.81.$$

(3)   Estimate the overall fuel savings, F_T, using equation (13)

$$F_T = 1-(1-0.583)(0.81/0.86) = 0.113.$$

The fuel savings are therefore 11.3%.

It can be seen that the fuel savings are approximately twice those achieved by heat recovery alone (5.8%). Note that if an alternative fuel was used it is necessary to use the boiler efficiency graph (Fig. 5) appropriate to that fuel.

Instead of using the above graphical method, the fuel savings could have been calculated by means of a computer program incorporating boiler efficiency calculations for a number of fuels, heat recovery data and flue damper calculations. Input data would be the same as the graphical method.

ACTUAL FUEL SAVINGS

The actual fuel savings that have been obtained in practice have been found to be in the approximate range of 6-16% according to the boiler and operating conditions prevailing. These fuel savings were determined in long duration tests, covering periods of time from 3 months to a full heating year. Fuel savings were deduced after making allowances for differences in degree-days during comparable monitoring periods.

The fuel savings were found to be within 10% of those predicted by the 'fuel-savings model' and in nearly all cases were under rather than over-predicted.

CONCLUSIONS

1.   The addition of a flue damper in a conventional economiser leads to enhanced fuel savings due to a reduction in boiler standby heat losses. Total fuel savings are in the range of 6-16% compared with 4-6% for heat recovery alone.

2.   The economiser with its potential for high fuel savings will be a cost-effective means of reducing fuel costs, with pay-back periods typically about half those of conventional economisers.

3.   Heat recovery data combined with the boiler cooling model can predict fuel savings with sufficient accuracy to enable the payback period and fuel savings to be assessed for any potential economiser installation.

4.   Standby fuel savings depend on the boiler operating time and number of burner cycles; heat recovery depends on boiler exhaust temperature and excess air level.

REFERENCES

1.   British Standards No. 845, Acceptance Tests for Industrial
      Type Boilers and Steam Generators, (1972, revised 1984).

NOMENCLATURE

CV	Fuel calorific value.
F	Fraction of original fuel saved.
m	Boiler fuel consumption.
n	Burner cycles per hour.
N	Number of hours per day boiler operates.
q	Boiler cooling rate, heat saved by heat recovery.
q(t)	Boiler cooling rate as a function of time.
Q	Total heat saved/heat input to water.
t	Time.
T	Standby time.
η	Boiler efficiency.
τ	Time constant of boiler.

Subscripts

1,2	With, without flue damper.
s1,s2	Steady state without, with damper.
F,SB	Boiler firing, standing-by.
b,c,d,p	Firing, cycle, delay, purging (time).
dl	Damper.
s	Saving.

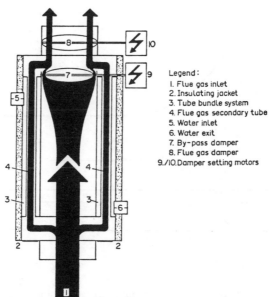

Legend:
1. Flue gas inlet
2. Insulating jacket
3. Tube bundle system
4. Flue gas secondary tube
5. Water inlet
6. Water exit
7. By-pass damper
8. Flue gas damper
9./10. Damper setting motors

Fig. 1:   The Economiser complete with flue dampers.

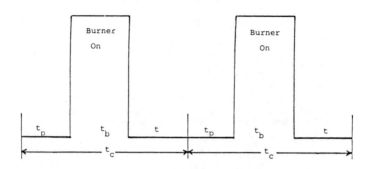

Burner Cycling - No Damper

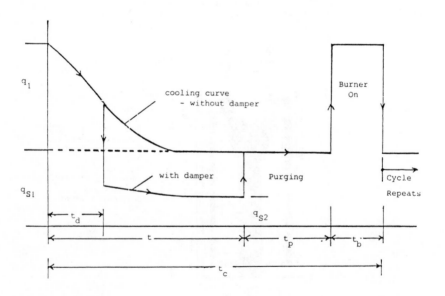

Cooling Behaviour with and without Damper

Fig. 2: Boiler cooling behaviour with and without flue damper.

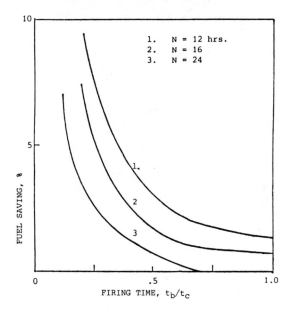

Fig. 3: The effect of boiler operating time on standby savings.

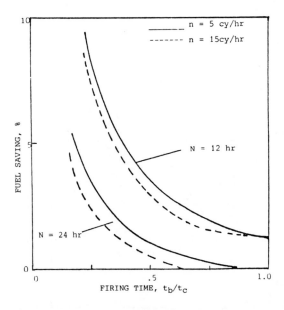

Fig. 4: The influence of burner cycles on standby savings.

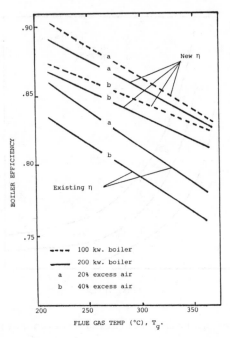

Fig. 5: Boiler efficiency resulting from economiser heat recovery.

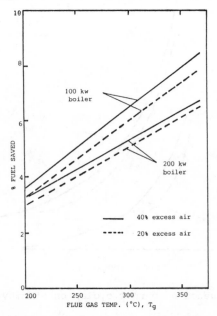

Fig. 6: Boiler fuel savings due to direct heat recovery.

THE BRITISH GAS HIGH TEMPERATURE DIRECT CONTACT WATER HEATER

M.J. Baker*

Over recent year many gas fired appliances have been
developed to replace centralised steam systems for
industrial process use with ever increasing
efficiency. However, to achieve efficiencies
approaching 100%, a heater has been developed
using direct contact between the combustion
products and process water in order to obtain the
necessary heat transfer rates to produce a compact
economical design.

1. INTRODUCTION

The field of industrial low temperature process and space heating is dominated
by the steam boiler. Excluding transport, steam raising is the largest energy
user in the U.K. and in manufacturing industry more than 60% of all fuel
consumed is used to raise steam. There are, of course, many reasons for the
continuing widespread use of centrally generated steam in low temperature
processes. Steam can be generated in boilers fired by a variety of fuels,
giving users a high degree of flexibility in fuel choice. Steam heating can
often be fitted to a process readily and at low cost and it is easy to
control. Any required process temperature can be achieved by selection of the
appropriate steam pressure and heat is transferred at virtually constant
temperature which may be a valuable characteristic in some applications.
Finally, steam boilers are generally very reliable and have long working
lives.

This is a formidable array of technical and operational advantages but it
is increasingly outweighed by the only significant disadvantage of centrally
generated steam, low overall efficiency. As manufacturing industry is
subjected to increased competition in the U.K. and overseas market places
users are being compelled to examine the most cost-effective ways in which
their heating requirements can be provided. The area of low temperature
process heating is one of particular interest as there is much evidence to
indicate that steam heating systems operate at a typical gross overall
efficiency of 40-50%. There is some scope for improvement through paying
attention to plant design and control and there is also potential for saving
fuel through the reduction or elimination of heat losses from the steam
distribution system. The major opportunity lies, however, in the concept of
decentralisation, a process of replacing centralised steam heating by

*British Gas Plc, Midlands Research Station, Solihull, West Midlands B91 2JW

by appliances which produce heat directly at the point of application. This approach eliminates many of the losses inherent in centralised systems and yields overall efficiencies in excess of 80%, offering the possibility of fuel savings of around 50% in many cases. The concept of decentralisation can be applied to a range of processes including space heating and drying. The Midlands Research Station of British Gas has carried out much work in support of decentralisation for these processes but has, in recent years, placed particular emphasis on industrial liquid heating which is a very large heating load distributed across many sectors, the most important ones being chemicals, textiles, food and drink and general engineering.

The principle of direct gas firing for liquid heating is, of course, well established, equipment such as undertank bar burners and natural draught immersion tubes having been in use for many years. Bar burners are inexpensive, simple and easy to install but have an efficiency of only 25% although this can be increased to 50% by ducting the combustion products through a jacket around the tank. Natural draught immersion tube burners fire directly into a heat exchanger immersed in the tank and have efficiencies of about 70% but their low firing intensity, or heat input per unit cross-sectional area of tube, necessitates the use of large tubes which are often difficult to accommodate.

More recently, British Gas has developed a range of immersion tube burners having much higher firing intensities (Figure 1). These compact units use small bore tubes or a variety of multitubular heat exchangers (Figure 2) and have an efficiency of around 85%. If efficiency is to be further increased the combustion products must be cooled to below their dew point of 55°C, in order to recover the latent heat of the water vapour which they contain, formed by combustion of natural gas. The use in industrial process heating of conventional heat exchangers for this duty is virtually precluded by their size and cost and therefore the most appropriate technique is that in which heating takes place by direct contact between combustion products and liquid.

2. DIRECT CONTACT HEATING

Direct contact heating is also a well established concept but traditional techniques have significant disadvantages which have prevented widespread use.

The submerged combustor, shown in Figure 3, has a burner located above the liquid surface, the combustion products being directed downwards through a dip tube which extends some distance into the liquid. The products enter the liquid as a large number of small bubbles, presenting a large heat transfer surface area. Figure 4 shows that submerged combustion is highly efficient at low liquid temperatures but efficiency falls rapidly as temperature increases due to an increasing proportion of the heat input being used to evaporate water from the bath. Analysis shows that at 50°C about 94% of the gross heat input appears as sensible heat in the bath, 5% as latent heat of water vapour and 1% as the heat content of the dry combustion products. In contrast, at 89°C, the "submerged combustion boiling point", no sensible heat is transferred to the bath, 86.8% of the heat input being used for evaporation whilst 11% appears as latent heat of water vapour in the flue gases and 2.2% as heat content of the dry gases. In view of its limitations submerged combustion must be seen as a specialised technique mainly confined to the heating of highly corrosive liquids and such applications as concentration of dilute solutions.

The efficiency of a submerged combustor can be increased by the addition of a direct contact recuperator, as shown in Figure 5. Such recuperators can also be used to recover heat from the flue products of boilers, dryers and other appliances and are accepted waste heat recovery devices. The principle is also used to produce self-contained direct contact heaters, as shown in Figure 6 but, as with submerged combustion, these heaters have a practical upper limit of 60-65°C on water delivery temperature if an acceptable efficiency is to be achieved.

3. THE HIGH TEMPERATURE DIRECT CONTACT HEATER

British Gas's objective in developing its direct contact heater was to produce a device having the high efficiency of direct contact heating without suffering the limitation on water delivery temperature, characteristic of traditional direct contact techniques. The specification laid down therefore required.

(a) An efficiency of around 95% on a gross calorific value basis in most applications.

(b) The ability to deliver water at high temperatures, approaching 100°C if necessary, whilst still operating at an efficiency of at least 90%.

(c) A low total combustion air pressure drop to allow the use of a packaged burner, generally the most cost effective way of providing heat inputs in the required range.

The heater designed to fulfill this specification is shown in Figure 7. It uses a combination of direct and indirect heating to satisfy the dual requirement for high efficiency at high temperatures and operates in the following way.

The burner fires into a comparatively short large diameter (to reduce pressure drop) firetube immersed within the base tank. The high temperature differential between the products of combustion and the water in this indirect section allows the heat content of the combustion products to be reduced by about 50% before they enter the base of the tower section which contains a series of perforated trays. Cold water enters at the top of the heater and is uniformly distributed across the surface of the first of these trays before flowing down through the tower where it removes both sensible and latent heat from the combustion products. The water is collected in the base tank where it is heated to final delivery temperature.

An initial programme of laboratory test work proved the principles of the new heater and it was decided that further development could be undertaken through field trials.

4. FIELD TRIAL PHASE

British Gas often prefers to introduce its developments to the marketplace through licensed manufacturers. It is considered advantageous for manufacturers to provide input to new developments at an early stage in commercialisation, the preferred route to obtaining this input being a joint development agreement under which British Gas and its development partner contribute their own particular expertise to the project. The direct contact heater was the subject of such an agreement.

Field trials play a vital role in developing new equipment as they are the only way in which the performance of the device can be assessed in its normal working environment. Where an item of equipment is covered by a joint development agreement it is normal practice for trials to be carried out and funded jointly and this was the case with the direct contact heater trials, the three trial units being constructed by Thurley International Ltd to specifications prepared by the Midlands Research Station.

4.1 Holmes Halls Tanners

The leather industry provides a potential market for high efficiency liquid heating equipment as tanneries use large quantities of warm water in hide washing and processing. Accordingly, two of the field trial heaters were installed in tanneries.

Holmes Halls Tanners in Hull, the trial site for the first heater to be constructed, is, with a 20% market share, the largest tannery in the U.K. On this site hot water was originally provided by gas fired steam boilers and calorifiers but the boilers were reaching the end of their working lives and the Holmes Halls management were considering the options available for their replacement. The projected cost savings from the installation of direct contact heating were attractive and it was agreed that a heater should be constructed to supply the requirements of the hide preparation area of the factory. This area required around 30,000 l/h at 45°C which was provided by a unit having a heat input of 1.5MW. Later experiments on the site showed that the high water flow rate yielded somewhat higher heat transfer rates in the tower section than had been expected, allowing heat input to be 1.8MW without adversely affecting efficiency. The heater supplies a storage tank of 55,000 l capacity which allows the system to cope with very high peak demands without the necessity for a unit of correspondingly high heat input. The heater operates virtually continuously throughout the day, showing that it is correctly sized.

The system at Holmes Halls has now been in operation for two years and only one major problem has been encountered. The site uses borehole water which has high levels of nitrates and chlorides and this lead to rapid corrosion of the original galvanised mild steel base tank. The tank was replaced with a stainless steel unit matching the tower section and corrosion is no longer a cause for concern.

During the trial period the performance of the heater was monitored by British Gas and an average overall efficiency of 98% recorded. This corresponds to a flue gas exit temperature of 20°C and has resulted in a reduction in fuel costs for this area of the factory of 45%. Holmes Halls have recently demonstrated their satisfaction with this performance by purchasing two further heaters.

4.2 J. Clarke

The second heater to be constructed was also for use in a tannery, Joseph Clarke of Milborne Port, Dorset. This heater is virtually identical to that at Holmes Halls and is rated at 1.5MW. The duty is rather different however as the requirement is 18000 1/h at 70°C, the water being used for intermediate tanning and final dyeing. No storage vessel is used at Clarkes, the heater supplying the process directly and functioning on-demand. The installation therefore differs in a number of respects from that at Holmes Hall and allowed the performance of the heater at higher delivery temperatures to be assessed. As expected, overall efficiency was reduced, an average figure of 92% being obtained during the trial period. It is thought, however, that improvements to the water distribution system and method of tray location within the tower would allow this to be increased. Even so, the reduction in running cost of the system compared with the original oil fired boilers enabled the capital cost of the heater to be recovered in less than one year, a similar figure to that yielded by the Holmes Halls unit. The heater at Clarkes has been in operation for about 18 months and reliability has been excellent.

4.3 Cooksons

Cooksons of Wallsend, Tyne and Wear, produce zircon and zirconium oxide powders for use in many applications including ceramic products, optical polishes and certain types of glass fibre. As part of the process the raw slurry produced by grinding, which is at 70°C and has a specific gravity of 2.3, is diluted with hot water to reduce its temperature and specific gravity to 55°C and 1.5 respectively. This hot water was originally provided by gas fired steam boilers but the duty has now been assumed by a direct contact heater rated at 360 kW which supplies 6000 1/h at 50°C. As at Clarkes, no storage is used, the heater supplying the process directly. Overall efficiency has been monitored at 96% and Cooksons claim that the use of the direct contact heater has reduced their fuel cost per tonne of finished product by 76%.

5. FURTHER WORK

As intended, in addition to providing a good deal of valuable information on heater performance in a variety of applications, the field trials also highlighted some aspects of heater design and construction which would benefit from further attention. In parallel with the latter stages of the trials a programme of laboratory tests was therefore begun to investigate these aspects. This included the design of the water distribution system and the method of location of the perforated trays. The trials had also shown the importance of sealing the edges of the trays against the tower to avoid the channeling of combustion products up the inside walls and consequent creation of hot spots. Various methods of sealing have been investigated and a suitable one identified.

The design of the heater has also been studied with a view to reducing its capital cost. This study has enabled material thicknesses and methods of construction to be optimised for minimum cost whilst maintaining the necessary structural rigidity. It has also revealed the possibility of eliminating some components used in the original heaters.

As shown in Figure 8, the trial heaters originally incorporated a "buffer" storage tank. It was initially thought that the extreme difficulty of balancing water inflow and outflow on this design of heater made this tank an essential component but the cost of the tank, its associated pipework and the series of level switches necessary to control the system made it desirable to investigate the possibility of an alternative configuration. Laboratory experiments showed that the tank could be eliminated by allowing the water in a slightly larger heater base tank to overflow directly into an oversized, heavy duty, self-priming centrifugal pump for transfer to the process or a storage tank as required. To confirm the practicality of this approach the heater at Cooksons was modified accordingly and has now been operating reliably in this mode for three months. Other work carried out subsequent to the field trials could be described as application engineering. British Gas foresees a range of applications for the direct contact heater including the textile and food and drink industries and also in heating swimming pools.

6. CURRENT STATUS

The direct contact heater has reached commercialisation. British Gas's original partner Thurley International Ltd., is no longer involved with the development and installation of the high temperature direct contact water heater although they have retained an interest in larger scale direct contact process water heaters. Two additional licensees have been appointed, Nordsea Gas Technology and the Aerogen Company Ltd., to manufacture and market the development. The heater is available from the manufacturers in a range of

sizes with heat inputs of up to 3 MW and both companies have begun to make commercial sales. Laboratory test work and mathematical modelling have enabled a design procedure to be produced which will allow one-off heaters to be readily designed for particular applications and should ultimately lead to further optimisation of the manufacturer's standard ranges.

7. CONCLUSIONS

The three field trials carried out on this novel heater were all highly successful and although the development is still in the early stages of commercialisation indications are that this performance will be repeated by heaters sold by the licensed manufacturers. Overall thermal efficiencies of up to 98% were recorded by the trial heaters and although this exceptional figure was obtained on a heater with a comparatively low delivery temperature efficiencies of around 95% can be expected on many installations. This should result in most cases in fuel cost savings in excess of 50% and extremely attractive capital payback periods.

British Gas and the manufacturers are now satisfied that the heater is fully developed and can be sold with confidence. In future therefore effort will be concentrated on further reduction in capital cost and on proving the heater in other markets.

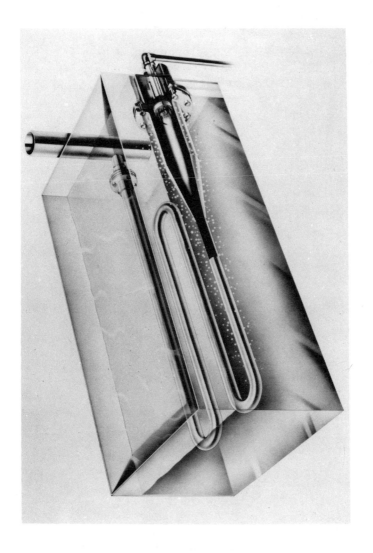

FIG. 1. M.R.S. SMALL BORE IMMERSION TUBE BURNER

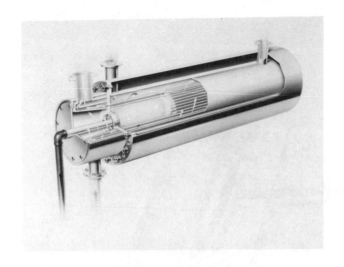

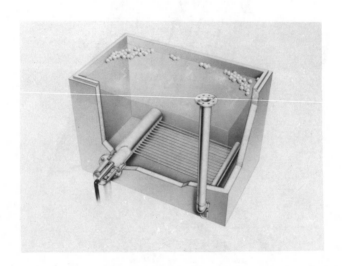

FIG. 2.   MULTITUBE HEAT EXCHANGERS

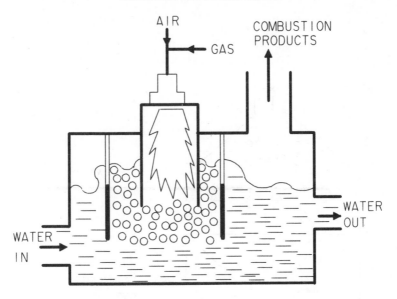

FIG. 3.   A TYPICAL SUBMERGED COMBUSTION SYSTEM

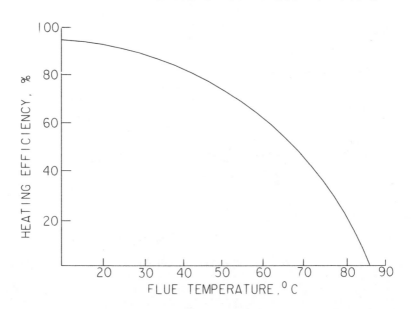

FIG. 4.   HEATING EFFICIENCY AGAINST FLUE TEMPERATURE
FOR DIRECT CONTACT HEATING

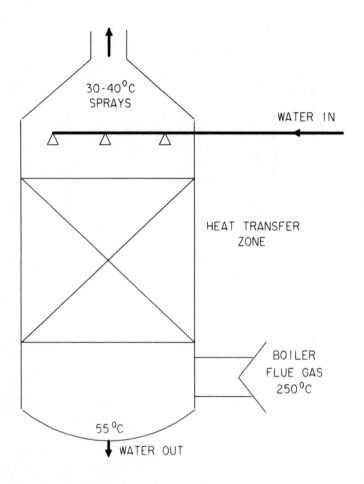

FIG. 5. SPRAY RECUPERATOR

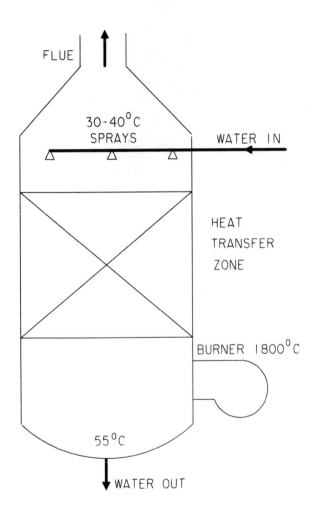

FIG. 6.   SPRAY HEATER

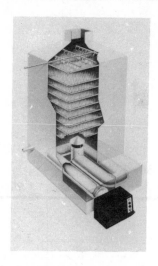

FIG. 7.   THE HIGH TEMPERATURE DIRECT CONTACT
          WATER HEATER

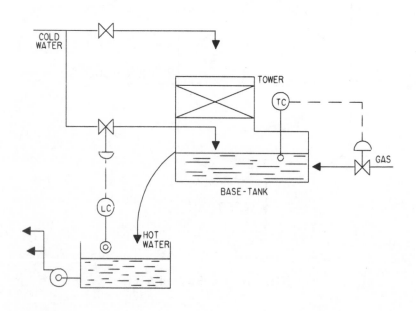

FIG. 8.   TEMPERATURE CONTROL, VARIABLE FLOW
          AND PUMPED OUTFLOW

IMPROVED ENERGY UTILISATION USING VORTEX COLLECTOR POCKET TECHNOLOGY FOR
CLASSIFICATION OF MATERIALS

*A. Mohammed-Ali, *M. Biffin, and *N. Syred

Classification of many materials is a very energy
intensive process, this paper describes the development of
a new type of low energy use, simple classifier which has
potential applications in many areas. The concept is
based on recent work at Cardiff on novel design of cyclone
dust separators, Fig. 1. Energy savings can be considerable, especially in terms of reduced power input to
milling/crushing equipment and/or power required to drive
the rotor in rotating blade classifiers. The paper will
discuss work carried out on a small laboratory prototype
with high dust loadings up to $\dot{m}_{dust}/\dot{m}_{air}$ = 0.85. The work
clearly shows the classification potential and also the
ability to remove the fine product efficiently using VCP
with adjustable aperture and a central collector.

INTRODUCTION

With the recent energy crisis there has been increasing demand for better
utilisation of energy in the process industries. One important area is the
increasing demands for finer powders with well controlled size characteristics.
Classifiers have an important role here in areas as diverse as the production
of cement, pharmaceuticals, pulverised coal, recovery of waste products,
separation of particles of different densities. Large classifiers (3) can
consume large quantities of energy and there is a need for new designs of
simple, compact, low cost units. The work described here originated in recent
work at Cardiff directed at producing new designs of cyclone dust
separators (1),(2). One of the new concepts introduced was that of the vortex
collector pocket, Fig 1, in which the flow in the main cyclone chamber is used
to drive, a weak subsidiary vortex in the vortex collector pocket (henceforth
VCP). Particles on entering the main cyclone chamber are forced into a thin
boundary layer region close to the outer wall, and due to inertia effects pass
directly into the VCP where they are rapidly decelerated due to friction and
drop to the base of the VCP to be collected. No net gas flow enters the VCP
whilst several VCP's may be dispersed on a cyclone chamber to give enhanced
dust separation.

Recent work has shown that the VCP concept can be modified to give an
extremely sharp 'cut' to the grade efficiency curve, one of the most important
characteristics of a classifier. This is achieved by altering the width of the
VCP aperture. Conventional cyclone dust separators do not produce a
particularly sharp grade efficiency curve and hence are usually considered to
be unsuitable for use as classifiers, except in very rudimentary cases.

* Department of Mechanical Engineering & Energy Studies,
  University College Cardiff.

Thus the VCP concept cannot only be used as a classifier, but can be combined in a single unit, with VCP's of different designs to give a simple, compact classifier/separator with very low energy utilisation.

These units have not been just considered as a competitor to conventional rotating blade classifiers, but also as a low cost modification to existing separation processes, allowing better, more consistent, products to be produced. One possible application described involves the conversion of a large cyclone separator to such a configuration, so that large sizes of coal can be preferentially separated for crushing, whilst the finer material passes directly into the process, thus reducing the formation of superfine material.

The paper considers two types of unit, the first one is based on the unit, shown in Fig 2 and comprises an outer VCP with classification, a second VCP and a central collector to recover fine material. The second type of unit is based on the conversion of conventional cyclones and involves replacing the conventional long cone with a flat bottomed section with a VCP (for classification) and a central collector, Fig 3.

The VCP concept has proved to be useful for classification by virtue of the fact that the cut size may be altered by varying the aperture between the VCP and the main cyclone chamber, as illustrated in Figs 6a,b,c. Reducing the width of the aperture reduces the quantity of fine material entering the VCP, thus allowing the cut size to be easily altered. Secondary air addition before the VCP has also proved to be useful by reducing the numbers of fine particles in the outer boundary layer, thus improving the shape of the characteristic.

Energy savings can be considerable; an efficient classifier on a cement plant can, for instance, give increases in production of order 30%. The reason for this is that the sharp 'cut' of the classification curve ensures that only coarse particles are recycled to the mill for crushing, thus reducing crushing of fine particles and hence energy consumption. The increase in production of order 30% means that energy costs/ton production are also reduced by 30%. Similarly the elimination of the rotor on conventional cement classifiers can also lead to very substantial extra energy savings. Typically the motor used to drive the rotor on a cement classifier absorbs two to three kWh/ton of cement produced thus needing 100 to 150 kW electrical power input to drive the rotor for a 50 ton/hr production rate.

The rotor on conventional cement classifiers is used for cut adjustment; this function can in part be replaced by adjustment of the VCP apperture which demands negligible energy input.

Similarly the application of the classification technology described in this paper can lead to considerable energy savings in coal washing, probably of order 40 to 50%, especially when it is applied to a system where little or no classification is at present used.

COMPACT CYCLONE CLASSIFIERS USING VCP TECHNOLOGY

The two designs of classifiers described in this paper are illustrated in Figs. 2 and 3. The first design is based on that of the Mk6c cyclone dust separator (1) and was tested at two scales, 30 and 60 mm exit diameter.

The aperture of the outer VCP could be varied by means of a simple sliding plate, whilst secondary air could also be added (on the 60 mm unit) before the VCP to improve the shape of the cut.

The second design of unit is based on the conversion of a standard Stairmand design of high efficiency cyclone whereby the long vertical cone is replaced by a VCP and a central collector, the exit diameter on the pilot scale unit being 150 mm.

The units tested in the laboratory were typically 10 to 20% of the size of full scale units and thus consideration of required cut sizes on large units indicated that scaling criteria for cut sizes on the model need to be derived in order to be able to scale laboratory/pilot scale results to a full size system. These scaling criteria were derived by consideration of available literature on the modelling and scaling of the performance of cyclone dust separators (4-8). Several different approaches are possible including that of Koch and Leith (4), Leith and Licht (5), Stern (7) and Dietz (8). All these models consider forces acting on an individual particle (in the Stokes Law Regime) and the balance between centrifugal and viscous forces. The more sophisticated models derive an expression for the grade efficiency curve of form

$$\eta = 1 - \exp f \left| \left[\left(\bar{t}_s \frac{W_{in}^2}{\mu} d_p^2 \rho_{dust} \right) g(n) \right] \right|$$

where $\bar{t}_s$ is the average gas residence time
ρ_{dust} is the dust density
d_p is the particle diameter
μ is the dynamic viscosity of the gas
W_{in} is the tangential inlet velocity

Such expressions can be remarkably effective but experience shows (9) that expressions have to be matched to individual designs of unit by adjustment of empirical constants in the expressions. Factors such as the shape factor of a dust are also important. The expressions are always derived for spherical particles and the authors have thus found that when designing large units via experimental data gathered from small scale units it is better to scale experimentally derived data via the use of cyclone scaling parameters which are available in the literature.

One such parameter which the authors have found useful is that due to Stern (7) and discussed by Tengbergen (6). This cyclone parameter, Z, is defined by the equation

$$Z = \frac{d_p^2 \, W_{in} \, \rho_{dust}}{\mu \, D_o} = \text{constant}$$

where D_o is the main body diameter of the cyclone.

This equation is only really applicable for particles complying with Stokes Law, i.e. $d < 100$ μm and $\rho_{dust} \cong 1000$ kg/m^3. This, however, covers the range of prime interest. Thus a plot of Z against separation efficiency for a given particle size band gives a curve which can be scaled for the same inlet dust concentration. In terms of scaling this means that a grade efficiency curve obtained on a small scale pilot model can be scaled to large sizes using this cyclone parameter Z. In practice then a grade efficiency curve as shown in Fig. 5 can be scaled to a different sized unit or operating temperature by multiplying the particle size scale by

$$\sqrt{\frac{(\mu \, D_o) \text{ new unit}}{(\mu \, D_o) \text{ model}}}$$

providing a similar inlet velocity and dust loading is used. Similar comments apply to fines collection efficiency curves as shown in Fig. 6.

Quite often different grinds of material need to be used on a pilot scale model so as to realistically model the behaviour of a much larger unit. An important factor is that of dust loading, i.e. $\dot{m}_{dust}/\dot{m}_{air}$ (3). This is especially important and experience shows that it is essential to test laboratory/pilot scale models at realistic values. Large variation in this parameter can drastically alter the cut sizes obtained as well as the shape of the fine grade efficiency curve. Most commercial classifiers now operate with values of $\dot{m}_{dust}/\dot{m}_{air} > 0.5$ (3) and sometimes up to 2. In this work an attempt was made to obtain as much data as possible for $\dot{m}_{dust}/\dot{m}_{air} > 0.5$.

PURPOSE DESIGNED CLASSIFIERS

Initial work was directed at purpose designed classifiers based on the design of Fig. 2. Pressure drop measurements on a 30 mm exhaust diameter unit clearly show the effects of dust loading, Fig. 4, where reductions in pressure drop of up to 40 to 50% are shown for mass loading of about 1:1, thus indicating a drop of tangential velocity level of about 20%. The shape of the resulting grade efficiency curve is quite good, Fig. 5, (VCP aperture fully open) with a cut size between 20-30 μm for $\dot{m}_d/\dot{m}_a = 0.7$. The tail of the curve (for VCP1) needs some improving, terminating at around the 20% level, i.e. too much fine product is returned to the process with the large material. VCP2 does not perform well, possibly its performance is too close to that of VCP1. The central collector performs very well indeed, giving a good overall separation performance.

The aperture of the VCP1 was then adjusted to 88% and 75% open and the performance of VCP1 re-evaluated and compared to that that with VCP1 fully open, Figs 6a-c. This time the data for VCP1 is rearranged to give the fine grade collection efficiency, i.e. the curve of the material rejected by VCP1 and finally collected as product; it is in fact close to the inverse of the curve for VCP1 shown in Fig. 5.

With the aperture fully open, Fig. 6a, the shape of the curve so obtained is quite acceptable, although the loss of fine material to the rejects can be clearly seen, the d_{50} cut being about 10 μm. When VCP1's aperture is closed to 8% open there is a marked change in the shape of the curve, Fig. 6b, recovery of fine product is enhanced, whilst the curve obtains a better form. The d_{50} cut is now about 35 μm to 40 μm. When VCP1's aperture is closed to 75% open, Fig. 6c, there is again an improvement in the form of the curve with a d_{50} cut of about 50 μm.

These promising results had been obtained at mean dust to air loadings of about 0.8 and it was thus then decided to proceed to a larger model of 60 mm exhaust diameter to investigate the effects of scale changes.

This larger unit was provided with a secondary air inlet to redisperse fine dust from the outer boundary layer into the main section of the cyclone chamber. It was also felt that the relatively large volume of the unit would enable better dust dispersion to be made. Materials tested included Fullers earth and Pozzillan PFA material. This unit was designed to operate with a dust capacity of 500 kg/hr at a pressure drop of 5,900 N/m² (610 mm wg).

Two typical results from this unit are shown in Figs. 7a and 7b with a 50% open aperture for Pozzillan PFA material. It is clear that the shape of the curve has improved with the secondary air, with a d_{50} cut of about 50 μm, Fig.

7b. Without secondary air the d_{50} cut is reduced to about 30 μm albeit at the expense of a slightly less steep curve, Fig. 7a. Recovery of fine products for d < 15 to 20 μm is improved somewhat. It appears, in fact, that the d_{50} cut point can also probably, in part, be controlled by throttling the level of the secondary air, as well as the VCP aperture. The indications are that there is an optimum width of VCP aperture (between about 88% and 50% shut) which gives the best shape of curve for a classifier.

Comparison of the results from the 30 and 60 mm units shows that the 60 mm unit is giving better results in terms of classification performance, ie shape or form of the curve, recovery of fine products, rejection of coarse material back to the process. This is in accord with experience with large rotating blade classifiers and is probably due to changes in particulate loading as the size of a unit is scaled up. Such units are usually scaled up via the use of a loss coefficient and a formula of the form (9).

$$\Delta P = \xi \tfrac{1}{2} \rho_g v^2$$

where ΔP is the system pressure drop
  ρ_g is the gas density
  ξ is an Euler number or loss coefficient
  v is a characteristic velocity.

In practice the characteristic velocity is kept constant for scale up and a linear scale factor for all dimensions is used, thus giving similar system pressure drop for both sizes of unit. Hence inlet, exhaust and surface areas all increase as a square law with increase in linear scale factor. Thus the internal volume to internal surface area ratio of a classifier also increases linearly with scale. Thus dust in large classifiers/separators for a similar $\dot{m}_{dust}/\dot{m}_{air}$ ratio is always better distributed with a lower propensity for effects such as agglomeration, deposition and formation of shifting dunes of material on the baseplate. Moreover, boundary layers are thinner, allowing less room for build up of high concentrations of particulate material in these boundary regions.

CONVERSION OF CYCLONE DUST SEPARATORS TO CLASSIFIER CONFIGURATION

After the success of this initial programme of work on purpose designed classifiers attention was directed at an industrial problem arising from the use of conventional cyclones. This is illustrated in Fig. 9, which shows a cyclone train installed after a coal dryer. The first stage cyclone removes coarse material which is fed to a crusher whilst the secondary cyclones remove fine material which is fed directly to the process. Unfortunately the first stage cyclone also removes much fine material which on crushing generates large quantities of 'superfine' material which can cause other problems in the process.

The work described in this section discusses experimental modelling work directed turning the first stage cyclone into a classifier/ separator, Fig 3. The long conical section is replaced by a flat bottomed section, containing one VCP and a central collector. The VCP again had an adjustable aperture for classification and the central collector is designed to collect fine material. Only material from VCP1 is fed to the crusher, the rest passing directly to the process. The model was 1/11 of the size of the full scale system and was run at a similar pressure drop and $\dot{m}_{coal}/\dot{m}_{gas}$ loading. The full scale system required that no significant quantities of material greater than 1 mm should enter the product thus requiring a realistic cut size of about 500 microns. Scaled to the model this means a cut of $\dfrac{500}{\sqrt{11}}$ microns or about 150 microns is required. In actual practice for different coals and operating conditions adjustability around this figure is required.

Some typical model results are shown in Figs 8a and b for aperture widths of 13% and 20% which gave d_{50} cut sizes of about 100 to 130 microns thus fulfilling the original objectives of the programme. The central collector removes virtually all the remaining material. Sharpening up the entry to the central collector improved the classification performance of the outer VCP.

CONCLUSIONS

The work described in this paper has shown it is possible to produce systems which incorporate both classifications and separator function simultaneously. The use of Vortex Collector Pockets (VCP's) has been shown to be important, classification being obtained by adjustment of the VCP aperture. The use of secondary air just before the VCP can help to prevent fine particles being entrained into the VCP. Two systems have been described, the first being a developed/integrated classifier/separator, the second being a conversion of a standard Stairmand cyclone to flat bottomed configuration. As discussed earlier the importance of this work lies in the following features:-

a) The ability to generate very sharp cuts, plus fine material removal without the need for a rotor. This can lead to high levels of energy saving by reducing mill utilisation/crushing or increasing production, remembering the high energy usage of mill equipment.

b) The possibility of eliminating the rotor from many designs of classifiers thus giving energy savings of, for instance, 2 to 3 kWh/ton for cement.

c) The ability to cheaply classify many products recovered by cyclones is an advantage which can be exploited to give useful energy recovery. For instance fines from coal fired boilers can be recovered by one of these units, acting as a clean up system in the first instance. Experience shows that carbon is concentrated in the product collected in the VCP, whilst ash is primarily collected in the central collector. Thus the product from the VCP can be refired to improve the combustion efficiency of the boiler and reduce energy consumption, possibly by 4 to 8%.

ACKNOWLEDGEMENTS

The authors thank the following bodies for their support in this programme of work.

1) The Coal Research Establishment, Cheltenham.
2) Coal Processing Consultants, Wingerworth, Chesterfield.
3) The Science & Engineering Research Council.

Opinions expressed are those of the authors.

REFERENCES

1. Syred, N., Biffin, M., Dolbear, S., Wright, M. and Sage, P. EFCE Publication Series No. 52, p 17, Inst. Chemical Engineers, 1986.

2. Biffin, M., Syred, N. & Sage, P. "Enhanced Collection Efficiency for Cyclone Dust Separators", Chem. Eng. Res & Develop. Vol. 62, pp 261-265, 1984.

3. Furukawa, T., Onuma, E. and Misaka, T. "New Large-Scale Air Classifier O-Sepa, It's Principle and Operating Characteristics", International Symposium on Powder Technology (1981, Kyoto).

4. Koch, W.H. and Leith, D. Chem. Eng. No. 7, p 81, November 1977.

5. Leith, D. and Licht, W. A.I.Ch.E Symposium Series, Vol. 68, No. 126, p 126, 1972.

6. Ebbenhorst Tengbergen, M.J. Ingenier, Vol. 77, No. 2, p W1, January 1965.

7. Stern, A.C., Caplan, K.J., Bach, P.D., "Cyclone Dust Collectors", American Petroleum Institute, New York, 1955.

8. Dietz, P.W. A.I.Ch.E.J., Vol. 27, p 88, 1981.

9. Gupta, A.K., Lilley, D.G., Syred, N. "Swirl Flows", Abacus Press, Tunbridge Wells, Kent, U.K., 1984.

ENERGY MANAGEMENT IN PALM OIL MILLS

A.F. Shafii

The palm oil milling industry utilises waste solid
fuels mainly palm fibre and shells to fire its boilers
for thermal and motive power generation. A study
described in this paper has shown that owing to the
new type of planting material being processed and
weevil assisted pollination practices, the energy
management scenario in a typical mill has drastically
changed. The implications to the design of new mills
and process operations are discussed.

INTRODUCTION

The Malaysian palm oil industry is the world's largest served by 272 mills
and 36 operating refineries producing 4.2 million tonnes of palm oil and
1.2 million tonnes of palm kernels in 1985. In addition, the milling
process also produced an estimated 2.52 million tonnes of palm mesocarp
fibre, 1.44 tonnes of palm shells and 4.14 million tonnes of Empty Fruit
Bunches (EFB) or bunch stalks as wastes.

Currently, all mills utilise fibre and shells as boiler fuel for thermal
and motive power generation. Over the years, the acceptable engineering
practice is to run a given mill beyond a certain critical throughput of
Fresh Fruit Bunches (FFB) per hour so that a condition of positive fuel/
steam balance is achieved. Typically, the rule of thumb has been to design
a power plant that would raise at least 0.5 tonne of steam per tonne of
FFB processed. Turbine turbo-alternator units are then employed to provide
mill and estate domestic power requirements. On average, about 32 kg of
steam is used to generate 1 kW of power. Clearly, incidences of badly
designed mills are those suffering from acute fuel/steam balances. In such
cases, external fuel (fibre/shell import) or, in its worst case, the bulky
and high moisture EFB are utilised.

Mill steam/power generation has also been an increasingly important issue
as most mill boiler exhausts are not able to meet the new Department of
Environment (DOE) atmospheric emission standards of Ringlemann #2 smoke
and 0.4 gm/Nm^3 particulates. Further, the era of new Dura x Pisifera
(D x P) planting material and weevil (*Elaeidobius kamerunicus*) pollinated

HMPB Engineering Services Scheme,P.O.Box 207,42700 Banting,Selangor,Malaysia.

FFB have drastically changed the fruit morphology. D x P planting have resulted in FFB with less shell (endocarp) content giving less calorific values while the weevil has given rise to bigger and compact FFB which very often require three-peak steam sterilisation for good subsequent stripping of fruitlets. The increased demand of steam per tonne of FFB was acknowledged but there was no updated design data available on the new status. In view of the pressing situation, a detail study of energy availability and steam distribution was conducted in an operating 25 tonne/hr mill to provide a bottom-line understanding of existing situations. This paper describes the main findings of the study and discusses their implications to the design of new mills and process operations.

BASIC FUEL CHARACTERISTICS

In practice, palm waste solid fuels have highly variable fuel characteristics governed by its moisture and residual oil contents. Previously reported figures for calorific values and proximate analysis are few and limited to essentially two early studies [1,2]. A more recent finding [3] also included compositions from an ultimate analysis and the chemical components of ash produced by the combustion of palm waste fuels. These are summarised in Tables 1 and 2 respectively.

Palm waste fuels are therefore characterised by high volatile matter, low fixed carbon and relatively high ash contents. Ultimate compositions of C, H, and O are in the typical range for cellulosic fuels while the S content is relatively low. Although fibre and shell ashes are characterised by higher silicate contents, in practice, when firing EFB, more clinkers are usually observed due to the presence of extraneous foreign matter (sand, laterite, etc.) accompanying the bunches. For a well designed mill, there is always sufficient fuel from fibre and shell. EFB are then incinerated, slowly, in large excess air, to mazimise the potash yield which is then used as a 3:2 substitute for the Murite of Potash fertiliser source in oil palm estate applications.

ESTIMATION OF COMBUSTION TEMPERATURES AND FLUE GAS HEAT LOSSES

Since there are no reported measurements of combustion temperatures and flue gas heat losses, it is decided to estimate these from basic principles with the following assumptions :

·   complete carbon combustion takes place and heat contributions from S and N in fuel material being small, are neglected.

·   average gross calorific values (GCV, kJ/kg)*, from literature, are used. Thus,

$$GCV_{shell} = 20552 > GCV_{fibre} = 19714 > GCV_{EFB} = 18813$$

·   for flue gas heat losses determinations, 1-2 % losses are allocated to unburnt fuels and radiation.

* dry basis, averaged over figures including GCV obtained in this study.

There are various ways to present the results of this exercise. To compare with typical industrial fuels, it is decided to plot, for each palm waste fuel and a typical fibre/shell mix,

- CO_2, O_2 content in flue gases (on dry basis) as a function of % excess combustion air used

- flue gas heat losses (as a percentage of fuel heat input) at different stack temperatures as a function of % excess combustion air used

- combustion or furnace temperatures as a function of % excess combustion air used

The results are shown in Figures 1 through 4 where, to reflect on practical situations, the following fuel moisture contents have been assigned :

EFB	@	63.0	% moisture
Fibre	@	37.6	% moisture
Shell	@	16.4	% moisture

Oil content in each of these fuels will alter the individual GCV. Since the GCV of dry fuel used are the averages of reported literature values, their reliability should be taken with the reported oil contents (dry basis), viz.,

EFB	@	0	to	5	% oil
Fibre	@	0	to	10.4	% oil
Shell	@	0	to	1	% oil

FUEL AVAILABILITY AND STEAM DISTRIBUTION IN AN OPERATING MILL

This study was conducted over 25 processing days in a 25 tonnes FFB per hour mill. The mill has two 15,200 kg/h water-tube boilers where at any one time, only one is in operation. The boilers are designed to generate steam at 17.6 kg/cm^2 and 50 ^{o}C superheat.

The measurement of fuel availability was conducted on a daily basis since the total amount of input FFB from different supplying estates were known. Fibre and shell flowrates were measured via manual sampling techniques. A sampling interval of 30 minutes and sampling times of 15 seconds (fibre stream) and 30 seconds (shell stream) were used after some statistical experimentation.

Steam distribution was measured by placing orifice plates in the main superheat steam header, steam lines to the turbine, two sterilisers, factory, and the steam make-up to the back-pressure vessel as illustrated by Figure 5. Recording, including line temperatures, pressures and mill power consumption meters, were conducted at intervals of 20 minutes. As flow totalisers were available, the average steam flows over the period were used.

Calorific value measurements of fibre and shell were conducted on the half-hourly fuel samples collected. The samples were analysed for their moisture and oil contents, and GCV using the Isothermal Method : BS 1016 Part 5 1967. Although EFB were not used to fire the boilers during the study, they were also subjected to the above analyses.

Results

The calorific values of fibre, shell and EFB were found to be in the range reported in literature except that the values obtained in this work covered the variability over a wide range of oil and moisture contents. The average net calorific values (NCV) obtained were :

EFB	@ 63.0 % moisture, 5.9 % oil (dry basis)	-	5723 kJ/kg
Fibre	@ 37.6 % moisture, 10.4 % oil (dry basis)	-	11627 kJ/kg
Shell	@ 16.4 % moisture	-	16685 kJ/kg

The study showed that over each processing day the mill was in positive fuel/steam and power balance. Combined fibre/shell availability averaged at 23 % of FFB and boiler consumption rate was around 16.7 % of FFB. Steam production was determined to be 3.9 kg steam per kg of fuel where the latter averaged .68:1 (W/w) shell/fibre at a combined NCV of 13517 kJ/kg. From these results, the boiler gave an average thermal efficiency of 71.5 %.

Steam flows were found to be highly variables as typically shown in Figure 6 for two different processing days. The steam losses (exhausted to the atmosphere) from the Back-Pressure Vessel were unexpectedly high. As Day 2 indicates, twice as much steam may be lost if (as it was the case) for example, one of the steriliser stations was down for some technical reasons (FFB cage derailment, repairs, etc.).

Over the study period during which the input FFB material were determined to be mainly of D x P species (7 to 13 % D x P below 5 years and 59 to 66 % above 5 years, from planting) the average steam consumption was found to be about 650 kg/tonne FFB processed.

DISCUSSION

Even on a dry basis, NCV of palm waste fuel may vary considerably over a given processing day. This is due to differing residual oil content in the fibre and EFB. In this study, fibre oil content ranges from 7.5 to 18.3 % (dry basis) while for EFB it was 4.8 to 7.0 %. As expected, no residual oil was detected in the palm shells. In practice, however, some unrecovered palm kernels (oil content 57.8 %, GCV = 28492 kJ/kg, both on dry basis) do remain unseparated from broken shells that go into firing the boilers.

In the past, mill designers (4) worked on the experience that each kg of fibre/shell mixture produces about 3.6 kg of steam and that 28 kg of steam would be required to generate 1 kW of power via the turbine turbo-alternator unit. This is based on 15 % FFB Fuel availability, at 13965 kJ/kg, and that the efficiency of thermal conversion is 75 %. Thus, this study has shown that although fuel availability has increased by 8 %, the effective calorific value entering the furnace seemed to have decreased slightly and this may be explained by the fact that there is now less shell in a unit weight of combined fuel derived from the new D x P planting material.

A more significant observation is the large increase in steam consumption per tonne of FFB processed. While previously, designers worked with 500 kg steam/tonne FFB, it appears that at least 30 % more may now be required. This increase in overall steam requirement is readily explained by the fact that FFB today are weevil-pollinated. The bunches tend to be bigger and more compact making them more difficult to steam-sterilise. Usually, triple-peak (steam pressure cycles) sterilisation have to be used in order to maintain an acceptable level of unstripped bunches at the mill. As the palms get older (each palm tree has a production life of 25 years), the FFB will get even bigger and as a result, engineers would rather use a conservative figure of 750 kg steam/tonne FFB now in designing new mills.

Perhaps, the most unexpected finding in the study was the quantitative observation that a typical palm oil mill may be losing considerable amounts of low pressure steam (3.1 kg/cm^2g) to the atmosphere due to the inefficient design of the steriliser station. Many mills in the country were designed with two sterilisers and these are designed such that only one steriliser may draw in steam from the Back-Pressure Vessel at a time. Only when the steriliser unit has reached its final holding pressure (2.8 kg/cm^2 g for 45 minutes), may another sterilisation process be started. The entire operation therefore needs careful sequencing in response to steam availability and the need to run above a critical FFB throughput so that surplus fuel is always available at the boiler platform.

The trend now is to design a mill with at least three steriliser units. This would allow at least one fully loaded steriliser to be on queue while one is drawing steam and the remainder being unloaded or being filled. Any excess steam in the Back-Pressure Vessel may now be automatically channelled to the 'waiting' steriliser for preheating. Experience has shown that with good process control, the Back-Pressure Vessel hardly loses steam through its exhaust valves and that the actual sterilisation process has registered faster peak build-ups and less steam condensation especially at the early stages of the cycle.

CONCLUSION

A detail fuel availability and steam distribution study in an operating palm oil mill has confirmed that with the new type of planting material and weevil-assisted pollination practices adopted by oil palm growers, the mill energy management situation has drastically changed. The implications of this scenario on mill design and sterilisation operations have been discussed.

ACKNOWLEDGEMENTS

The author wishes to thank Harrisons Malaysian Plantations Berhad for the permission and cooperation to publish this paper.

REFERENCES

1. Research on Production and Storage of Palm Oil (The Mongana Report) IRSIA (1955).

2. Maycock, J.H., The Burning of Oil Palm Refuse for Steam Generation (Palm Oil Research Institute of Malaysia, Private Communication).

3. Shafii, A.F. and K.Y. Boon, Instrumentation and Control Strategies for Industrial Boilers, Proc. of ENSEARCH Conf. on Instrumentation and The Environment, Kuala Lumpur Regent, 1-2 May (1986).

4. Hammond, K.L., Lecture Notes on Power Station Design., Malaysian Oil Palm Growers Council Course for Mill Engineers (1982).

Table 1 Palm Waste Fuel Analysis (Wt. Percent Moisture Free Basis)

Proximate Analysis	Empty Fruit Bunch	Fibre	Shell
Volatile Matter	75.7	72.8	76.3
Fixed Carbon (by difference)	17.0	18.8	20.0
Ash	7.3	8.4	3.2
Total	100.00	100.00	100.00

Ultimate Analysis			
Hydrogen	6.3	6.0	6.3
Carbon	48.8	47.2	52.4
Sulphur	0.2	0.3	0.2
Nitrogen	0.7	1.4	0.6
Oxygen (by difference)	36.7	36.7	37.3
Ash	7.3	8.4	3.2
Total	100.00	100.00	100.00

Table 2 Ash Composition (Wt. Percent)

	Empty Bunch	Fibre	Shell
SiO_2	34.7	63.2	65.4
Al_2O_3	1.2	4.5	2.1
Fe_2O_3	1.8	3.9	3.3
CaO	3.3	7.2	3.1
MgO	2.9	3.8	3.2
Na_2O	0.8	0.8	0.5
K_2O	40.1	9.0	12.7
TiO_2	0.1	0.2	0.1
P_2O_5	2.5	2.8	3.3
SO_3	8.0	2.8	3.2
CO_2	0.1	2.2	3.8
Total	99.5	100.4	100.7

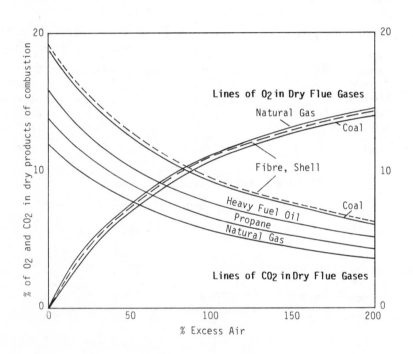

Figure 1   O$_2$ and CO$_2$ Content in Flue Gases
when Firing Different Fuels

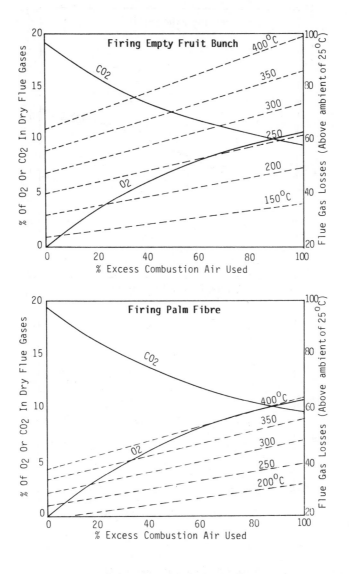

Figure 2 Flue Gas Heat Losses

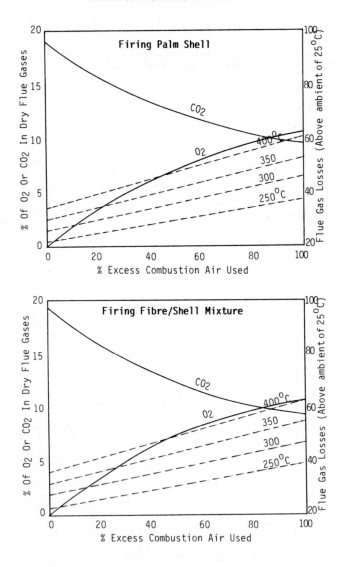

Figure 3   Flue Gas Heat Losses

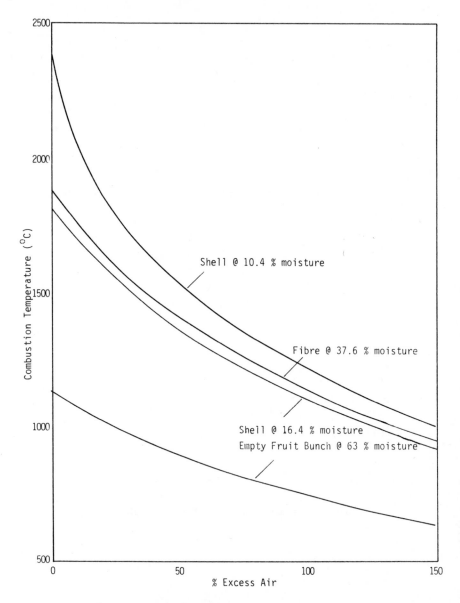

Figure 4 Estimated Combustion Temperatures When Firing
Different Palm Waste Fuels

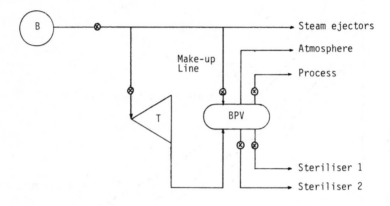

B   - Boiler
T   - Turbine
BPV - Back Pressure Vessel
⊗   - Orifice Meter

Figure 5 Schematic layout of the boiler plant

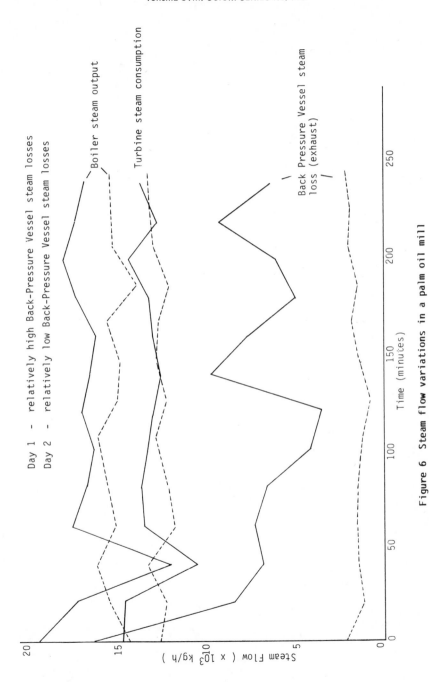

Day 1 - relatively high Back-Pressure Vessel steam losses
Day 2 - relatively low Back-Pressure Vessel steam losses

Boiler steam output

Turbine steam consumption

Back Pressure Vessel steam loss (exhaust)

Steam Flow (x 10³ kg/h)

Time (minutes)

Figure 6 Steam flow variations in a palm oil mill

BURNING RICE HUSK IN A HORIZONTAL CYCLONE FURNACE : A CASE STUDY

Dr.Santanu Chakrabarti*, P.Chakrabarti**,Dr.S.Saha***, Syamal Datta.****

India's annual husk production has an energy content of 200 X 10^{15} Joules, although this energy is used ineffici- ently. Energy-needs of a rice-mill for parboiling and dry- ing can be amply met by the heat generated by cyclonic combustion of rice-husk produced by the mill itself.

This paper describes the design development of a plant drying 15 tonnes of paddy by husk-fired cyclone furnace operating with a thermal efficiency higher than that of conventional husk-fired furnaces in India. Payback period of above six years is likely to come down with further design developments described in the paper.

INTRODUCTION

After a crop of paddy is harvested, it is threshed and prepared for milling. In the pre-milling process of parboiling (the process that differentiates between raw rice and parboiled rice), batch-loads of paddy are soaked in steam and water for a few hours and then dried. Usually furnaces fired by coal and/or husk generate the required steam and also the hot air for drying. In India, 40% of the total husk produced per year is consumed as a fuel in parboiling 45-50% of the total rice crop.

Parboiling thus places a higher demand on the energy-bill of a rice-mill and the mill-owner has to find the optimum mix of fuel for his mill. For example a rice-mill handling 4 ton rice per hour needs approximately 64.2 X 10^6 KJ for parboiling and drying and 125 KW for other auxiliary services like milling.

Husk as an Energy-Source

Husks are the largest milling by-products of rice. Though husk content of paddy varies form 14% to 20% by weight, an average of 18% yields about 14 X 10^6 tonnes husk per annum in India alone. The sheer volume of this waste product makes its disposal a serious problem, though its use as a supplementary fuel has been continuing for several years in rice-mills all over the world. A look at the cost-table (T-1) reveals that it is the cheapest available fuel. At the current rate of prices (December '86), a tonne of husk replaces 285 litres of furnace oil (44000 KJ/KG), monetary equivalent thus being five times the actual value of husk (See Table 2).

Contd. ...2

* &****: Projects & Design Dept., Wellman Incandescent India Ltd. Cal.- 71
** Consigne Engineers, Calcutta- 68.
*** Mechanical Engineering Dept., Jadavpur University, Calcutta- 700 032.

Husk Characteristics and Husk Burning

However husk does not easily release its total heat content, as oil or even coal does, and hence it becomes imperative to find the best means of burning rice-husk in a furnace.

In view of the husk characteristics (See Table 3) such as low bulk-density, high abrasiveness, flame-retarding as well as self-extinguishing nature in combustion, different methods of husk burning have been tried viz. atmospheric pile burning, incineration, briquet-burning and furnace-burning. Out of these, furnace burning has a large number of variations :-

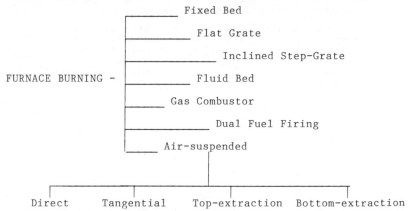

| Fixed Bed |
| Flat Grate |
| Inclined Step-Grate |
FURNACE BURNING – | Fluid Bed |
| Gas Combustor |
| Dual Fuel Firing |
| Air-suspended |

Direct   Tangential   Top-extraction   Bottom-extraction

Cyclone Burning of Husk

Laboratory studies have indicated that thermal efficiency of cyclone burning of husk (tangentially-fired air-suspended) is greater than that of grate-firing. No comparison has yet been made between efficiencies of a fluid-bed husk fired furnace and cyclone furnace. Ash disposal and temperature uniformity are found to be better in a cyclone furnace. Unfortunately very little data are available on direct continuous flow dryers running on cyclone furnaces. In 1978, a model unit was designed, built and tested at Indian Institute of Technology, Kharagpur, for steam raising. The furnace part was a cylindrical chamber (1.10 m dia X 2.39 m long) which consumed 110 Kg/hr. of husk and released a maximum of 70% heat, with an air flow rate of 12.74 m^3/minute.

We decided to follow this model to find a solution for a specific problem and used the existing theoretical frame work in designing and developing an innovative commercial model.

THE PROBLEM

A rice-mill in Sainthia (88°E, 24°N), West Bengal has been sundrying 20 to 30 tons of parboiled paddy on its yard (Photo 1) employing 30-40 labourers on a daily wage basis. Drying is thus a function of not only the labour availability but also of the incident solar radiation of 4 kwh/m² day (global, annual average), humidity (65-85%), temperature (10-40°) and air flow.

To reduce this dependence, as well as to increase the milling output, an efficient mechanical dryer coupled to a husk-fired furnace is designed. For achieving the maximum rate of heat release, air-suspension burning in cyclone is employed. A batch-load of 15 tons of paddy containing 30% moisture is to be dried to 14% in about 4 hours, excluding handling time (for loading of parboiled paddy in the elevator feed-hopper at the ground-level and discharging of dried paddy from the top chute of the elevator) to produce 12 tons dried paddy. Hence the design exercise has two parts viz. to determine the quantity of heat required to dry the batch-load and to calculate the air-volume needed to carry the thermal energy at 125±5°C to the dryer. The first part of the exercise establishes the furnace size, while the second ascertains the dryer dimensions (See figure 1).

FUNDAMENTAL PRINCIPLES OF DESIGN

a) Process should ensure thorough mixing of husk and air in proportions meeting combustion requirements.

b) Sufficient air velocities to be provided in the fuel conveying system as well as in combustion air lines to ensure cyclone formation in ignition and combustion zones.

c) Husk particles must have adequate residence time to ensure complete burning; presence of a bluff-body for recirculation of combustion products is a must.

d) A pressure-balance is to be maintained between the entry and the exit of the furnace so that the husk does neither 'flash-back' towards the entry door, nor does it 'blow off' towards the exit end. Simultaneous maintenance of the forward swirling motion is important.

Design Steps

The design of the husk-fired cyclone furnace with a dryer follows from the following design information :-

(i) Complete fuel analysis (ii) Weight of wet batch-load of parboiled paddy (iii) Initial and final moisture content of paddy (iv) Ambient temperature and desired temperature of drying.

After obtaining these data, the design exercise is carried out in six steps :-

a) Thermal duty of the dryer is ascertained considering that a batch load of 15 tons of parboiled paddy containing 30% moisuture is to be dried to a final moisture content of 14% in 4 hours. Once the thermal duty is calculated on the basis of the heat to paddy and moisture, casing losses and flue losses rate of heat generation is found out.

b) Combustion chamber volume is now determined with the empirical relation that heat release rate in a cyclone combustor varies from 5.2 to 9.3 MW/M³. The total heat release also gives an idea about the rate of husk consumption per hour assuming calorific value of husk to be 12,600 KJ/Kg.

c) Air-volume for drying the paddy in 4 hours (net) is calculated from an empirical relation that 680 NM³/ minute of air at 120°C is required to dry 10^T paddy from 30% to 14% in 3.5 hours. This is checked later by a conventional heat balance between the hot combustion products from the combustion chamber at 800°C when mixed with the required quantity of dilution air. This helps establish the blower-capacity as well as the process air chamber size.

d) Combustion blower selection is made from a knowledge of stoichiometric air requirement for husk @ 3.94 m³/Kg calculated from an ultimate analysis of the fuel and checked against empirical relations available. An excess air level of 50% is allowed, but 30% of the blower capacity is to be diverted to the fuel conveying line, also helping the initial ingnition.

e) Ducts are sized with a knowledge of allowable air velocities in the different lines viz. 15-20 m/sec for the combustion air, 10-15 m/sec for the husk-conveying line, and 10-12 m/sec for the dilution air as well as for the process air to dryer.

Nozzle-dimensions for husk conveying are determined on the principle of energy conservation : Velocity of entry (V_0) must be sufficient for husk particle to complete a turn so that the swirl is maintained (refer to figure 2).

$$\text{K.E. at A} = \text{P.E. at B (neglecting frictional losses)}$$

$$\tfrac{1}{2}\,\dot{m}\,V_0^2 = \dot{m}\,g\,d_0 \qquad \text{Where } \dot{m} \text{ is the rate of mass flow.}$$

$$\text{or } V_0 = \sqrt{2\,gd_0} \qquad \text{where } d_0 \text{ is the diameter of the entry-chamber.}$$

Assuming a coefficient of friction μ between husk and chamber -wall, corrected velocity $V_0 = \sqrt{2\,gd_0\,(1 + \mu\pi)}$ - this is the minimum sustaining velocity for the cyclone.

342

To ensure the formation of torroidal recirculation zones for the combustion products in swirls, value of the Swirl Number 'S' is kept above 0.6 (S is defined as the ratio of axial flux of the angular momentum to the axial flux of the linear momentum times the chamber radius). A strong swirl would persist for a hundred or more pipe diameters though the linear flow tends to be damped by its motion along the pipe.

f) The dryer dimensions are determined from a knowledge of the holding capacity of 15 tons (design batch-load) and the bulk-density of wet parboiled paddy. Cross-flow velocity of the hot mix through 100 ports of varying sizes in the dryer is fixed around 15 m/sec.

DESCRIPTION OF EQUIPMENT

A view of the furnace-dryer assembly is shown in photo 2. Detailed designs of the individual components are not shown here, only brief descriptions are given (refer to figure 1 and photo 3). Circled numbers relate to figure 1.

Husk Feeding Device ①

Husk is manually fed into a hopper. From the hopper, husk is introduced in an air-line② via a husk-metering device. Thus air-borne husk tangentially enters the ignition chamber through a rectangular nozzle. (refer to fig. 3)

Ignition Chamber of the Furnace ③

In this primary chamber, incoming husk is initially ignited from an external source; an intermediate air nozzle supplies primary air also helping to carry the swirl generated over to the next chamber. As husk does not sustain ignition below 450°C, temperature in this chamber is to be kept around 500°C. The chamber is insulated with refractory bricks and a 3mm thick stainless steel cladding is used to take care of the eroding stream of husk.

Combustion Chamber of the Furnace ④

Combustion chamber is a refractory lined cylindrical chamber where cyclonic combustion takes place. Nozzles supplying combustion air are arranged tangentially along the length ensuring that the swirl is also maintained. A re-entry thorat ④a is provided at the end of this chamber; this acts as a bluff-body ensuring ample recirculation of the burning particles. Lighter products of combustion escape through the throat into the next chamber.

Process-Air Chamber of the Furnace ⑤

It is a lagged air-mixing chamber where hot products of combustion from the combustion chamber at around 800°C mix with a tangentially entering dilution air stream from the process air-blower. The sudden expansion of the combustion gases into this larger chamber causes the ash-particles lose their momentum and fall to the bottom of the chamber, while the hot mixture at about 125° travels upwards to the dryer. Ash accumulates at the bottom and is collected periodically from the ash-trap by operating a manual flap-type damper.

Hot Air Plenum ⑥

It is the vertical chamber that takes in the hot mix to the main dryer ⑦ from the process air chamber. A multiple flap-type damper is provided at the middle of the entry to the dryer to regulate the hot flow to the main dryer.

Air Handling System

It consists of a combustion air blower ⑧ and a process-air blower ⑨ along with ducts, dampers and nozzles. The combustion blower also feeds the husk conveying line ② and the intermediate air-nozzle ②a in the ignition chamber. Seventy percent of the combustion air is fed through a cylindrical air drum into second chamber via air-ducts ending tangentially inside the chamber in specially-shaped nozzles (see photo 3). The process air blower tangentially throws the dilution air in the final chamber through a rectangular duct. A control air-bleed line ⑩ is tapped from the delivery line to the hot air plenum, so that the drying temperature can be controlled by operating a damper when required.

Dryer ⑦

It is a rectangular column-type drying chamber handling wet parpoiled paddy and hot air. Triangular baffles are welded on its inner panels in a staggered manner so that the downcoming stream of paddy is deflected and falls slowly. During this journey the stream meets hot air entering from ports on one side of the dryer body. This cross-flow through entry ports on one side and exit ports on the other gradually dries the paddy which returns to the same path via the elevator ⑪ . There are five motor-operated equispaced rotors at the bottom of the drying chamber distributing the partly dried paddy into the elevator feed-hopper and the cycle continues.

Elevator ⑪

Wet parboiled paddy is loaded at the bottom feed hopper of the spaced-bucket type chain-elevator. It is then taken up and discharged in the top charging hopper. Once drying is complete, the downcoming stream is deflected to the mill house through a separate chute.

RESULTS

PARAMETERS	TRIAL-I	II	III	IV	V
Weight of Parboiled Paddy Fed in (Tonnes)	14.7	15	15	15.88	14
Initial Moisture Percentage	30	30	30	30	30
Weight of Dried Output (Tonnes)	12.1	12.3	12.3	12.93	11.75
Final Moisture Percentage	15	14.66	14.6	14	13.9
Time taken for Drying only (Hours).	4	4.25	4	4.25	4
Husk consumed (Total in kg.)	520	535	520	550	490
Ash Collected (Kg.)	100	108	90	115	95

DISCUSSION OF RESULTS

From the available set of results, an average thermal efficiency of 85% is indicated. Analysis of initial ash samples indicated total absence of volatile matter and traces of unburnt carbon. Ash-samples from step-grate furnaces for similar duty were found to contain ten times higher percentage of unburnt carbon i.e. partially charred husk.

Husk consumption figures for mechanical drying of paddy with our dryer were 15% lower than the figure quoted in the only reported work on horizontal husk fired cyclone furnace.

A look at the up-dated table (T - 5) of comparative costs of processing paddy by different fuels highlights the potential of husk as a fuel for the future, particularly for the rice-mills.

A straight-forward calculation on pay-back period could not be done in absence of systematic records with the millowner. However an exercise with simplifying assumptions indicate that the capital expenditure on this furnace-dryer assembly in a rice-mill is paid back in less than six years at the current rate of prices, when compared to the cost of sun-drying on yards. When compared to a steam heat-exchanging type dryer, the extra investment is justified within 1.5 years.

SCOPES OF FURTHER WORK

Extensive experimentations are required

a) with different varieties of husk to ascertain their behaviour during burning as well as heat release rates under different seasonal and climatological conditions;

b) to assess performance of the dryer with higher batch-loads and to see what percentage of moisture is removed each hour under different loads;

c) to determine the effects of preheating the conveying air as well as the combustion air with a part of the hot flue, if possible;

d) to assess the effectiveness of different dampers on the combustion air line on the nature of combustion in the furnace;

e) to assess the applicability and effectiveness of automatic temperature control of the drying air;

f) to judge the relative merits and demerits of a vertical cyclone furnace (VCF) and a horizontal furnace;

g) to assess the possibility of automatic de-ashing;

h) to see if husk consumption rate decreases by insulating the hot air plenum.

CONCLUSIONS

Whatever may be the teething troubles of the particular plant, and the scopes of future developments, an optimum design of this furnace and drier is operating successfully at Sainthia Rice and Oil Mills, Kandi Road, Sainthia, Birbhum (W.B.).

Agricultural production in India is being geared up to meet increasing demands of food-grains by the exploding population. It is therefore imperative that mill output is increased and/or processing time in mills is reduced. Our design, described in this article, is a step towards tackling the problem of energy conservation as well as of waste utiliation, in order to provide scopes of improving the quality of life in the rice-eating countries. After all, these contain almost 50% of the global population.

ACKNOWLEDGEMENT

The authors are grateful to M/s. RDB Engineering Works, Calcutta for providing them the opportunity for design and experimentation and to M/s. Wellman Incandescent India Ltd., Calcutta- 71 for allowing to publish this paper. The first author in particular expresses his indebtedness to his ex-professor Dr. I.E. Smith of Cranfield Institute of Technology, Bedford and ex-colleague Councillor Dr. P.D. Fleming of Milton Keyness, both of whom helped him with papers, comments and relevant literatures.

T A B L E S

Table 1. Energy Cost Table

	Coal	Oil	Husk
Net Calorific Value (KJ/Kg)	25100	44000	12560
Price(Rs.)	1000/tonne	3.90/Kg	2.00/20 Kg bag
Apparent Price (Paise/MJ)	3.98	8.85	0.80

Table 2. Percentage Composition and Calorific Value of Different Fuels

FUEL	H	N	S	O	C	ASH	C.V. (KJ/Kg)
STRAW	5.0	0.5		38.0	31.0	4.75	4550
RICE HUSK (IR-8 Variety)	5.0	2.0	0.1	35.4	40.2	17.3	12560
SAWDUST	10.2	55.2		0.8	20.7	0.9	14770
BAGASSE	6.0			46.0	45.0	3.0	18840
WOOD	6.0	0.3		43.0	50.0	0.7	19800
PEAT	5.7	1.5		31.6	54.2	7.0	22100
ANTHRACITE	1.8	0.7	1.0	2.1	94.4		34270
FURNACE OIL	9.0		4.0		87.0		43960

Table 3. Physical Properties of Rice Husk

Bulk density	:	128.3 Kg/m^3
Porosity	:	82.7%
Angle of Repose	:	47. 52
Coefficient of Friction	:	0.63

Table 4. Proximate Analysis of Rice Husk

Volatile Matter	:	65-67%
Fixed Carbon	:	12-15%
Ash	:	15-18%
Moisture	:	3-4%

Table 5. Comparative Cost of Processing Paddy by Various Fuels. (ref. 16 paper 3)

Fuel	Cost (Rs./Kg)	Parboiling		Mechanical Drying		Total
		Fuel needed per tonne paddy (Kg)	Cost Rs.	Fuel needed per tonne paddy (Kg)	Cost Rs.	Cost of parboiling and drying Rs.
Diesel	4.30	16.4	70.52	15.6	67.08	137.60
Kerosene	2.88	16.3	46.86	15.5	44.56	91.42
Furnace Oil	3.90	17.1	66.69	34.4	134.16	200.85
Coal	1.00	45.5	45.50	163	163	208.50
Electricity	0.70/Kwh	147Kwh	102.90	158Kwh	110.60	213.50
Rice Husk	0.10	140Kg	14.00	152	15.20	29.20

REFERENCE

1. Renewable Energy in Action - A booklet published by C.A.S.E. Govt. of India, Technology Bhavan, New Delhi-16, 1982.

2. India, The Energy Scene - Presentation by The Indian National Committee at the 12th Congress of the World Energy Conference, New Delhi, 1983.

3. Don't waste Rice Husk - Burn it - S.K. Bhide, Economic Times, Calcutta, P6, 1st October, 1986.

4. Rice-husk Conversion to Energy - E.C. Beagle, Published by FAO of the United Nations, 1978.

5. Rice Milling Industry in India - Paper presented in the Seminar on Role of Rice Milling in Ensuring Better Availability of Rice & its By-products organised by Bengal Rice Mills Association, Calcutta-1981.

6. Method of Paddy Processing : NGC Iyengar and G.Rajendra, ibid.

7. Energy Technology Hand book - (Ed) D.M. Considine, Mc Graw Hill Book Co., P 9-80, 1977.

8. North American Combustion Handbook - Published by N.A. Mfg. Co., Cleveland (2nd ed), 1983.

9. BS-1042 Part-I : 1964.

10. Influence of Swirl Chamber Geometry on Turbulence in Multifuel Engineers - M.N. Elkotab et al in Combustion in Engineering, I. Mech.E. proceeding, c 55/83, 1983.

11. "Some Aspects of Fluidised Bed Combustion of Paddy Husk" - S.C. Bhattacharyya et. al, P 307-316, Applied Energy (16), 1984.

12. "Design Considerations for Calculating Fluidised Bed Combustors" - P. Basu in J. of the Institute of Energy, P 179-183, December, 1986.

13. Personal Correspondence with J.G. Hare of Gull Air Limited, Hampshire, U.K. (1985).

14. Personal Correspondence with Prof. I.E. Smith of Cranfield Institute of Technology, Bedford, U.K. 1985.

15. Ideas and Innovations - Simple Husk Fired Furnaces - R. Singh, Science To-day, December, 1980.

16. Paper No. 1,2,3,4,5,6,8,12,13,15,17,18 :-

    Proceedings of the National Workshop on Rice Husk for Energy : Organised by National Productivity Council, New Delhi, August, 1982.

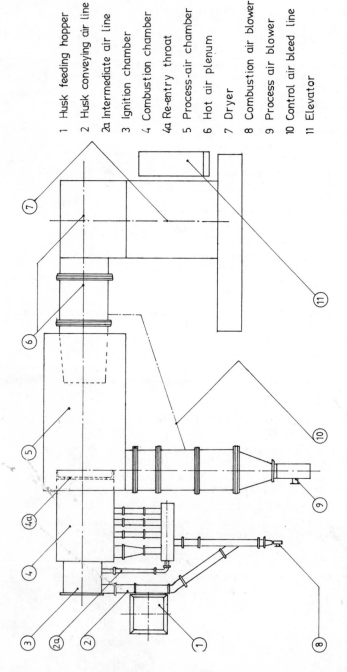

1 Husk feeding hopper
2 Husk conveying air line
2a Intermediate air line
3 Ignition chamber
4 Combustion chamber
4a Re-entry throat
5 Process-air chamber
6 Hot air plenum
7 Dryer
8 Combustion air blower
9 Process air blower
10 Control air bleed line
11 Elevator

FIG.-1 OVERALL PLAN OF THE CYCLONIC
HUSK FIRED FURNACE & PADDY DRYER

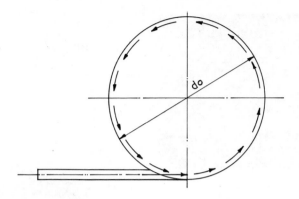

FIG. 2   Assessing minimum particle velocity for swirl
maintenance   inside a cylindrical chamber

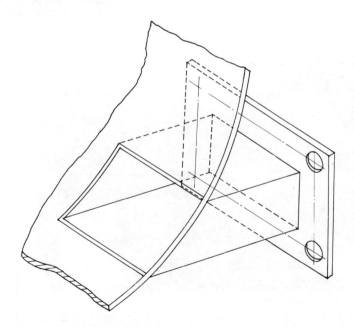

FIG. 3   Tangential entry in a cylindrical chamber
through a rectangular nozzle

Photo 1. ↑ Sun-drying of paddy on the mill-yard

Photo 2 → End-view of the furnace-dryer assembly

Photo 3 A partial view of the cyclone furnace showing ①②③④ [refer to fig-1]

GASIFICATION AND PYROLYSIS OF PEAT IN A FLUIDISED BED

B.M. Gibbs* and W. Nimmo*

Peat was gasified autothermally using a partial combust-
ion/pyrolysis process whereby the heat required to promote pyrolysis and gasification reactions was produced in
the fluid bed by burning a portion of the peat fuel. The
use of an inert fluidised bed of sand as a carrier of
heat resulted in rapid and efficient transfer of heat
from zones of combustion to zones of reduction and pyrolysis. The influence of a number of process variables
on the gasification efficiency and fuel gas quality were
studied including the effects of temperature, fuel feed
rate and bed depth. The results indicate that to produce a fuel gas of C.V. greater than 4 MJ/Nm^3, temperatures in excess of 750° degrees C are required with a
peat feed rate greater than 2.75 times stoichiometric.
The process is also benefitted by bed depths (static)
greater than 20 cm.

INTRODUCTION

Some manufacturers of boiler equipment have recognised a need in
industry for small scale pyrolytic incinerators and gasifiers
where solid fuels such as coal, peat and municipal or industrial
wastes can be processed to give a low calorific value fuel gas.
The gas could be piped economically over short distances in the
locality of the gasifier to provide energy for space heating etc.
The small gasifier units already in production generally take the
form of fixed or fluidised beds with the fluidised bed gasifiers
using technology gained from experience with combustor construction.

Gasification tests on peat have been successfully carried out in
conventional commercial gasifiers (1-2) to produce low, medium
and S.N.G. quality fuel gases. Fluidised bed processes in the
form of the Winkler gasifier compete well with other gasifier
types, namely fixed and entrained beds, in the production of low
and medium C.V. gases using air and O_2 respectively.

There will be a continuing need for high utility gaseous fuels in
the domestic and industrial sectors of the market as natural resources are depleted. In order to meet the overall requirements
of U.K. consumers, British Gas are basing future production of
substitute natural gas on coal using the British Gas/Lurgi slag-
ging gasifier which is at present in advanced staged of develop
ment.

*Department of Fuel and Energy, University of Leeds, Leeds LS2.9JT.

By the turn of the century, fuel gas consumed in the U.K. will largely be produced from solid fuel, mainly coal, supplemented by small gasifier units supplying factories in the immediate locality. There is no doubt that peat would provide a useful fuel for use in small fluidised bed gasifiers particularly in areas where peat is locally available, thereby minimising costs of transportation.

GASIFICATION OF PEAT IN FLUIDISED BEDS - STRATEGY

Gasification processes producing fuel gases of low (< 6 MJ/Nm³) or medium (10-15 MJ/Nm³) calorific value have the potential of operation autothermically, by achieving a thermodynamic balance between the endothermic gasification reactions and the exothermic combustion reactions.

Fluidised beds exhibit good heat and material transport properties and generally operate with an almost uniform temperature profile through the bed. This means that heat generated from burning peat particles is rapidly transferred to adjacent sand particles, which in turn, through solids circulation, distribute heat throughout the bed where heat absorbing steam-char, CO_2 - char and pyrolysis reactions are taking place. The use of an inert material such as sand as a heat carrier overcomes solids fluidisation and recirculation problems (3-4) and helps to prevent the formation of hot-spots in 'dead' regions, which could lead to peat ash sintering, resulting in defluidisation and reactor failure.

In a well fluidised bed of sand, fuel particles of a wide size range can be accommodated, therefore it is inevitable that a portion of the fuel input will be elutriated since the flow of fluidising air is determined by the mean particle size of sand which has a higher density than peat char. This represents a loss to the process which could be recouped if the fines are returned to the bed, after separation from the product gas flow.

When a peat particle is fed onto the surface of a vigorously bubbling fluidised bed of hot sand (800°C) it undergoes physical and chemical changes associated with a rapid increase in temperature (5). The stages of devolatilisation can be described as follows:

1. At temperatures up to 300°C physically bound water evaporates and decarboxylation reactions release CO_2 CO and water.

2. At high temperatures (up to 600°C) organic material is volatilised which includes tar vapours and low molecular weight incondensibles. (i.e. CH_4, C_2H_6, etc.).

3. At temperatures in excess of 600°C, any residual oxygen and hydrogen in the coke or char will be evolved by the breaking of higher-energy bonds.

The ultimate result of the rapid distillation of volatiles from peat is the production of

1. Condensible tar vapours and water.
2. Incondensible gases (both inert and combustible).
3. A char residue, low in volatiles.

The incondensible gas products of devolatilisation will mix with combustion and gasification products produced when peat is gasified using air or oxygen. As the peat particle is heated after introduction to the fluid bed, it enters the solids recirculation pattern. As it passes from top to bottom it leaves a zone essentially reducing in character and enters a region of oxidation above the distributor where the bulk of combustion takes place. The extent to which oxygen penetrates the bed depends largely on the bed-char loading. It is clear that high bed-char loadings will restrict the oxidation reactions to a region close to the distributor whereas light char loading may in the extreme case allow oxygen to leave the fluid-bed and subsequently become a part of the fuel gas. It is also possible under these conditions for valuable light hydrocarbon products to be burnt in the freeboard region of the reactor representing a loss to the process.

EXPERIMENTAL INVESTIGATION

Objectives

The aim of the series of tests performed on the air-gasification of peat was primarily to assess the effects of the two main process variables, namely, temperature and fuel feed rate, on the quality of the fuel gas produced. A secondary aim was to characterise the effects of these variables on gas yields and process efficiencies.

Fuel Description and Preparation

The peat used in the tests was supplied by Bord na Mona of Southern Ireland in the form of sods which were reduced in a jaw crusher to give the size distribution shown in Table 1.

TABLE 1

μm	% by weight
< 6000	100
< 3353	66.4
< 1676	37.7
< 1003	25.4
< 699	19.4
< 500	15.29
< 295	10.01
< 210	6.34
< 178	6.17

The crushed peat was allowed to dry at room temperature for 24 hours which reduced the physically bound moisture to an average figure of 17% (by weight). The ultimate and proximate analyses of the fuel on a dry basis are given in Table 2.

TABLE 2

Ultimate Analysis (dry basis)		Proximate Analysis (dry basis)	
C	54.2% (by weight)	Volatile matter	64.4%
H	4.95%	Ash	5.2%
N	2.02%	Fixed carbon	30.4%
S	0.3%		
O	33.3%		
Ash	5.2%		

In this form the handling characteristics were very good and no problems were encountered in feeding or metering using a vibratory feeding system.

EXPERIMENTAL EQUIPMENT AND OPERATION

Fluidised Bed Gasification Unit

The tests on peat gasification with air were carried out on a laboratory scale fluidised bed unit of 15.4 cm i/d shown schematically in Figure 1.

The apparatus consisted of three main sections, namely:

1. Preheater.
2. Reactor and freeboard.
3. Fuel feeding system.

The preheater unit was used to heat the reactor by warming the fluidising air passing through it. The desired reactor temperature was achieved by burning peat on the bed. The fluidising velocity employed was determined by the sand particle size distribution, therefore to achieve good fluidisation, a velocity of 0.25 m/s was selected corresponding to greater than four times the minimum required for fluidisation. The temperature of the gasifier bed was controlled by regulating the heat loss through the reactor wall by rearrangement of insulating material. This method proved satisfactory on the scale of reactor used in the experiments. On larger sized equipment, temperature control may be achieved more conveniently and efficiently by using water cooled heat transfer tubes whereby the reactor temperature could be controlled and steam generated at the same time.

The feeding system consisted of a hopper-fed vibratory feeder isolated from the reactor by a rotary sealing value. This meant that product gas and other vapours could not enter the feeding system which was slightly pressurised with N_2. The peat feed rate was varied by altering the frequency of the vibrator which permitted reasonably accurate calibration to be obtained. The feeder calibration was checked several times during an

experimental test. After metering the fuel passed down a side-
arm and entered the reactor at a position 10-20 cm above the
fluidised bed depending on the height of bed employed.

Gas Analysis

The composition of the product gas was determined by gas
chromatography and the compositions presented in this paper are
the average of a minimum of 4 samples. Gas samples were extracted from the product flow in the cyclone exit pipe, as indicated
in Figure 1. Samples were collected in sealed syringes and gas
analysis was performed on the same day as the experimental test
to minimise the risk of air ingression or hydrogen escape. In
general the O_2 content of gas samples was less than 1% (by
volume) and typically less than 0.5% indicating that syringe sealing was effective.

EXPERIMENTAL RESULTS

General

Results presented in this section were obtained from the
gasification of crushed peat in a 15.4 cm i/d fluidised bed. Information has been obtained concerning the effect of temperature,
peat feed rate and bed depth on fuel gas quality and yields. In
all the tests performed, the superficial air velocity through the
bed was 0.25 m/s at 800°C. This velocity was chosen to cut down
losses through elutriation, but still maintains a well-fluidised
bed of sand. The moisture content of the fuel was 16.9% by
weight.

Effect of Temperature

The influence of reactor bed temperature on C.V. and yields
of product fuel gas, are shown in Figures 2 and 3. Results in
Figure 2 shown for a number of different inert bed depths (static
before fluidisation). The general trend is to a better quality
fuel gas at higher temperatures and a fuel gas of gross C.V.
> 4 MJ/Nm³ can be obtained at temperatures in excess of 750°C at
bed depths (static) greater than 20 cms. At temperatures, up to
about 875°C, the reactor could be operated without any preheat
and control was achieved by adjusting heat loss through the reactor wall by removal of sections of the insulating material.

The effect of temperature on the yield of individual components in the product gas is shown in Fig. 3. It is clear that
the combustion content of the product gas, i.e. CO, H_2 and light
hydrocarbons, increases with temperature, whereas CO_2 decreases.
The results show that the yield of gas is increasing with the
possible reduction of yields of the other two products namely char
and tar vapour. At higher temperatures, oil and tar vapours will
tend to crack more readily and the rates of gas interchange from
char particles to surrounding bulk gas will be enhanced. This
means that the char yield will be less since oils and tar vapours
will not have the same tendency to stagnate around fuel particles
which is the case of lower temperatures.

Higher temperatures also favour the char-steam/CO_2 reactions resulting in enhanced yields of CO and H_2. The bulk of the steam originates from the evaporation of fuel moisture which reacts on contact with hot char particles to form CO and H_2.

Effect of Peat Feed Rate

The rate of indtroduction of peat into the fluidised bed was varied over a wide range of fuel factors (i.e. multiples of the feed required for complete combustion with air). Fuel feed rates in the region of 1.5 to 4.5 times stoichiometric (S) were employed and some results are shown in Figures 4, 5 and 6. Figure 4 shows the effect of fuel reed rate on the fuel gas C.V. for a number of bed depths at temperatures of around 800°C. The general trend is to an improved quality gas at higher fuel feed rates. Over the fuel feed range under investigation, doubling the feed rate from 2 to 4 can improve the C.V. of the gas by 1 to 2 MJ/Nm^3 which represents a significant increase in quality for low C.V. fuel gases. There is, however, a price to be paid for this improvement in the quality of product gas. Figure 5 shows that the amount of heat returned in the fuel gas C.V. per kg of fuel feed decreases at higher feed rates and shows an average decrease in efficiency of around 6% for a twofold increase in feed rate. This extra loss may be due to increased tar production at higher feed rates whose C.V. is not reflected in the heating value of the product gas.

It has been shown that the loss in efficiency at higher fuel feed rates cannot be apportioned to increases in solids carry-over. Figure 6 shows that the percentage weight of char carried over remains virtually constant over the range of feed rates studied and a plot of elutriation rate vs. fuel feed rate returns a linear relationship.

The dependency of gas yields on fuel feed rate variation is shown in Figure 7, for a bed depth of 30 cm (static) in the temperature range 780-820°C. The total yield of gas decreased although the yield of the combustible components H_2, CO and C_nH_m increase at higher feed rates. The most significant increase is in the yields of light hydrocarbons (C_1-C_2) due to increased cracking activity at the bed surface.

CONCLUSIONS

It is clear that peat has the potential of being a useful fuel for the production of low C.V. fuel gas, using air, in a relatively simple fluidised bed reactor. The problems associated with possible peat ash fusion can be eliminated by using an inert bed material, such as sand, as a heat transfer medium, reducing the risk of hot spots occurring.

By operating the reactor at temperatures around 800°C, the process can behave autothermally, giving a fuel gas of C.V. approaching 5 MJ/Nm^3 (Gross) with a typical composition as follows:-
CO 12.4%, H_2 9.6%, CH_4 2.3%, C_nH_m 1.4%, CO_2 14.3%, balance N_2 (dry basis).

REFERENCES

1. Leppamaki, E., Asplund, D., and Ekman, E., Gasification of
   Peat - A Literature Review, Symposium Papers, Management
   Assessment of Peat as an Energy Resource, July, 1979,
   Arlington, Virginia.
2. Punwani, D.V., Synthetic Fuels and Peat, ibid.
3. Salo, D., Filen, H. and Asplund, D., Gasification of Milled
   Peat on a Fluidised Bed, 6th International Peat Congress,
   Duluth, Minnesota, U.S.A., August, 1980.
4. MacDougall, D., Production of Water-Gas from Milled Peat in a
   Fluidised Bed, International Peat Symposium, Dublin, July,
   1954.
5. Fuchsman, C.H., Peat, Industrial Chemistry and Technology,
   Academic Press, 1980.

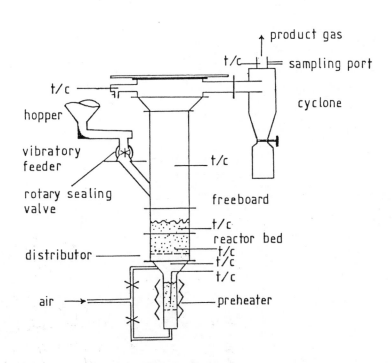

Fig. 1.   Schematic diagram of fluidised bed gasifier.

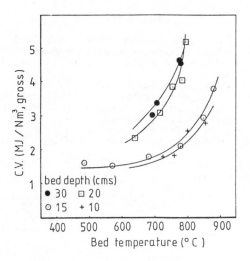

Fig. 2.   The influence of bed temperature on fuel gas calorific value.

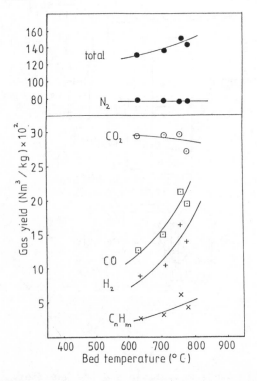

Fig. 3.   The effect of bed temperature on individual and total gas yields.

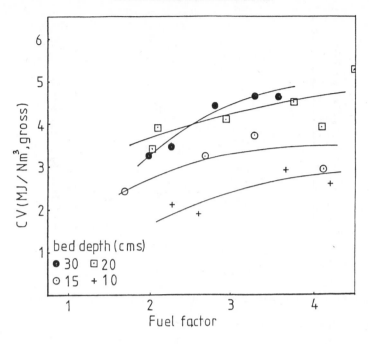

Fig. 4.   Influence of peat feed rate on fuel gas calorific value.

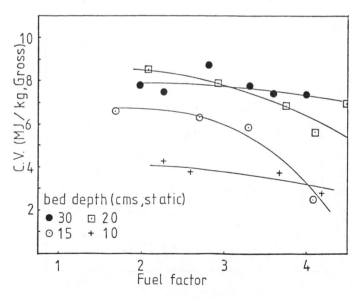

Fig. 5.   Effect of peat feed rate on specific C.V. of the fuel gas.

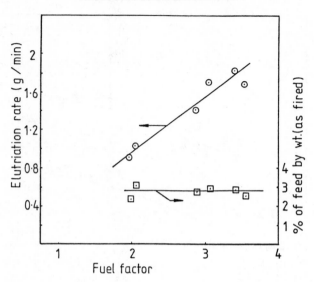

Fig. 6.   Effect of peat feed rate on solids elutriation rate.

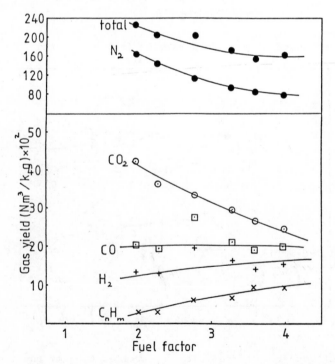

Fig. 7.   Effect of peat feed rate on individual and total gas yield.

FUNDING OF INNOVATIVE PROJECTS - TAPPING RESOURCES IN THE UK AND EUROPE

D.A. Reay*

The fields of energy efficiency and process innovation are ones in which participants can receive substantial funding from a variety of sources. This paper presents information on schemes managed by the UK Energy Efficiency Office and the European Commission. Some guidance to potential applicants is given, based in part on the Author's work for the Energy Technology Support Unit at Harwell and the Non-Nuclear Energy R&D Programme administered from Brussels.

INTRODUCTION

The fields of energy efficiency and process innovation are ones in which participants in industry, research laboratories and educational establishments can receive substantial funding from a variety of sources. The UK government, principally via the Dept. of Energy and the Dept. of Trade & Industry, is able to support industry in a number of ways, either as a manufacturer or energy-efficient or energy-conserving products, or as a user of energy.

The European Commission, operating principally from Brussels, has many programmes supporting innovation. These are directed both at energy saving and at improving industrial competitiveness - often similar or complementary routes can be used to meet these aims. The BRITE and EURAM programmes, together with EUREKA, the Non-Nuclear energy R&D and the Energy Demonstration Programmes, are among the major initiatives taken by the EC.

In this paper it is intended to present data on 4 of the schemes which could benefit companies concerned with process energy efficiency. The Author will also give advice on proposal preparation, and show some examples of projects which have been recently funded.

* David Reay & Associates, PO Box 25, Whitley Bay, Tyne & Wear NE26 1QT

THE UK DEPARTMENT OF ENERGY

The D.En. operates a number of schemes which assist users and in some cases suppliers of energy-consuming plant to improve efficiency. Innovation is not a precondition in all of these, which are listed in Table 1. Those involving research, development and demonstration and administered by the Energy Efficiency Office (EEO), in conjunction with the Energy Technology Support Unit at Harwell (ETSU) are, however, directed at innovative approaches to energy efficiency, either in terms of the application or the technology employed.

TABLE 1   UK Department of Energy:
Schemes for Industrial Support to
Reduce Energy Costs*

Energy Efficiency Surveys:

>   Short Survey
>   Extended Survey
>   Combined Heat and Power Feasibility Study

Energy Efficiency Research & Development

Energy Efficiency Demonstration Scheme

Coal-Firing Scheme

Monitoring and Targeting

* Correct as of December 1986

2.1   R&D Projects

2.1.1 Programme Aim & Nature of Projects Supported

The aim of the R&D programme in energy efficiency is to "encourage innovation in energy-efficient systems in industrial plant and operation and in all classes of buildings". An emphasis is put on cost-effectiveness when assessing projects, and this theme runs through all the programmes.

There are a number of basic criteria which should be met before funding by the EEO can be considered. Bearing in mind that the "raison d'etre" for the scheme is the saving of cost and energy on a national basis, preference is given to projects which:

(i)   could lead to significant national energy efficiency benefits

(ii)   have a potential for commercial as well as technical success, and

(iii)   will be carried out by organisations which have the capacity for either exploiting the results or successfully transferring them to other organisations who can.

With regard to the projects themselves, guidelines also exist for the types of R&D which can be carried out. As well as support for prototype equipment, feasibility and other studies can be supported where these could contribute to the evaluation of concepts which may lead to substantial savings in the longer term. In particular, projects which could lead to one or more of the following benefits could be supported:

- increased efficiency in energy usage
- reduced capital costs for energy-saving plant
- extended markets for energy efficiency measures
- provision of exploitable knowledge relevant to energy efficiency

2.1.2 Financial Support Available

Funding of these R&D projects can take one of two forms. A grant of up to 25% of the cost of the R&D may be made available. Alternatively, up to 50% of the cost of the project could be provided on a shared cost basis. In order to recoup some of its investment, a levy would be placed on sales (assuming that the project was successful).

Unlike the demonstration projects scheme discussed later, support for R&D tends to favour companies who are suppliers of equipment or systems, rather than users. This does not, however, preclude a potential user being involved at a stage in the R&D work so that it may readily exploit the technology. Links within the R&D project or, more logically, via a demonstration project, would ensure that the user also benefitted from government assistance.

2.1.3 Advice for Potential Applicants

The EEO publishes brochures describing in detail the Scheme and the criteria on which applications are judged. They also contain information on the format of the application, and some advice on how to satisfy the requirements of the assessors, (normally a committee made up of representatives from industry, commerce, and government).

It is important to note one difference between the UK and CEC programmes. Brussels normally sets a deadline for proposal submission under annual or, more commonly, 3-4 year programmes. The UK government schemes in general accept submissions at any time, although in both cases the proposer should allow ample time between first contact with the funding authority and notification of the success, or otherwise of the application. As with many other programmes, retrospective funding is officially discouraged, and if project implementation is urgent, other funding routes may be preferable. Alternatively, government could be asked to participate at a later stage in the project.

R&D project proposals, in common with those for UK Demonstration Projects, are guided through the committee stages by a project officer at ETSU (or BRECSU) with responsibility for the sector or technology with which the proposal is conerned. Contact with the appropriate officer will be arranged at an early stage, and the help he will give is generally substantial. Although unsolicited proposals can be submitted, it is inevitable that the project officer will become involved. The sooner this takes place

the better - at worst it will save the potential proposer a lot
of unnecessary data collection and proposal composition.

An important section of any proposal is that dealing with the
case for support.   As well as the more obvious data dealing with
the local impact on the host's process or product line, national
(or European in the case of the EC) interests have to be taken
into account.   These are normally reflected in terms of the potential national energy savings, which generally have to be in excess
of a certain figure, and/or the financial value of these savings.

2.1.4. Project Example - Process Integration Study

Within the above programme, a number of projects covering the
use of process integration (PI) have been funded.   One such study
was carried out at the Tioxide plc plant at Seal Sands, Teesside,
(Ref.1).   The study was performed by the ICI process integration
consultancy service which, at the time, existed as a separate
unit, (now integrated with that at Harwell).

The execution of the project involved the collection of process
stream data, (at the site there were 27 process streams and 3
utility streams, these being steam, cooling water and brine).
This permitted ICI to calculate the "grand composite curve" -
the graphical representation of all the process streams requiring
heating or cooling.   These are illustrated in Table 2, together
with the target energy consumptions which the study predicted
could be achieved by adjusting the composite curves for each of
the major operations to minimise energy consumption.

TABLE 2   Minimum Heating and Cooling Requirements

	Heating	Cooling
	(Comparative Values)	
Total site as a single entity		
Present Consumption	1000	1046
Energy Target	592	638
Individual plants		
Chlorination plant		
Present consumption	63	301
Energy target	00	238
Oxidation plant		
Present consumption	24	90
Energy target	0	66
Chlorine recovery plant		
Present consumption	200	142
Energy target	154	96
Wet treatment plant		
Present consumption	713	513
Energy target	553	353
Total energy target for the four individual plants	707	753

In this application of PI, each major plant on the site was treated separately, as operating flexibility precluded large scale transfer of process heat from one part of the site to another. This accounts for the two targets identified in the Table.

Among the changes suggested was a modification to the process in the chlorine recovery plant. This is illustrated in Fig.1, and gives a potential saving of 31%, which at the time was worth £200,000 per annum.

2.2   The Energy Efficiency Demonstration Scheme

2.2.1   Programme aim & nature of projects supported

The Energy Efficiency Demonstration Scheme (EEDS) has as its aim the encouragement of more widespread use of technologies which contribute towards energy efficiency. This is done in two ways - by demonstrating both the technical and economic effectiveness of the technology in a working situation and by encouraging replication.

The EEDS covers industrial processes and most types of buildings, and in addition to projects concerned with the installation of equipment, the application of new materials or design techniques are eligible.

A project involves 3 principal activities which, in the case of, for example, use of a novel process heat exchanger for heat recovery, would be:

(i)   Installation of the heat exchanger on the process.

(ii)   Monitoring of the energy used by the process both prior to and following installation of the heat exchanger.

(iii)   Promotion of the project by various means and media to encourage other appropriate potential users to adopt the heat exchanger - i.e. replication.

The use of a novel heat exchanger falls within the category of projects covering new or improved technologies which enable energy to be used more efficiently. Additionally, new applications of established energy - efficient technologies may be supported. In the context of mechanical vapour recompression (MVR) systems, an application on an evaporator in the chemicals industry is receiving support in spite of the existance of MVR installations in the food and drink sector. This reflects upon the different nature of the fluid being concentrated, and the increased risk due to uncertainties regarding corrosion and system component design. Other projects supported under this scheme in the chemicals sector are listed in Table 3, (Ref.2).

2.2.2   Financial support available

Financial support can take one of two basic forms:

TABLE 3   Some Recent EEDS Projects in the Chemical Industry

Area	Project	Host Company
Process heating	Heat recovery on a catalytic reformer	ICI Mond
Process heating	Heat recovery for use in distillation	ICI Petrochemicals & Plastics
Distillation	Structured packing on an distillation column	Phillips Imperial Petroleum
Evapora-tion	Mechanical vapour recompression	Staveley Chemicals
Drying	Spray dryer heat recovery	ABM Chemicals
Heat recovery	Heat recovery from a heavily contaminated vapour steam	Tioxide UK
Utilities	Cold de-aeration of boiler feedwater	BP Chemicals
Utilities	Solvent recovery using liquid nitrogen	Xidex Corporation

(i)   Up to 25% of the installed capital cost of the innovative system, plus up to 100% of the monitoring costs (including additional necessary instrument-ation and the services of the independent monitoring organisation).

(ii)   A shared-cost contract under which the EEO can provide up to 50% of the installed capital cost, plus of course full monitoring costs. However in this case a proportion of the EEO costs, up to their full value plus interest, could be recover-able, depending upon the success of the project.

In addition, where the proposed project contains a considerable degree of uncertainty, the EEO may request a feasibility or design study, for which support up to its full cost may be provided.

Although the R&D scheme described above is a more likely route for funding, some scope exists within the EEDS for equipment manu-facturers to carry out development work which would lead to a demon-stration project.

2.2.3   Advice for potential applicants

In common with support for R&D, the documentation available to assist potential applicants is comprehensive. This is complemented by a back-up service, normally operated from ETSU for industrial projects, the principal role of which is to help the applicant progress his proposal through the assessment and, hopefully, con-tractural procedures. This must rank as one of the most "user friendly" services available from central government.

ETSU is able to give an indication of the areas within individual industrial sectors where projects are required. Early in 1987 this was done for the chemicals industry, and the data are given in Table 4, (Ref.2). Other opportunities are, of course, available, but the data in the Table are based on a detailed strategy study of the industry performed during 1986. This identified unit operations where scope still existed for energy efficiency improvements which fell within the ETSU remit. It is therefore likely that a degree of preference will be given to the projects listed.

TABLE 4   New Projects Required in the Chemical Industry

Process heating	Batch process heat recovery*
Process heating	Calciner heat recovery
Distillation	Mechanical vapour recompression
Distillation	Structured packing on pressure distillation column
Drying	Novel rotary dryer designs*
Drying	Improved band dryer operation
Drying	Humidity control
Heat recovery	Innovative heat recovery schemes*
Utilities	Heat recovery below oil-fired boiler exhaust dewpoint

*More than one project envisaged

As with other programmes, considerable emphasis is given to the "replication potential". In the case of the EEDS this is currently expressed as the value of the potential national energy savings resulting from realistic replication of the technology being demonstrated. These savings should be in excess of £1 million per annum. ETSU assists in identifying the replication potential, but it is in the client's interest to make ETSU aware of any opportunities known to him.

2.2.4   Project example - structured packing in a crude oil atmospheric distillation column

Typical of the projects funded recently in the chemicals sector is that at Phillips Imperial Petroleum Ltd. (PIP). PIP at Teesside has received support under the scheme to install structured packing in an atmospheric crude distillation unit. It is predicted that the replacement of 11 existing trays with structured packing will double the number of theoretical stages in these sections. This will benefit the user in two ways - improved yield and energy saving. The location of the packing is shown in Fig.2.

The refinery handles about 5 million tonnes oil equivalent per annum, and the target savings of 97,000 GJ give an anticipated payback of 3 months. With a total investment cost of £270,000. it can be seen that a significant part of the payback in this project results from an increased yield of the more valuable middle distillates. It is anticipated that a substantial reduction in stripping steam will also result.

The project is being monitored by Thermal Developments Ltd. This company is quantifying the value of the benefits, assessing the ease of installation, and using radioisotopes to study liquid distribution in the column. This work will be completed by December 1987.

THE EUROPEAN COMMISSION PROGRAMMES

There are several programmes managed by the EC which have relevance to energy use in processes. In some, notably BRITE, EURAM and EUREKA, the energy interest tends to be peripheral, but nevertheless of some interest. However, two schemes which to some extent parallel those operated by the UK D.En. are solely related to energy. These are:

(i)   The Non-Nuclear Energy R&D Programme - managed by DGXII

(ii)   The Energy Conservation Demonstration Programme - managed by DGXVII.

(DG's are the Directorate Generals responsible for administering different areas of Community policy and legislation - roughly equivalent to a UK government department).

3.1   Non-Nuclear Energy R&D Programme

This programme, which involves calls for proposals on a variety of topics on a large scale every 3 to 4 years, with occasional intermittent more restricted calls, covers several topics. These include renewable energy sources, the better exploration and exploitation of liquid, gaseous and solid fuels and energy conservation R&D. This section concentrates on this last subprogramme.

The current (3rd) programme commenced in 1985, and is scheduled for completion in 1988. A 4th programme is currently being planned. There are 3 principal objectives of the energy conservation subprogramme:

(i)   The promotion of energy conservation in the Member States by means of new and improved energy technologies, processes and products.

(ii)   The development of new and improved energy technologies which will additionally lead to a reduction in pollution.

(iii)   The encouragement of industrial involvement in the above by financial incentives, publicity and example.

Approximately 100 projects are currently being supported, and funds of up to 26.5 MECU have been made available. (In early 1987 1 ECU = £0.70).

3.1.1   Proposal procedures

A strict deadline exists for submissions, and the selection procedure is exhaustive, involving a number of assessment hurdles. Recently the EC has set out the evaluation procedure in some calls for proposals, and in the most recent call in this programme it was as follows:

- receipt, registration & acknowledgement
- distribution to experts appointed by the EC for assessment
- consideration by Management & Co-ordinating Advisory Committee
- final conslusions by the EC
- notification of decision to proposers
- submission of more detailed administrative and financial form by proposers
- contract negotiations
- signature of contract.

During implementation of the project, formal progress and final reports are normally specified. These are assessed by the experts. Project review meetings are held in conjunction with other contractors at least once during the course of a programme, and a full conference is likely to review the programme. These may be in Brussels or at a centre elsewhere in Europe.

The EC and/or its representatives (eg a consultant employed as an expert) may visit the site of the work during its progress. This is likely to occur at least once during the course of the contract.

With regard to financing, up to 50% of the cost of the R&D project can be provided by the EC. In the case of Universities and Polytechnics, marginal cost funding is provided. In practice this pays the full salary of the researcher, (not his supervisor), and up to 40% overheads.

There is, in some programmes, a necessity to link with laboratories/ companies in other Member States before funding for a project can be made available. This is not the case in this programme, but it is advisable to attempt to formulate an international project if possible. The EC may suggest that you co-ordinate activities with another contractor, as is happening with the expert system project I later use to illustrate the scheme. I suspect the trend will be towards more projects involving several contractors, as in BRITE.

3.1.2   Advice for potential applicants

The documentation associated with the programmes of the EC tend to spell out in some considerable detail the topics which will be regarded as acceptable. For example, in the case of heat exchangers for heat recovery, high temperature corrosion resistance may be cited as a particular goal, but the development of units for

371

operation below the acid dewpoint may not be mentioned. It is imperative that the proposer takes note of the specific requirements, as requests for support in the second category would be unlikely to pass the first screening. (Such proposals may be directed by the EC to other programmes where they may be more applicable). These requirements have been arrived at following detailed discussions involving the EC, its experts, and possibly subcontracted organisations whose role it may be to assess European capabilities in a specific area of technology.

Implicit in this is the desirability to "make your mark" with the Commission or one of its advisers at an early stage, preferably before a programme is announced. This can be done at seminars occasionally organised by bodies promoting European collaboration, via the D.En. who have representatives on the steering committee, via the experts, or directly with the Commission. It is most effective, at some stage, to include a visit to the offices of the appropriate personnel in Brussels.

The proposal form gives precise information on the data required from the proposer. Accurate costing is essential at this stage as, while any cost escalation may receive sympathy, it is highly unlikely to extend as far as financial bailing-out!

It is worth noting that the technical activities, once agreed, form part of the contract. It is therefore useful to all concerned if the technical programme is spelt out in some detail and in a form such that "milestones" and specific tasks can be readily identified. A bar chart of course complements this.

Implementation of the results of the R&D is a requirement if the contractor is to retain rights and/or funding received from the Commission. A peiod of 3 years is normally allowed between termination of the R&D contract and commencement of exploitation. The Commission may thereafter offer rights on a non-exclusive licence basis to others, or demand reimbursement. Renegotiation is always of course a preferred third option, insofar as timescale for exploitation is concerned.

Finally, language difficulties in general do not occur when dealing with the Commission. Documentation is in English and all the personnel with whom you are likely to come into contact will speak English well.

3.1.3   Project example - the development of expert systems for heat exchangers

The subject of Expert Systems comes within the area of computing developments known as intelligent knowledge-based systems (IKBS). The broad area of IKBS has received attention within the EC ESPRIT programme, but no Community-wide initiative had been taken for the application of expert systems, and the opportunity to include projects in the non-nuclear energy R&D programme is of considerable importance.

The principal objective of this project is to develop expert systems linked to data bases to assist users in the specification of heat recovery equipment. More specifically, the system being developed at NEI IRD with the assistance of Pergamon BPCC will relate to heat exchangers. Financial support from the EC is matched by inputs from NEI, ETSU (representing the UK Dept. of Energy) and Pergamon BPCC.

Note that in this particular project funding has been obtained from the EC and the EEO. Normally this will be unlikely to exceed 50% of the total project cost, but if the technology is of particular interest to ETSU, for example, support by way of a form of "buy-in" may lead to a small increase above the 50% limit.

A parallel project is being undertaken by Comprimo BV in the Netherlands, concentrating on a system for heat pumps in buildings. Close collaboration is taking place between Comprimo and IRD.

The project involves a number of stages leading to the testing of the expert system and preparation of packages. These include minimum hardware specification, software specification and development, rule bases for equipment selection and preparation of the data bases.

The project is administered by the Commission with the assistance of a Project Officer in Brussels and an expert. The expert assesses the progress of the work, makes recommendations as and when necessary, and attends some progress meetings.

3.2   Commission Energy Demonstration Programme

3.2.1   Programme aim & nature of projects supported

The aims of the demonstration projects supported by the European Commission are broadly similar to those in the UK. However, the level of support available and the governing rules differ. Also, the areas covered by the EC programme are wider-ranging, particularly in the fields of alternative energy and solid fuels processing. The topics included in the programme announced in 1986 are listed in Table 5.

It is worth noting that within each category certain projects may be ineligible for support. For example, sports centres and hospitals were among those excluded from the "Energy savings in buildings" sector in 1986. Considerable help is given to the potential proposer in this way in the call put out by the Commission, in common with that of the R&D programme.

As in the UK, there is a requirement to ensure that the project demonstrates an innovative technique, process or product, or involves a new application of a known technique. Proposals may be submitted by users or producers of equipment. However as the onus to stimulate replication is to some extent on the party being funded, it is more likely that the supplier will take the lead in an EC project. In the UK the EEO takes on itself the task of promoting successful projects to encourage replication.

If the project is likely to be implemented without EC support, the criterion for funding related to technical and/or economic risk disappears, and the case for support would be difficult to justify.

3.2.2   Financial support available

Financial support may cover the whole project, or specific stages of it.   In the 1986 call, an upper limit of 40% of the eligible cost was put on support available from the EC.   The eligible cost must include the cost of evaluating the results and all report preparation.

Note: In the case of Commission-funded demonstration projects, no requirement has to date existed for the use of independent monitoring agents.   Reliance is put on the host site and/or equipment supplier to assess the energy savings resulting from the installation or modifications.   You may well, of course be subjected to a financial audit by EC accountants!

A degree of flexibility in funding is possible due to the policy of the UK government to consider "topping up" suitable projects. It should be emphasised that this is never meant as a contribution to bale out an under-costed project, but is made where the project may have particular local interest, or could complement the UK demonstration strategy.
Retrospective funding is not possible, except in some cases related to new phases of projects already receiving financial support.

The total level of funding allocated to the 1986 scheme was about 90 MECU.   It was recognised that only in exceptional cases would the allocation to individual projects be likely to exceed 1.5 MECU, and most projects would be supported at a level well below 1 MECU. (This, nevertheless is often greater than the average funding for projects within the UK scheme, although it should be noted that the nature and scope of the projects differ).

TABLE 5   Sectors included in the 1986 European
Commission Demonstration Programme

Energy saving in industry
Energy saving in building
Energy saving in transport
Energy saving in energy industry
Solar energy
Biomass and energy from waste
Geothermal energy
Hydro-electric power
Wind energy
Use of electrical energy and heat
Use of solid fuels
Liquifaction and gasification of solid fuels

Repayment of Commission funding is no longer requested.   In previous programmes the Commission attempted to recoup part of its investment by negotiating repayment as either a levy on sales or as a proportion of the value of the energy saved per annum.

CONCLUSIONS

Four of the major UK and European Commission programmes which could provide support for innovative projects involving energy efficiency have been described. There are significant opportunities for the process industries in the UK to make good use of these funds.

REFERENCES

1. Clayton, R.W. Cost reductions on a titanium dioxide plant identified by a process integration study at Tioxide UK Ltd. EEO R&D Report RD/6, August 1985.

2. Mercer, A.C. Chemical sector gets boost for fuel saving projects. Process Engineering, Feb. 1987.

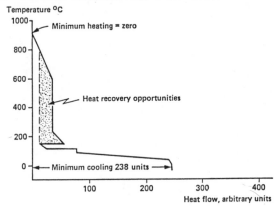

Fig. 1. Grand Composite Curve for Chlorination Plant.

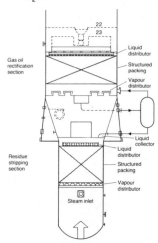

Fig. 2. Location of Structured Packing at PIP.

COAL FIRED APPLIANCES FOR PROCESS HEATING

A.R. Butt, C.J. Bower, R.C. Green, N.P. Paterson, J.J. Gale *

ABSTRACT

Two coal fired fluidised bed systems are being developed
for the generation of hot gas; a spouted bed gasifier
and a twin-bed combustor/pyrolyser. Fuel conversion
efficiency of the gasifier has been raised at a
laboratory scale to meet the target of 90%w and
demonstration at 0.5 t/hr scale is planned. A 3.5 MW_t
twin-bed combustor/pyrolyser fired water tube boiler has
been designed, built and evaluated. Emissions of NO_x
and SO_2 from both processes can be reduced to acceptable
levels but further development of particulate removal
systems is required.

1. INTRODUCTION

The UK market for process heating currently represents about 50 m
tonnes coal equivalent, of which about 9 m tonnes is supplied by coal;
the remainder is provided by natural gas and oil. Whilst the prices of
these fuels are currently competitive with coal, reserves are more
limited and eventual price rises are inevitable. Before the advent of
cheap natural gas and oil, coal supplied a large proportion of the
process industry energy demand for both direct and indirect firing. In
both cases, fixed bed gasifiers were generally employed to produce a
low calorific value gas. However, for many applications in the process
sector (capacity up to 60 MW_{th}) a restricted coal specification,
capital and operating costs associated with fixed bed gasifiers reduce
the competitive advantage of coal. In addition, atmospheric pollution
is increasingly a debated topic with regard to emissions of
particulates, hydrogen chloride and oxides of nitrogen and sulphur[1].
In recognition of the potential market for coal in the process heating
sector, and the need to develop environmentally acceptable coal fired
equipment, the Coal Research Establishment (CRE) of the British Coal
Corporation is currently developing coal fired fluidised bed systems
for the generation of hot gas. This paper describes the state of
development of two such processes, viz: a spouted bed gasifier, and a
twin-bed combustor/pyrolyser.

* CRE, British Coal Corporation, Cheltenham, Glos

The gasifier is aimed at applications requiring an essentially 'clean' hot gas at temperatures up to 1600°C, eg brick manufacture. Higher gas temperatures (up to 2000°C) should be possible by oxygen enrichment of the fluidising gas to the gasifier, eg for glass melting. The fuel gas can be distributed to individual burners on a furnace, and to several furnaces on an industrial site. The twin-bed combustor/pyrolyser is aimed at close-coupled applications which can accept a hot 'dirty' gas at temperatures up to 1500°C.

2. GASIFICATION

2.1 Process Description

The gasifier is based on the use of a submerged spouted bed in which a proportion of the fluidising gas is introduced in the form of a jet at the apex of a conical base. The spouted bed promotes internal solids circulation which enables caking coals to be processed without agglomeration. In order to ensure bed mobility, additional fluidising gas (usually air/steam) is introduced through the conical section. The bed, generally in the form of char particles of size 90% less than 6 mm (for a 'singles' grade coal), is operated at atmospheric pressure and at temperatures in the range 950 to 980°C. Control is achieved at a given fluidising velocity by adjusting coal feed rate to maintain bed height and limiting the bed temperature if required by increasing the proportion of steam in the fluidising gases.

A 0.5 th^{-1} semi-commercial demonstration unit at CRE, incorporating gas cooling and cleaning equipment is shown in Figure 1 and schematically in Figure 2. The cooled gas (calorific value about 4 MJm^{-3}) is burned in a purpose designed burner. At present the gasifier system, referred to as the Industrial Fuel Gas process (IFG), has a fuel conversion efficiency of about 69% on a mass basis (about 63% on an energy basis).

2.2 Background of IFG Process

The initial development work (jointly funded by British Coal Corporation and the European Coal and Steel Community) was carried out at CRE on two laboratory scale gasifiers of diameter 0.15 and 0.3 m. In 1982, the process was successfully demonstrated on a converted 1.4 m diameter fluid bed reactor operating at 0.5 th^{-1} during a 750 hour trial period[2]. Subsequently, a 0.5 th^{-1} semi-commercial unit was installed at CRE during 1985 (with financial support from the European Economic Community) to provide design and scale-up data for process plant of capacity 0.5-3.5 th^{-1}. A licence has been granted to Otto-Simon Carves Ltd with a view to exploiting the technology worldwide[3].

2.3 Objectives of the Development Programme

The objectives of the ongoing development programme are to:

(i) Maximise the coal conversion to gas.

(ii) Raise the calorific value of the gas by oxygen enrichment of the fluidising gas, as an option for processes requiring high firing temperatures (eg refractories and glass).

(iii) To quantify and minimise emissions of particulates and gaseous pollutants.

(iv) To demonstrate the Industrial Fuel Gas process at a semi-commercial scale.

2.4 Fuel Conversion Efficiency

In common with many fluid bed gasification processes, fuel conversion to gas is low (about 70%). A conversion efficiency of at least 90% on a mass basis (80% on an energy basis) is considered to be necessary for the process to be economically attractive in the UK under the current economic climate. It should be noted that elsewhere in the world such as India or South Africa, coal is very much cheaper relative to oil or gas and a low conversion is acceptable. However, even there the attractiveness of the process is enhanced by maximising the conversion of coal to gas.

The low efficiency is attributed to the elutriation of fines from the bed containing about 80% carbon. Clearly it is necessary to reduce the elutriation rate, or to effectively utilise the fines by recycling or combustion in a secondary system.

2.4.1   Reduction of Elutriation

Mechanisms for the production of elutriable carbon fines from fluid bed gasifiers and combustors has been reviewed by Massimilla[4,5]. The primary mechanisms are coal fragmentation, attributable to thermal stresses during heating and devolatilisation, and mechanical and combustion/gasification related attrition. For the latter, the uneven progression of the reaction due to ash or fissures in the coal particles results in the production of surface asperities, which can be readily removed by interparticle collisions. These studies showed that the elutriation rate is primarily dependent on coal type, bed and feed size distribution, bed carbon content, and excess gas velocity, and is insensitive to bed temperature.

In order to alleviate coal handling problems, early test work was conducted on laboratory scale gasifiers using crushed coal (<2 mm size). The ability to burn coal of this size grading was considered an important advantage over fixed bed gasifiers, which have a limited capacity to process such material. However consideration of the elutriation problem and the desirability of using a commercial grade of coal prompted a series of tests using a larger grade of coal. Accordingly a 'singles' grade coal (size 12-25 mm) was assessed in a 0.3 m diameter laboratory scale gasifier. Results (Table 1) show that fuel conversion efficiencies were similar to those achieved with fine coal. A suggested explanation is the counterbalancing effects of increased size distribution (of the bed) and fluidising velocity. Despite the apparent lack of improvement in fuel utilisation, the higher fluidising velocities employed did increase specific throughput, which subsequently leads to a reduction in plant size and therefore capital cost. Consequently the 0.5 th[-1] gasifier was designed to process a 'singles' grade coal.

To date the effect of reducing the bed carbon content has been
assessed by using sand beds in a 0.3 m diameter gasifier (Table
1). Use of a sand bed also permitted operation at higher
temperatures (up to 1020°C) without sinter by preheating of the
fluidising air to 600°C (equivalent to about 7½% of the total
energy input). Lower preheat temperatures should be required for
commercial gasifiers with reduced heat losses due to refractory
insulation. Elutriation rate was reduced allowing an increase in
fuel conversion efficiency from 58 to 85% on a mass basis and 49
to 80% on an energy basis. Gas calorific value dropped from 4.3
to 3.6 MJm_n^{-3} which is a little low, but adequate for$_1$ many
potential applications. A fluidising velocity of 1 ms^{-1} was
employed with sand beds to minimise erosion of the stainless steel
reactor, which has the effect of reducing gas output and coal feed
rate. Increased throughputs will be attempted when the use of
sand beds is demonstrated on the 0.5 th^{-1} gasifier. Reduced
elutriation in the 0.3 m gasifier tests was attributed to a lower
carbon content in the bed and reduced fluidising velocity,
combined with enhanced gasification reactions due to the raised
operating temperature and additional heat input as air preheat.

2.4.2   Recycle of Elutriated Fines

In order to improve fuel conversion efficiency, fines collected in
the primary cyclone of CRE gasifier systems have been injected
into the spout region of the gasifier, using a steam-driven
ejector (Figure 2). Data from the 0.5 th^{-1} unit operating on a
char bed (Table 3) showed a significant improvement in fuel
conversion from 61 to 69% on a mass basis, attributed to
consumption of fines in the combustion zone at the spout.
However, conversion efficiency was still below the 90%w target
value because the char bed continues to generate a significant
elutriation rate.

The effect of recycling fines to a sand bed was investigated in a
similar way, but using a 0.3 m diameter gasifier (Table 1). Fuel
conversion efficiency was raised from 85 to 94% on a mass basis,
and from 80 to about 82% on an energy basis. Calorific value of
the fuel gas was little changed. Thus recycling fines into the
spout region of the gasifier when operating with a sand bed
allowed the 90% fuel conversion target (mass basis) to be
exceeded. Demonstration of this process configuration is planned
for the 0.5 th^{-1} unit.

2.4.3   Combustion of Elutriated Fines in a Secondary System

The possibility of raising fuel conversion efficiency by
effectively utilising the carbon content of elutriated fines in a
separate combustion system is currently being investigated. Such
a system could find application in association with either a char
or a sand bed. It is anticipated that the hot gases from the
combustor would be utilised either directly as part of the
gasifier fluidising medium or to provide a preheated air supply to
the gasifier spout. A survey of the various methods available for
burning low grade materials indicated that a cyclone combustor

operated in a non-slagging mode might lend itself to incorporation into the IFG process[6].

Principal features of the cyclone combustor are illustrated in Figure 3. Combustion air and fines are injected tangentially into a cylindrical combustion chamber. Ash is centrifuged towards the chamber walls and removed either in side-vortex collector pockets or through the conical base of the combustor. Combustion occurs at low temperatures (900 to 1000°C) and the cyclone provides a flue gas of relatively low dust content.

As a first stage towards the incorporation of a cyclone combustor into the gasification processes, a 5 kgh^{-1} test rig (Figure 3) has been commissioned at CRE. The cyclone consists of a refractory-lined chamber of 0.2 m diameter and 0.8 m height. Fines collected from the gasifier are metered and conveyed into the chamber with the combustion air. During start-up the combustor is preheated to the ignition temperature of the carbon (about 800°C) by burning a mixture of air and natural gas in the chamber.

Data (Table 4) show that combustion efficiencies of up to 85% were achieved. In order to prevent ash sintering to the side walls of the combustor it was necessary to replace approximately 10% of the combustion air with steam. Air is currently being investigated as an alternative to steam for wall cooling.

Providing the concept of the cyclone combustor can be successfully demonstrated, the unit will be incorporated into the gasification process, initially on a 0.3 m diameter laboratory scale gasifier. Following successful demonstration on the 0.3 m gasifier it is proposed to install a combustor on the 0.5 th^{-1} gasifier.

2.5 Use of Oxygen Enriched Fluidising Gas to Raise Fuel Gas Calorific Value

The effect of oxygen enrichment of the fluidising gas on fuel gas calorific value and fuel conversion efficiency has been investigated using char and sand beds in a 0.3 m diameter gasifier. It was fed with crushed coal (-2 mm) and primary cyclone fines were recycled to the bed (Table 5).

In tests using char beds the calorific value of the fuel gas was raised from 4.7 to 6.4 MJm_n^{-3} as the additional oxygen concentration of the fluidising gas was raised from 13.8 to 28.9%v. The corresponding calorific value without oxygen enrichment was 3.1 MJm_n^{-3}. Fuel conversion efficiency ranged from 71 to 77% on a mass basis, and 58 to 71% on an energy basis. However, the values appeared to stabilise at higher levels of oxygen addition indicating a limiting value. A test with 12.4%v added oxygen using a sand bed containing a low carbon content produced a fuel gas with a calorific value of 5.0 MJm_n^{-3}.

The enhanced calorific value achieved in these tests was attributed to reduced thermal load associated with heating nitrogen in the fluidising gas, permitting a larger proportion of the energy released in combustion reactions to be employed in gasification and steam conversion reactions. Demonstration of the use of oxygen enriched fluidising gas to raise calorific value is planned for the 0.5 th^{-1} unit, in association with sand beds.

2.6 Particulate Emissions

There is little information regarding the performance and applicability of dust cleaning equipment to atmospheric pressure fluidised bed gasification plant. In order to assess the suitability of the various techniques available, several systems have been installed on the 0.5 th^{-1} gasifier. The gas clean-up system illustrated in Figure 2 consists of recycle and high efficiency cyclones, a bag filter and a venturi scrubber. With the exception of the recycle cyclone, the performance of each system can be assessed independently in terms of the dust concentration in the fuel gas. A target value of 0.85 gm_n^{-3} was set to ensure that atmospheric emission limits will be met after combustion.

The performance of the secondary high efficiency cyclone is given in Table 6. The dust concentration in the fuel gas (12 gm_n^{-3}) was well above the target level and it is doubtful that the performance of the cyclone could be significantly improved. However, it should be noted that the data refers to operation with a char bed and it is anticipated that emission levels would be reduced by operating the gasifier with a sand bed.

A fixed throat venturi scrubber has been assessed on the 0.5 th^{-1} gasifier. The pressure drop across the scrubber was high compared with that of a cyclone or bagfilter, typically 10.1 kPa. In operation, the system proved unreliable and allowed toxic gases to escape. Poor design of the liquor handling system has prevented long term testing, and the unit is currently being redesigned.

To date, the most reliable high efficiency gas cleaning system has been the pulse-jet bagfilter. Preliminary studies using a single filterbag on a sidestream from a laboratory scale gasifier (0.15 m diameter) have shown that an outlet dust concentration of 0.086 gm_n^{-3} can be achieved (Table 7). This concentration is much higher than would be expected on a commercial bagfilter, but the work showed that operation on fuel gas might be viable. Subsequently a 36 element bagfilter has been installed on the 0.5 th^{-1} gasifier. After commissioning difficulties a short duration test on this filter (Table 8) suggested that a near stable pressure drop can be achieved, but that blinding of the bags with tar aerosols is a problem. Test work on this unit continues with plans to assess a limestone coating system as protection for the bags, and later to test the feasibility of feeding 'singles' grade coal into the gasifier bed rather than over it, in a bid to reduce the tar emission. If the blinding problem can be resolved then it should be possible to achieve dust concentrations in the fuel gas of less than 0.02 gm_n^{-3}, so meeting any anticipated client process or future atmospheric emission requirements.

2.7 Gaseous Pollutants

The concentrations of gaseous pollutants emitted from the gasifier using a coal containing 0.2%w chlorine and 2.0%w sulphur are shown in Table 9, and discussed below.

2.7.1 Hydrogen Chloride

Virtually all the chlorine in the coal is released as hydrogen chloride in the gasifier. A laboratory scale test programme has recently commenced to evaluate candidate retention agents. Candidates selected by literature search will be evaluated using a 0.3 litre packed bed reactor supplied with hydrogen chloride/nitrogen gas mixtures at temperatures up to 700°C. Those found to possess a high retention capacity will be evaluated further using different contacting methods on a sidestream of gas from a 0.15 m gasifier.

2.7.2 Sulphur

Approximately 70% of the sulphur in the coal is released from the fluid bed as hydrogen sulphide. Preliminary tests on the 0.5 th^{-1} gasifier using char beds have shown that about 90% of the sulphur can be retained by limestone addition to the bed at a calcium:sulphur molar ratio of 2:1 to give about 200 vppm of hydrogen sulphide in the fuel gas. The hydrogen sulphide in the gasifier reacts with the calcined stone to form calcium sulphide which is oxidised to sulphate within a few days of exposure to air. Sulphur in the fuel gas converts to about 100 vppm of sulphur dioxide in flue gas which should ensure compliance with any anticipated emission regulations. Confirmatory tests using char beds in the 0.5 th^{-1} gasifier are planned. The feasibility of sulphur retention using limestone associated with sand beds will also be tested.

2.7.3 Nitrogen

Emission levels of nitrogen oxides derived from fuel gas nitrogen are low, due to the low concentrations of ammonia and other nitrogenous compounds. Typical NO_x levels measured in the flare of the 0.5 th^{-1} gasifier are given in Table 9. The NO_x emission level of 140 vppm (280 mgm^{-3} of NO_2 referred to 6% oxygen) was comfortably within proposed EEC legislative limits. An experimental programme is investigating the combustion characteristics of low calorific value fuel gas, and includes development of low NO_x burners.

2.8 Demonstration of the Process

The 0.5 th^{-1} semi-commercial gasification unit at CRE has operated for a total of 500 hours. The availability of the process has exceeded 70%; increased availability is expected as minor mechanical reliability problems are overcome. The current EEC demonstration grant expired in April 1987. A new EEC contract has been awarded which will allow operation of the plant to continue up to November 1989. During this period, methods of improving fuel conversion by the use of sand beds and raising the fuel gas calorific value by oxygen enriched fluidising gas will be further assessed. In addition, it is planned to demonstrate combustion of elutriated fines in a cyclone combustor and prove gas cleaning systems adequate for emission and client process requirements.

3. THE DEVELOPMENT OF A TWIN-BED COMBUSTOR-PYROLYSER SYSTEM FOR THE GENERATION OF HIGH TEMPERATURE GASES

3.1 The 1 MWt Furnace System

3.1.1   General Features

Supporting studies leading to the design of the furnace have been presented previously[7]; the principal features (Figure 4) are summarised below. The refractory-lined combustion chamber consists of two adjacent fluidised beds with a dividing wall. The beds are operated respectively as a char combustor and pyrolyser. The lower section of the wall contains an array of apertures to facilitate interchange of material between the two beds. Inport standpipes are provided in order to retain fluidisation of material during passage through the ports and to regulate the solids circulation rates across the division wall.

3.1.2   Discussion of Performance and Operating Characteristic of the Twin-Bed Furnace

Steady-state operating conditions (initially for a 50mm thick division wall) were identified for 'singles' grade coal (Table 10) and demonstrated that hot gas temperatures up to 1420°C can be achieved in the twin-bed unit. A combustible gas of calorific value 3.3 MJm^{-3} (dry purge free at NTP) was formed in the pyrolyser. This gas, when mixed with the oxygen-rich exhaust from the char combustor, produced a hot gas at temperatures ranging from 1100°C to 1420°C. The oxygen content was 6.0%, equivalent to an excess air level of 40%. The overall combustion efficiency of the system was better than 98%. In a second series of tests conducted with smalls grade coal, for the same range of operating conditions, the corresponding measured values of hot gas temperature and excess air level were 1500°C and 25% respectively. A combustible gas of calorific value 3.3 MJm^{-3} (dry purge free at NTP) was formed in the pyrolyser. The reduced excess air and increased gas temperature levels were attributed to the burning of elutriated char particles in the furnace freeboard. In particular, the concentration of carbonaceous solids in the freeboard will depend upon the coal feed size distribution. The combustion efficiency, when burning 'smalls' grade fuel was in excess of 97%.

Whilst scale-up of the twin-bed furnace concept to semi-commercial application (up to 3 MW) is practicable using a single 50 mm partition wall, many industrial processes require capacities within the range 5-50 MW. For furnaces within the capacity range 5-15 MW, a thicker division wall offering improved mechanical stability is a pre-requisite. Larger units may be provided in the form of modular multiple-bed systems. As part of the test

programme, the twin-bed furnace was modified to incorporate a
150mm division wall, with the facility to fluidise the material
within individual transfer ports.

The close degree of control on char and sand transfer achieved by
regulation of the air flowrate to the in-port air distributors was
such that a combustible gas of calorific value 3.6 MJm^{-3} could be
produced in the pyrolyser. The higher gas yields can be
attributed to the increased residence time of the coal particles
in the pyrolyser and the associated higher coal feed rates
required to maintain the temperature of the adjacent char bed.
Higher gas temperatures (1450°C) at lower excess air levels (30%)
were attained. The combustion efficiency of the unit was greater
than 98%.

Fluidisation within the transfer port is also necessary to prevent
the generation of locally high temperatures at the walls of the
ports. The extent of fluidisation across the width and depth of
the transfer ports was inferred from measured temperature
gradients within the material being transferred. Typically
temperature gradients in the vertical plane were $3°C\ mm^{-1}$; the
mean temperature of the material flowing through the port did not
exceed 980°C (the bed temperature). There was no evidence of bed
material sintering in the transfer ports in the absence of in-port
fluidising air.

The calorific value of the combustible gas formed in the pyrolyser
could be enhanced by reducing the number of transfer ports or
reducing the air flowrate to the in-port distributors.
Consequently, the circulation rate of bed material and char was
reduced. In order to maintain a constant temperature in the char
bed it was necessary to reduce the air flowrate or increase the
coal flowrate to the pyrolyser bed. Both actions resulted in the
production of higher final gas temperatures with a lower excess
air level. An increase in coal feed rate will result in an
increased char transfer rate. Since the components of the
combustible gas (e.g. hydrogen and methane) are products of
pyrolysis, their yields would be expected to increase with
increasing coal feed rate.

3.1.3 NO_x and SO_2 Emissions

The test furnace studies have confirmed that simultaneous[8]
reductions in emission levels of NO_x and SO_2 can be achieved
(Table 10). NO_x emission (referred to 6% oxygen) can be reduced
from 580 vppm (single bed operation) to 228 vppm. Sulphur
retentions (by limestone addition to the bed) of 75% (single bed)
and 70% (twin bed) have been measured for a Ca:S molar ratio of 4,
Figure 5. The spent sorbent removed from the char-bed was found
to be free of calcium sulphide. There are no disposal problems.

3.2   Application and Scale-up of the Twin-Bed System

3.2.1   Background

As part of the collaborative programme, a 4,500 kgh^{-1} (2.8 MW_e)

steam boiler was chosen to host a twin-bed system based on the design features of that for the 1 MWt furnace. The boiler (Figure 6), one of the Senior Green 'A' frame range of factory-packaged units, was originally designed for use with a conventional bed. Following a period of operation at the manufacturers site (to provide steam for process and space heating) the unit was subsequently relocated at the Coal Research Establishment (British Coal) where it was modified to accept a twin-bed firing system.

The boiler is of the natural circulation type, with external unheated downcomers feeding into square-sectioned flat-bottomed headers from a single steam drum located above the combustion chamber. The boiler walls, water-cooled and of membrane construction, form part of the convection riser tube circuit. The unit is equipped with a Senior Green vertical multi-pass extended surface economiser. The relatively tall combustion chamber and generously-proportioned first boiler pass, combined with the absence of in-bed cooling tubes, favoured the incorporation of a twin-bed system.

3.2.2   The Twin-Bed Boiler

Principal features are shown in Figure 7.

The boiler consists of a combustion chamber (which accommodates the twin-bed firing system), the first boiler pass (serving as the hot gas mixing zone), convection banks and economiser. Figure 7 shows that the dividing wall separating the combustor from the pyrolyser bed spanned the width of the combustion chamber.

The wall, constructed from 150mm thick interlocking refractory concrete blocks was supported on a water-cooled lintel; the upper end of the latter was attached to the boiler headers. The lower section contained an array of ports to facilitate the interchange of suspended coal and sand between adjacent beds. In-port standpipes were incorporated to ensure continued fluidisation of material during their passage from one bed to the other.

To prevent corrosion of the water cooled containment walls, the pyrolyser bed was refractory-lined. Localised areas of the tube surface were left exposed to the reducing gas in order that an assessment of corrosion (if present) could be made. Anti-erosion shelves were fitted to those parts of the char combustor side walls in contact with the bed. The initial bed depth was 150mm; the bed was formed by silica sand of mean size 0.75mm.

Fluidising air is supplied separately to the char and pyrolyser beds by forced draught fans of capacity 3,500 m^3h^{-1} and 1,500 m^3h^{-1} respectively. The air enters the beds via a sparge pipe type distributor in the form of ten 80mm bore horizontal tubes (six in the char bed and four in the pyrolyser bed). The tubes in the char bed each carry twenty-four screw-connected vertical standpipes on an 85mm square pitch; the tubes in the pyrolyser bed each carry thirty-eight standpipes on a 50mm triangular pitch.

Singles grade coal (size range 12-25mm) is admitted to each bed by a variable-speed screw feeder. The coal discharges directly from the feeder on to the upper surface of the fluidised bed through circular ports located in the boiler side walls.

The combustible gas generated in the pyrolyser is transported to the mixing zone via a circular duct constructed from stainless steel in order to withstand corrosion. Because of the unknown nature of the aerodynamic flow and mixing patterns prevailing in the mixing chamber, it was considered appropriate to have the facility to effect changes in the geometry of the mixing chamber (e.g. the installation of baffles or nozzles) without the need for major modification to the unit. The stainless steel transfer duct was designed to readily incorporate such changes. Oxygen-rich gases (from the char bed) enters the mixing zone through the screen tube wall (Figure 7) separating the combustion chamber from the mixing zone. Hot gases leaving this zone enter the boiler convection passes and economiser and are subsequently cleaned by a cyclone collector before discharge to the stack. The boiler is instrumented for measurements of air and coal flow rates, above-bed and exhaust gas composition, pressures and temperatures.

3.3 Start-up

3.3.1 General Features

Pre-heated gas can be supplied to both beds (or the char bed alone) by means of a gas burner mounted in the combustion air duct. The gas temperature is limited to about 700°C in order to prevent thermal distortion of the distributors. At a bed temperature of 600°C, coal is admitted. When the fluidised bed temperatures reach 750°C (sufficiently high to sustain coal combustion) the gas burner is extinguished. Subsequent control of the fluidised bed temperature is achieved by regulating the speed of the coal feeders. In particular, the bed temperature is monitored and fed to a P.I.D. controller which signals an appropriate change to the speed of the coal-feed screw.

3.3.2 Part-bed Start-up (Cold Pyrolyser from hot char bed)

With the char bed operating at a temperature of 900°C (on coal), the pyrolyser is pre-heated to 450°C by the circulation of sand between adjacent beds. The circulation rate of material is regulated by varying the air flowrate to the in-port standpipes. On attaining a pyrolyser bed temperature of 450°C, the coal feed rate to this bed is increased to raise the bed temperature to 900°C.

3.3.3 Twin-bed Start-up

For initial operation in the twin-bed mode, substoichiometric conditions in the pyrolyser are achieved by reducing the air flowrate and increasing the coal feed rate to this bed. There is an ensuing increase in temperatures of the adjacent char bed due to transfer of char from the pyrolyser. On attaining a pre-set temperature, the coal feed to the char bed is stopped; the char bed temperature is subsequently controlled by varying the air flowrate to the bed.

3.4 The Test Programme

Principal objectives of the on-going test programme are to:

   i)   Demonstrate part-bed start-up and twin-bed operation on the larger scale unit.

  ii)   Validate mathematical models used for scale-up and identify refinements to the boiler plant for incorporation in future designs.

 iii)   Develop a low-cost control system allowing fully automatic start-up and load following.

  iv)   Assess boiler performance for a range of operating conditions and with different coals, with particular attention being given to the fouling characteristics of the latter.

   v)   Establish the effect of operating conditions on NO_x and SO_2 emissions (with and without the addition of limestone).

3.5 Results and Discussion

3.5.1 Single Bed Operation

During March 1986, the unit was commissioned in the single-bed mode. Performance data (Table 13) showed a maximum output (for a 150mm deep bed) of 1600 kgh^{-1} of steam (40% MCR) at 120% excess air. The reduced operating excess air level was attributed to the enhanced heat transfer rates to the shelved containment walls (Section 3.2). Measured combustion efficiencies were within the range 96% to 98% for singles coal of characteristics given in Tables 11 and 12. Boiler efficiency was not less than 80% and could be increased to 82% by increasing the static bed depth to 300mm and reducing the excess air level to 90%. The refractory-lining in the pyrolyser chamber restricted the boiler operation to relatively high levels of excess air and hence precluded operation at MCR.

3.5.2 Assessment of Solids Circulation Rates

To assess the effect of scale-up on solids circulation rates between the interconnected beds, and its dependence on operating conditions, a series of part-bed start-up tests were conducted, including an assessment of the rate of heating in the cold bed at a range of hot-bed fluidising velocities and in-port air flowrates. Assuming, (i) a well-mixed bed and (ii) negligible amounts of coal were transferred with the circulating sand, then the solids circulation rate across the division wall can be determined from the following heat balance:

$$\dot{M}_s C_s (T_2 - T_1) = \dot{M}_g (C_g T_1 - C_a T_a) + \tilde{M}_b C_s \frac{dT}{dt} 1$$

These assumptions were considered appropriate (since they were valid for the test furnace) when the cold pyrolyser is heated from the hot char bed.

Figure 8 shows the start-up characteristics (temperature v time) for the pyrolyser bed and dependence on the number of transfer ports in operation. The corresponding solid circulation rates (calculated by heat balance) showed good agreement with predicted values from cold model studies[8] thereby validating the scaling relationships which were used.

With regard to start-up of the cold char bed from the hot pyrolyser, the mathematical model was further developed to incorporate the following features:

(i) the lateral mixing properties of the bed and coal particles in the larger char chamber

(ii) the higher transfer rates of coal from the pyrolyser to the char bed

The refined model for predicting the start-up characteristics of the char bed gave good agreement with experimental data.

3.5.3   Boiler Operation in the Twin-Bed Mode

Following the commissioning of the boiler in the single-bed mode and the assessment of the simultaneous heat and solids circulation rates between beds, operating parameters were chosen to give stable substoichiometric conditions in the pyrolyser. Preliminary tests (of short duration) were conducted to confirm steady-state operating parameters when burning a singles grade coal (Tables 11 and 12). Initially, the oxygen concentration in the hot gas mixing zone was lower than anticipated ($<4\%$). This was attributed to the previously-established reduced excess air levels above the char bed. In order to compensate for this reduced oxygen availability (without reducing the mixed gas temperature) the pyrolyser bed area was reduced by 10%. Subsequent performance data is summarised in Table 13. Generally, the operating characteristics resembled those for the twin-bed furnace[8]. Under optimum firing conditions, a combustible gas of calorific value 3.3 MJm^{-3} was formed in the pyrolyser. This gas, when mixed with the oxygen-rich gas from the char combustor, gave hot gas temperatures in excess of 900°C in the water-cooled mixing zone. The corresponding oxygen concentration (7.0%) was equivalent to an excess air level of 40%. The overall combustion efficiency of the system was 95%; boiler efficiency was not lower than 81%. During twin-bed operation, it was evident that combustion in the mixing zone was incomplete; further refinements were required, to restrict smoke emission to an acceptable level. To date, test periods have been too short to allow detailed assessments of corrosion of the exposed surfaces in the pyrolyser bed and of the ash-fouling propensity in the mixing zone and convection passes.

4. CONCLUSIONS

The British Coal Corporation is developing two coal fired fluidised bed systems for the generation of hot gas, viz: a spouted bed gasifier, and a twin bed combustor/pyrolyser.

A major objective of the gasification programme has been to raise fuel conversion efficiency of the process from about 60%w for a char bed fed with crushed coal (<2 mm) to a target value of 90%w. The poor conversion has been attributed to elutriation of fines from the fluid bed, containing about 80% carbon. Recycling of elutriated fines to the spout region of the 0.5 th^{-1} semi-commercial demonstration gasifier raised conversion efficiency to 69%w. Development work employing a 0.3 m diameter laboratory scale gasifier has suggested that feeding 'singles' grade (12-25 mm) coal to a sand bed containing a low carbon content, operating at 1020°C and fluidising velocity of 1 ms^{-1} with recycle of elutriated fines and preheating of the fluidising air, can raise fuel conversion to 94%w. Utilisation of elutriated fines by combustion in a secondary system and enhancement of fuel gas calorific value by oxygen enrichment of the fluidising gas have also been investigated.

Progress has been made in reducing emissions of particulates, hydrogen chloride, oxides of sulphur and nitrogen, but development of the particulate removal systems is required. Emissions of NO_x and SO_2 can be reduced to acceptable levels.

Demonstration of the use of sandbeds, oxygen enriched fluidising gas and ex-bed fines combustion (if required) are planned on the 0.5 th^{-1} semi-commercial demonstration gasifier at CRE.

Data obtained from a 1 MWt test furnace have been used in the design of 3.5 MWt twin-bed combustor/pyrolyser fired water tube boiler. Subsequent testwork on the twin-bed boiler (burning a singles grade coal) has demonstrated both the validity of the principle and the mathematical models used for scale-up purposes. In particular, the production of a low calorific value gas (typically, of calorific value 2.8-3.3 MJm^{-3}) has been demonstrated. Subsequent combustion of this gas with the oxygen-rich gas from the char bed has produced gas temperatures (in the water-cooled mixing zone) in excess of bed temperature. Start-up of one bed from another can be readily achieved; the temperature of the cold pyrolyser reaches that of the char bed within 12 minutes from cold.

Twin-bed test furnace studies have confirmed that simultaneous reductions in emission levels of NO_x and SO_2 can be achieved. NO_x emissions (referred to 6% oxygen) can be reduced from 580 vppm (single bed operation) to 228 vppm. Sulphur retentions of 75% (single-bed) and 70% (twin-bed) were measured for a CaS molar ratio of 4. The spent sorbent removed from the char bed was free of calcium sulphide.

5.   NOTATION AND DEFINITIONS

C_a :   Specific Heat of Fluidising Gas, $kJkg^{-1}K^{-1}$
C_s :   Specific Heat of Solids, $kJkg^{-1}K^{-1}$
M_b :   Mass of Pyrolyser (cold) Bed Material, kg
M_b :   Solids Circulation Rate, kgs^{-1}
M_s :   Mass Flowrate of Fluidising Gas, kgs^{-1}
T_2 :   Temperature of Hot Char Bed, °K
T_1 :   Temperature of Cold Pyrolyser Bed, °K
T_a :   Temperature of fluidising gas entering cold bed, K

Boiler Efficiency = Percentage of energy input (coal calorific value)
                   converted into useful energy as steam output

Combustion Efficiency = Percentage (wt) of carbon in feed coal
                       converted into gaseous carbon in fuel/flue gas

Fuel Conversion efficiency = Percentage of feed coal converted into
                            fuel gas, on a mass or energy basis

6.   ACKNOWLEDGEMENTS

The authors wish to thank the British Coal Corporation and Senior Green Ltd. for permission to publish this paper and, in particular, Dr. T.J. Peirce and Dr. M.St.J. Arnold for assistance in its preparation. The views expressed are those of the Authors and not necessarily those of the British Coal Corporation, Otto Simon Carves Ltd, or Senior Green Ltd.

7.   REFERENCES

1.   53rd National Society for Clean Air Annual Conference, Blackpool, 27-30 October 1986.

2.   Green, R.C., Paterson, N.P., Summerfield, I.R. 'Demonstration of Fluidised Bed Gasification for Industrial Applications', Energy World, July 1984, pp 7-10.

3.   Smith, A., 'Industrial Gasification Comes of Age', Coal Trans, Sept/Oct 1986, pp 64-65.

4.   Massimilla, L., etal, 'Carbon Attrition During the Fluidised Combustion and Gasification of Coal', DoE Quarterly Technical Report No DOE/PC40796-7, US Dept of Energy, Washington DC (1983).

5.   Chirone, R., etal, 'Char Attrition During the Batch Fluidised Bed Combustion of a Coal', A.I. Chem E. Journal Vol 31, No 5 (1985), pp 812-820.

6.   Sahatimehr, A., 'Combustion of Poor Quality Fuel Using a Multi-inlet Cyclone Combustor', PhD thesis, University College, Cardiff, 1983.

7.   Butt, A.R. and Peirce, T.J. 'The Generation of High Temperature Gas from Fluidised Bed Furnaces', Proc. of 8th Int. Cong. on F.B.C., Vol III, pp1435-1442, July 1985.

8.   Butt, A.R., Peirce, T.J. and Payne, E.C. 'The Development of a
      Two-Stage Fluidised Bed System for the Generation of High Temperature
      Gases', Proc. of the 8th Members' Conference of the I.F.R.F. Ijmuiden,
      April 1986.

TABLE 1

THE EFFECT OF COAL SIZE BED MATERIAL AND RECYCLE OF ELUTRIATION FINES
ON PERFORMANCE OF A 0.3 M DIAMETER GASIFIER

Operating Conditions

Coal		Daw Mill	Baddesley	Baddesley	Baddesley
Coal Size	mm	-2	12-25	12-25	12-25
Nominal Bed Material		Char	Char	Sand	Sand
Primary Fines Recycle		No	No	No	Yes
Fluidising Gas - Air	%v	100	85	86	86
- Steam	%v	-	15	14	14
Fluidising Velocity	ms^{-1}	0.8	2.0	1.0	1.0
Excess Fluidising Velocity	ms^{-1}	0.7	1.1	0.7	0.7
Coal Feed Rate	kgh^{-1}	24	61	16	14.6
Bed Height	m	2	2	2	2
Bed Temperature	°C	950	950	1020	1010
Fluidising Air Preheat Temperature	°C	0	0	600	600

Process Performance

Gas Output	$m^{-3}h^{-1}$ (wet)	96	160	75	78
Gas Calorific Value (gross @ NTP)	MJm^{-3} (wet)	3.7	4.3	3.6	3.7
Bed Purge Rate	kgh^{-1}	5	2.4	0	0
Primary Cyclone Catch	kgh^{-1}	4.7	15.0	1.7	Recycled
Secondary Cyclone Catch	kgh^{-1}	0.7	3.3	0.3	0.6
Dust Content of Fuel Gas	kgh^{-1}	0.6	2.0	0.3	0.4
Elutriation Rate	% coal fed	25	33	14	-
Fuel Conversion - Mass Basis	%	57	58	85	94
- energy basis	%	50	49	80	82

Notes: 1. See Table 2 for representative analyses.

          2. Heat loses from the 0.3 m gasifier were compensated with electrical surface heaters operating at a fixed heat rate for all runs. Small differences in lagging between runs and heat loss from the primary fines recycle system are sources of inconsistency in fuel conversion efficiency quoted on an energy basis.

          3. The sand employed in all tests reported was sized 0.5 - 1.0 mm and classified as Lower Green (Leighton Buzzard).

TABLE 2

REPRESENTATIVE CHEMICAL ANALYSES OF COALS EMPLOYED IN GASIFIER TESTS

Coal		Daw Mill	Baddesley	Seamoor
Moisture	%ad	6.0	7.3	6.5
Ash	%ad	4.8	4.3	7.1
Carbon	%db	76.5	76.9	76.1
Hydrogen	%db	4.5	5.45	4.95
Nitrogen	%db	1.35	1.22	1.56
Sulphur	%db	1.6	2.0	1.8
Chlorine	%db	0.22	0.20	0.30
Volatiles	%daf	53.3	51.2	ND
Calorific Value	$MJkg^{-1}daf$	40.2	33.47	34.12
BS Swelling No		1.0	1.0	1.0
Ash Fusion (Deformation Temperature) °C		1190	1200	ND

Notes: 1.   Abbreviations: ad = as determined
                              db = dry basis
                              daf = dry ash free basis
                              ND = not determined

         2.   Oxygen content can be estimated by difference

TABLE 3

THE EFFECT OF RECYCLING PRIMARY CYCLONE FINES TO THE GASIFIER ON THE PERFORMANCE
OF THE 0.5 TONNE/HOUR GASIFICATION DEMONSTRATION PLANT

Operating Conditions		Without Fines Recycle	With Fines Recycle
Mode			
Coal		Baddesley	Baddesley
Coal Size	mm	12-25	12-25
Nominal Bed Material		Char	Char
Fluidising Gas – Air	%v	86	90
– Steam	%v	14	10
Fluidising Velocity	ms^{-1}	2.0	2.0
Excess Fluidising Velocity	ms^{-1}	1.1	1.1
Coal Feed Rate	kgh^{-1}	482	452
Bed Height	m	2.0	2.0
Bed Temperature	°C	971	973

Process Performance			
Gas Output	$m_n^{-3}h^{-1}$ (wet)	1512	1508
Gas Calorific Value (gross @ NTP)	MJm_n^{-3} (wet)	3.5	3.5
Bed Purge Rate	kgh^{-1}	0	0
Primary Cyclone Catch	kgh^{-1}	140	38*
Secondary Cyclone Catch	kgh^{-1}	40	86
Dust Content of Fuel Gas ex 2° Cyclone	kgh^{-1}	11	15
Fuel Conversion – Mass Basis	%	61	69
– Energy Basis	%	54	63

Notes: 1.   See Table 2 for representative coal analyses

2.*   Limited consumption of the recycled fines allowed the recycle rate to increase
beyond the capacity of the recycle system, so the excess fines were removed as
a purge stream.

TABLE 4

INITIAL RESULTS FROM CYCLONE COMBUSTOR TEST RIG

Fines Feed Rate	kgh^{-1}	4 - 6
Excess Air Level	%	30 - 50
Steam Addition		10% of Fluidising Gas
Combustor Wall Temperature	°C	970 - 1170
Dust Content of Flue Gas	gm_n^{-3}	1.5
Ash Off-take	kgh^{-1}	0.8 - 1.5
Combustion Efficiency	%	70 - 85

TABLE 5

THE PERFORMANCE OF THE 0.3 M DIAMETER GASIFIER EMPLOYING OXYGEN ENRICHED FLUIDISING GASES

Operating Conditions

Coal		Daw Mill						Daw Mill
Coal Size	mm	<2						<2
Nominal Bed Material		Char						Sand
Primary Fines Recycle		Yes						Yes
Fluidising Gas - Added Oxygen	%v	0	13.8	16.9	20.9	25.6	28.9	12.4
- Air	%v	90	75.8	60.2	44.9	29.5	19.8	45.9
- Steam	%v	10	10.3	22.8	34.1	45.0	51.3	41.6
- Total Oxygen	%v	18.8	29.7	29.5	30.3	31.8	33.0	22.0
Fluidising Velocity	ms^{-1}	0.8						0.8
Coal Feed Rate	kgh^{-1}	14.9	27.4	28.6	29.0	29.9	32.3	19.2
Bed Height	m	2						2
Bed Temperature	°C	926	980	976	981	978	980	1020
Fluidising Air Preheat Temperature	°C	0						0

Process Performance

Gas Output	$m_n^3h^{-1}$ (wet)	76	78	82	89	86	88	85
Fuel Gas Composition CO_2	%v (wet)	8.6	12.4	14.4	17.2	17.6	17.3	15.1
CO	%v (wet)	12.6	22.8	22.8	24.9	24.5	25.0	21.0
CH_4	%v (wet)	0.5	0.7	0.8	1.1	1.2	1.2	1.1
H_2	%v (wet)	11.4	14.5	18.0	22.6	23.3	24.6	19.8
N_2	%v (wet)	57.7	40.1	31.5	16.8	14.3	9.6	20.2
H_2O	%v	9.2	9.5	12.5	17.4	19.1	22.3	22.8
Gas Calorific Value (gross @ NTP)	MJm_n^{-3}	3.1	4.7	5.2	6.1	6.2	6.4	5.0
Bed Purge Rate	kgh^{-1}	1.7	2.9	2.2	1.7	1.6	1.6	0
Secondary Cyclone Catch	kgh^{-1}	3.1	3.3	3.1	2.7	3.3	3.3	2.6
Dust Content of Fuel Gas	kgh^{-1}	1.4	1.9	1.9	1.5	1.9	2.5	0.6
Fuel Conversion - Mass Basis %		67	71	75	80	77	77	83
- Energy Basis %		53	58	63	71	69	71	79

Notes: 1.   Nitrogen content of fuel gas is estimated by difference.
             2.   See notes 2 and 3 on Table 1

TABLE 6

PERFORMANCE OF THE SECONDARY CYCLONE ON THE 0.5 TONNES/HR GASIFIER

Operating Conditions		
Gas flow rate	m^3h^{-1}	1519
Gas inlet temperature	°C	230
Inlet dust mass flow rate	kgh^{-1}	133
Inlet dust size distribution	(Coulter)	

	Size grade (μm)	%w in grade
	+125	0.4
	63 - 125	2.9
	20.2 - 63	5.9
	10.1 - 20.2	26.8
	5 - 10.1	26.9
	2 - 5	24.6
	1 - 2	9.9
	- 1	2.6

Pressure drop	kPa	35

Performance		
Overall collection efficiency	%w	87
Outlet dust concentration	gm_n^{-3}	12

Note: Gasifier operating conditions: Coal: Baddesley
Char bed at 971°C
Coal feed rate: 476 kgh^{-1}
Fluidising velocity: 2ms^{-1} (9:1
Air:Steam)
Primary cyclone fines
recycled, less 29 kgh^{-1} purge

TABLE 7

OPERATING CONDITIONS AND RESULTS OF TEST ON 'HUYGLAS' FILTER MEDIA

Operating Conditions

Gas Flow Rate	$m_n^3s^{-1}$	0.8×10^{-3}
Inlet Dust Concentration*	gm_n^{-3}	1
Filtration Velocity	ms^{-1}	0.025
Filtration Temperature	°C	200
Filter Configuration		0.6 m diameter tube x 0.28 m long
Pulse Pressure	bar	5
Pulse Volume	m^3	1.75×10^{-3}
Total Tar Concentration in Gas#	gm_n^{-3}	0.68

Performance

Outlet Dust Concentration*	gm_n^{-3}	0.068
Dust Collection Efficiency	%w	> 93
Maximum Pressure Drop (stable)	kPa	2.1
Pulse Interval	min	5
Test Duration	h	186

Notes:

* Dust samples were taken isokinetically. Determination of the mass of outlet dust required chloroform washing to remove tars.

The fraction of tar in condensed form is not known

TABLE 8

OPERATING CONDITIONS AND RESULTS OF TESTS ON THE PULSE-JET BAGFILTER ON THE
0.5 TONNE/HR GASIFIER

		A	B	C
Coal Feed to Gasifier		Baddesley	Baddesley	Seymoor
Inlet Gas temperature to Bagfilter	$^\circ C$	215	167	205
Filtration Velocity	ms^{-1}	0.013	0.009	0.013
Total Tar in Gas	gm^{-3}	0.48	ND	0.63
Maximum Pressure Drop	kPa	10	10	10
Pulse Pressure	barg	6	6	6
Duration	msec	200	200	200
Frequency	h^{-1}	5–60	5–15	5–12
Average Dust Collection Rate	kgh^{-1}	112	78	58
Test Duration	h	23.75	24.1	26.2

Note: (1) Filter media : Huyglas.
       (2) Test B produced a stable trace of pressure drop variation
           versus time.
       (3) Extensive deposits of NH_3Cl were found inside the unit after
           the run, ascribed to the low operating temperature during
           test B.
       (4) Air permeability of samples of filter media taken before and
           after this test series were 18% and 10% of new,
           respectively. Microscopic examination indicates that this
           permeability loss is due to tar deposits.
       (5) The secondary cyclone was out of service during these tests,
           to increase the ratio of dust to tar.
       (6) The fraction of tar in condensed form is not known.

TABLE 9

CONCENTRATIONS OF HYDROGEN CHLORIDE AND COMPOUNDS OF SULPHUR AND
NITROGEN DETERMINED IN GAS STREAMS ON THE 0.5 TONNE/HR GASIFIER

GAS	MATERIAL	CONCENTRATION
Fuel	HCl	350 vppm
	H_2S	200 vppm
	CS_2	12 vppm
	COS	100 vppm
	NH_3	300 vppm
Flue	HCl	150 vppm
	SO_2	100 vppm (@ 6%O_2)
	NO_x	140 vppm (@ 6%O_2)

TABLE 10

TYPICAL PERFORMANCE DATA (1 MWt TEST FURNACE)

Mode of Operation	SINGLE BED							TWIN-BED				
Coal feed Rate kgh^{-1}	34	34.3	38.5	34.5	38.7	53.2	36	90	91	88	93	95
Ca:S Molar Ratio	0	1	1	2	2	3	3	0	1	2	3	4
Bed Temperatures °C												
Pyrolyser –	–	–	–	–	–	–	–	970	970	970	970	950
Char bed/single bed	850	850	950	850	950	850	950	950	950	950	950	950
Mixing Zone Temp. °C	–	–	–	–	–	–	–	1410	1450	1425	1485	1450
Gas Composition												
Pyrolyser												
%CO	–	–	–	–	–	–	–	13.0	13.5	14.0	14.2	14.3
%CO_2	–	–	–	–	–	–	–	6.1	6.0	6.2	5.8	6.2
%H	–	–	–	–	–	–	–	8.5	8.3	8.4	8.0	8.6
%CH_4	–	–	–	–	–	–	–	3.1	2.8	2.9	3.0	3.1
NO_x (Vppm)	–	–	–	–	–	–	–	ND	ND	ND	ND	ND
NH_3 (Vppm)	–	–	–	–	–	–	–	300	325	340	350	375
HCN (Vppm)	–	–	–	–	–	–	–	2000	1650	1500	1250	1200
Char bed/single bed												
%CO	6.3	6.4	6.5	6.7	6.9	7.0	6.5	6.5	6.5	6.5	6.4	6.5
%CO_2	14.0	14.0	13.8	13.8	13.0	13.0	13.4	13.4	13.4	13.4	13.4	13.5
NO_x (Vppm)*	557	560	585	558	590	564	587	590	585	590	600	590
Mixing Zone												
O_2	–	–	–	–	–	–	–	6.1	5.9	6.0	6.5	5.8
NO_x (Vppm)*	–	–	–	–	–	–	–	228	234	228	240	236
% SO_2 Retention	0	18	11	46	27	63	46	10.0	14.4	32.0	51.0	70.0

* Referred to 6%O_2 ND: Not Determined

TABLE 11

CHEMICAL ANALYSIS OF COAL FIRED IN COMBUSTOR/PYROLYSER

Analysis % (as fired)	Coal	
	Gedling W. singles	Calverton W. singles
Total Moisture	12.7	12.8
Ash	4.5	5.8
Volatile Matter	40.0	33.8
Carbon	67.5	66.3
Hydrogen	4.46	4.59
Oxygen	8.01	7.7
Sulphur	1.35	1.42
Nitrogen	1.45	1.40
Calorific Value (kJ/kg)	27,770	27,550

TABLE 12

SIZE ANALYSIS OF COALS FIRED IN COMBUSTOR/PYROLYSER

% retained on screen	Coal	
	Gedling W. singles	Calverton W. singles
+26.5	2.9	3.1
+19.0	44.4	24.8
+13.2	46.4	19.8
+6.7	5.6	20.1
+3.35	)	10.3
-3.35	) 0.7	21.9

TABLE 13

PRELIMINARY BOILER PERFORMANCE DATA

Mode of Operation	SINGLE BED		TWIN BED	
Coal feed Rate kgh^{-1}	137	207	226	282
Gas Composition				
Pyrolyser				
%CO	-	-	5.8	5.5
%CO$_2$	-	-	13.3	12.8
%H$_2$	-	-	8.8	8.8
%CH$_4$	-	-	3.7	3.5
Char Bed				
%CO$_2$	8.7	8.8	8.7	8.2
%O$_2$	11.2	11.0	11.2	11.6
Mixing Zone				
%CO$_2$	-	-	14.5	12.8
%O$_2$	-	-	3.0	5.5
Bed Temperature				
Char bed/single Bed °C	900	900	840	840
Pyrolyser bed °C	-	-	965	930
Boiler Pressure (bar)	7.63	7.36	6.6	6.5
% MCR	26	40	43	54
Static Bed Height (mm)	150	150	150	150

FIGURE 1: The 0.5 Tonne/h Semi-Commercial Gasification Plant at CRE

① COAL FEED LOCK HOPPERS   ⑤ FINES RECYCLE LEG   ⑨ REJECT SOLIDS COOLER   ⑬ GAS SCRUBER
② SCREW FEEDER   ⑥ START UP BURNER   ⑩ REJECT SOLIDS COOLER   ⑭ BAG FILTER
③ GASIFIER   ⑦ REJECT SOLIDS COOLER   ⑪ GAS COOLER   ⑮ GAS FLARE
④ PRIMARY CYCLONE   ⑧ REJECT SOLIDS COOLER   ⑫ SECONDARY CYCLONE

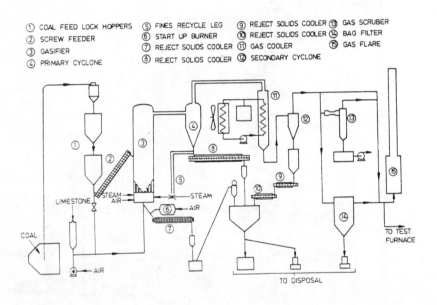

FIGURE 2: Process Flow Diagram of the 0.5 tonne/hr Gasification Plant

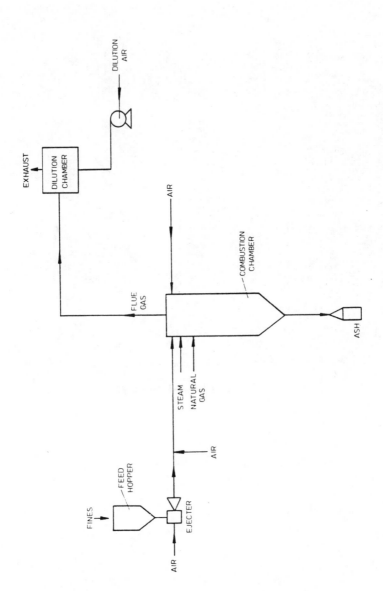

FIGURE 3   SCHEMATIC OF CYCLONE COMBUSTOR TEST RIG

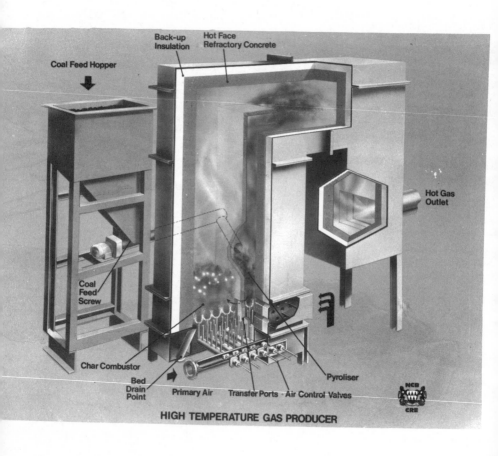

Coal Feed Hopper

Back-up Insulation

Hot Face Refractory Concrete

Hot Gas Outlet

Coal Feed Screw

Char Combustor

Pyroliser

Bed Drain Point

Primary Air

Transfer Ports

Air Control Valves

HIGH TEMPERATURE GAS PRODUCER

FIGURE 4: 1 MW$_t$ Twin-Bed Furnace

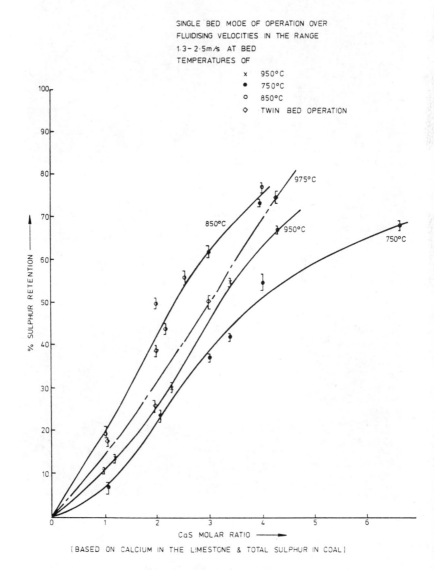

FIG. 5   SULPHUR RETENTION

FIGURE 6: 2.8 MW Senior Green Water Tube Boiler and Extended Surface
Economiser

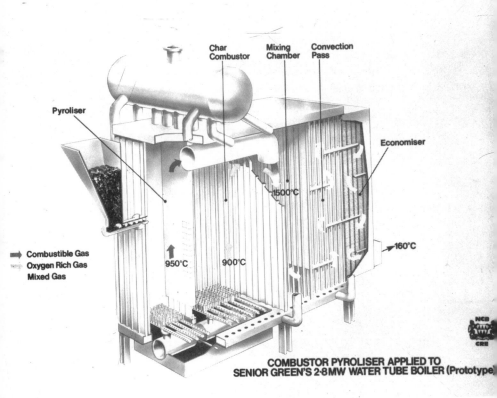

Pyroliser

Char Combustor

Mixing Chamber

Convection Pass

Economiser

1500°C

160°C

Combustible Gas
Oxygen Rich Gas
Mixed Gas

950°C

900°C

**COMBUSTOR PYROLISER APPLIED TO
SENIOR GREEN'S 2·8MW WATER TUBE BOILER (Prototype)**

FIGURE 7: Twin Bed Boiler

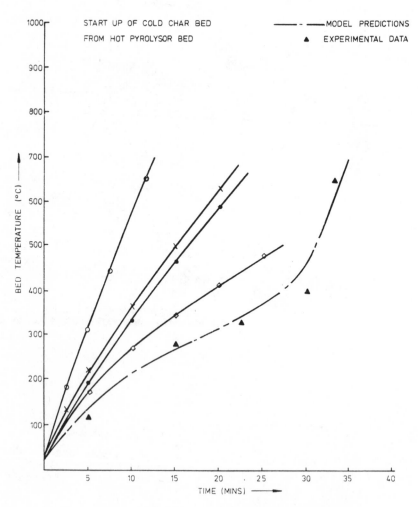

FIG 8. PART-BED START UP CHARACTERISTICS

CONTROLLING CORROSION IN LOW TEMPERATURE ENERGY UTILISATION

By

D C A Moore, W M Cox, D Gearey, CAPCIS/UMIST*,
R Littlejohn, CRE** D B Meadowcroft - CEGB***

ABSTRACT

The paper considers some of the practical problems associated with the recovery of low temperature heat in combustion systems and presents some of the results of a recently completed six-year investigation into the corrosion of flue gas handling plant. This work was sponsored by the Department of Trade and Industry in collaboration with the Central Electricity Generating Board, British Coal, Esso Engineering (Europe) Ltd, with CAPCIS/UMIST as main contractor. It was initiated through an awareness of the need to minimise degradation of constructional materials resulting from dewpoint corrosion and the possibility of improving the thermal efficiency of combustion plant by reducing back-end temperatures.

The programme comprised evaluation of plant experience and investigation of structural materials and protective coatings both in the laboratory and operational plant. An important feature of the programme was the application of advanced corrosion monitoring to continuous assessment of corrosion hazards in flue gas handling plant.

The outcome of the work has been to increase the understanding of dewpoint corrosion and the way it affects a wide range of combustion equipment. A number of important secondary factors have been identified which modify the simple condensation-corrosion mechanisms. It has been confirmed that a potential gain in thermal efficiency is available, even in large coal fired plant, subject to good housekeeping practices, as corrosion rates have been found to be severe only at temperatures below about 60 C contrary to the previously conceived acid dewpoint theory.

The paper also outlines the application of advanced corrosion monitoring techniques during the course of the project. These techniques can provide a continuous and immediate response to corrosion and by permitting confident operation at lower exhaust temperatures, facilitate a move towards more efficient and reliable operation of combustion systems.

---

*   CAPCIS-UMIST, Manchester, England
**   Consultant, late of British Coal - Coal Research Establishment, Stoke
      Orchard, England
*** CEGB-CERL, Leatherhead, England

INTRODUCTION

In the past twenty years there has been a considerable increase in the scale of combustion installations. The largest of these are power generation boilers where units commonly 60-100 MW in capacity were replaced by 500 or 660 MW units. In the same period the cost of oil rose from about $4 per barrel to over $30 per barrel in 1985/6. These two factors combined to concentrate attention on the expense of poor thermal efficiency and by 1986 the economic value of plant with increased thermal efficiency had become well recognised. While the more recent reduction in the oil price has led to a decrease in the margin to be saved and it is doubtful that energy prices will return to their previous high levels in the immediate future, it is equally certain that the earlier era of cheap energy has finished. Thus there is now an awareness of the benefits of good thermal recovery in many combustion installations which will continue.

In a 2000 MW power generation plant annual savings of over £1M have been quoted as a result of lowering the flue gas exhaust temperature by 15°C.[1]. New oil refinery installations, process heaters and cracking plant will have significantly improved thermal efficiency in comparison to those designed when heat was cheap. Even in medium capacity industrial plant and small domestic installations there is a marked trend towards increased efficiency.[2] The adoption of process heat optimisation engineering [3] also demonstrates progressive improvement. The effect of all these initiatives is to bring operating plant nearer to or below the critical temperature at which acid vapours in the flue gas stream begin to condense and attack the installation. Thus, the process engineering problem is to maximise useful heat recovery and, at the same time, by process and materials selection to ensure that equipment is not exposed to unacceptable rates of corrosion which might subsequently lead to loss of unit availability.

DEWPOINT INVESTIGATION

The problems of optimising thermal recovery were considered by the DTI Committee On Corrosion which met in the late 1970's. It was recognised that there was scope for improved combustion performance but the primary objection to exploiting it was the probability of sulphuric acid dewpoint corrosion. A collaborative project between the DTI, the National Coal Board (now British Coal), the CEGB and Esso Engineering Ltd was launched to investigate the phenomenon in detail. The work was conducted by the Corrosion and Protection Centre Industrial Services Unit at UMIST and comprised two parts - first, a fundamental study which examined the mechanisms of condensate attack including the effect of flue gas constituents such as HCl, and second, a materials evaluation which was designed to identify alloys or coatings which could withstand condensing flue gas conditions. Both laboratory combustion rig tests and in-plant field tests were conducted. Some of the work was reported at a conference held in 1985 [4], and a comprehensive final report of the programme was submitted to the sponsors in late 1986.

Eighty materials of which over fifty were alloys were tested under conditions of sulphuric and hydrochloric acid deposition. The majority of the highly alloyed materials were not found to be particularly resistant to corrosion in acid dewpoint conditions. The rates for nickel based alloys were comparable with bulk acid dissolution rates, indicating that the rate of acid deposition was not the rate determining step. Some low alloy steels showed a lower long term corrosion rate than mild steel, but would not be easily engineered for use in combustion plant.

Tests on organic coatings [5] showed that a fluoroelastomer gave good performance close to the acid dewpoint (-120°C). An American supplied coal tar epoxy with a specially developed curing agent also performed very well in the laboratory test rig. In site tests, cheaper polyester and isocyanate cured epoxy materials were also found to be adequate in certain conditions. Tests on inorganic coatings highlighted the difficulty of obtaining pore free, adherent coatings.

Several factors which influence the fundamental condensation/corrosion behaviour were also considered. These included:

    (1)   Kinetics of corrosion
    (2)   Pulverised fuel ash.
    (2)   Temperature cycling.
    (4)   Air in-leakage.
    (5)   Hydrogen chloride.
    (6)   Effects of cleaning systems.

Their roles may be summarised as follows:

Kinetics of Corrosion

Mild steel and most alloys, including nickel based alloys such as 625, give linear corrosion rates and in dust-free gases with typical SO_3 levels (10-20ppm). The corrosion rates of all are within a factor of five at metal temperatures ~20 deg.C below the dewpoint. Some low alloy steels can show protective kinetics. The mild steel corrosion rate is determined by the acid deposition rate while nickel based alloys corrode at rates similar to their bulk acid values.

Pulverised Fuel Ash

In the presence of PFA, further reactions are set up in competition with the acid/metal corrosion system. Ash particles are able to absorb, and/or reduce the acidity of, condensate which has been deposited on cool metal surfaces. This reduces the local amount of acid available to corrode the metal. Additionally, an ash-containing deposit may act as a highly effective thermal insulator, allowing the flue gas/deposit interface temperature to approach the flue gas temperature which will in turn reduce the acid condensation rate. Conversely, thick deposits are potentially hazardous due to the development of corrosion cells if the deposit remains wet and warm off-load longer than the surrounding surfaces.

Temperature Cycling

Intermittent operation of combustion plant increases the number of occasions when bonded corrosion products, built up during normal service, are disrupted. The disruption results from the hygroscopic nature of the deposits and corrosion products which 'sweat', i.e. absorb moisture from the air during shutdown periods. Further disruption occurs during the warm-up period when the plant is returned to service. These events tend to cause deposits to be detached leaving exposed areas which corrode rapidly until they re-form a protective layer or are re-covered with bonded deposits.

Air Ingress

Air leakage tends to cause localised reductions in metal temperature, increased aeration of the warm moist environment and an increased tendency to form agglomerate poultices. The poultices maintain the moist environment over extended periods, causing severe damage to ductwork.

Hydrogen Chloride

At metal temperatures below 60°C, the presence of HCl in the flue gas stream can lead to a marked increase in corrosion rate compared with that resulting from SO_3/H_2SO_4 alone because the surface is often below the dewpoint for HCl condensation.

Effect of Cleaning Procedures

Deposits on boiler surfaces and rotary heat exchangers are frequently removed by steam or air jets. This process is called 'sootblowing'. In cases of severe fouling, water washing may be used. Steam and water washing procedures can make a major contribution to attack by condensed acids, particularly at cold end air heater baskets. The contribution can be twofold: (i) moisture injection, producing dilute aggressive acid, and (ii) erosion. Protective scale as well as fouling is removed and fresh steel is exposed for the next cycle of corrosion.

It was concluded that although a theoretical estimation of the dewpoint temperature can be made, in practice, differences in fuel composition, excess air, ash characteristics and other operational variables mean that its value is limited in defining the limiting temperature for safe operation. It is necessary to measure corrosion behaviour on the specific material of interest in the actual plant as a function of temperature and to do this continuously while the plant is in normal operation. To satisfy this requirement, new techniques of electrochemical corrosion monitoring developed in the UMIST Corrosion and Protection Centre were applied by CAPCIS to the investigation of acid dewpoint corrosion. It was found that, contrary to the previously accepted view of acid dewpoint corrosion, based largely on data obtained in dust free test environments, active corrosion would commence at much lower metal temperatures, typically 60°C, than those suggested by simple measurements of the flue gas acid dewpoints. Furthermore, the majority of damage sustained by combustion equipment often occurs during short periods of aggressive corrosion rather than by continuous attack.

ELECTROCHEMICAL CORROSION MONITORING

The corrosion monitor comprises a controlled temperature electrochemical probe, electrochemical monitoring unit, temperature control unit and six channel computerised data acquisition unit. A typical layout of the system is shown in Figure 1. The probes are normally used mounted flush with the internal surface of the flue-duct. In addition, a weight loss probe containing six one-inch diameter coupons which are machined from the material of interest, each individually cooled, are exposed nearby for calibration. The electrochemical probe contains multiple elements of the test material which are linked via cable to the electrochemical monitoring unit. Four electrochemical techniques are employed simultaneously, these are electrochemical impedance monitoring (EIM), electrochemical potential noise (EPN), zero resistance ammetry (ZRA) and electrochemical current noise (ECN).

The electrochemical impedance monitor applies a sine wave voltage at two frequencies simultaneously. The resulting currents are measured and from the difference between the impedance responses, an estimate for the charge transfer resistance (R_{ct}) obtained. The corrosion rate is then obtained in an analogous manner to the polarisation resistance of linear polarisation resistance (LPRM) measurements, i.e. $I_{corr} = K/R_{ct}$ where K is a constant. The next set of electrodes is used to measure electrochemical potential noise ie. the small, low frequency variations in potential that occur between two nominally identical electrodes due to local micro-corrosion. The coupling current (zero resistance ammeter) module is connected to the third set of probe electrodes and measures the current flowing between the nominally identical electrodes. This coupling current is caused by variations in the relative areas of anodic and cathodic activity on the two electrodes. There is evidence that this current is proportional to the corrosion current. The electrochemical current noise module determines the low frequency (<1Hz) and low amplitude (<1µA) variations in the currents. This module and the potential noise module are both particularly sensitive to initiation of corrosion. The output from either of these modules gives characteristic traces if localised attack (pitting or crevice corrosion) is in progress and may be used to characterise the type of process occurring.

In summary, the four electrochemical techniques are complementary. The EIM and ZRA modules provide corrosion rate information with the EIM being suitable for more resistive environments, ie. less condensation. The electrochemical noise signals indicate whether localised or general attack is occurring and these signals may be processed to estimate corrosion rates, even low rates. However, all the rate estimations require calibration in the particular environment. One of the unique features of the system is its ability to monitor corrosion in the presence of thin films of condensate, such as are found in dewpoint corrosion environments.

The corrosion monitoring system has been described in detail elsewhere [6], but examples of typical data obtained in a selection of applications are presented here give an indication of its value:

Corrosion In Refinery Plant

The effects of an alteration in fuel composition on corrosion in the exhaust ducting of a oil/gas fired boiler are shown in Figure 2. The trace covers a period during which the fuel was changed from a light gas-oil to a residual fuel oil. The monitors record a small increase in corrosion of the probe material, in this example carbon steel. (ZRA and ECN signals increase, EIM decreases). The exhaust gas and probe temperature remained relatively constant, suggesting that the change in corrosion activity was related to differences between the fuels.

Effects of Cleaning Measures on Heat Exchanger Corrosion

The data shown in Figure 3 come from a carbon steel probe in a heat exchanger exposed to a gas turbine exhaust. For clarity each of the channels has been stored digitally and are displayed individually. The heat exchanger tubes were water washed once a week while the unit was off-line.

Prior to washing and with the unit off-line, the condensate on the tubes had dried out and corrosion activity had fallen to the low level shown in the first 1½ hours of the trace. The charge transfer resistance was high (EIM) while the coupling current (ZRA), the current noise (ECN) and the potential noise (EPN) were low. The steadiness of the latter indicates the absence of localised attack. The onset of washing, which may be identified by the sharp decrease in probe and gas temperatures, resulted in a marked increase in electrochemical activity which slowly diminished thereafter. This behaviour probably arose from initial wetting of the dried acid film on the tubes resulting in corrosion attack, followed by rinsing of the aggressive species from the probe surface. The sharp peaks in the ECN and EPN signals suggest that after washing localised off-line attack continued, at a low rate. The EPN signal also shows that localised activity continued for approximately one hour after the furnace was refired (gas temperature increase). The final thirty minutes of the trace indicate that uniform corrosion was increasing as the EIM signal decreased (lower charge transfer resistance), while the EPN and ECN signals became more steady.

Corrosion In Flue Gas Desulphurisation Plant

Figure 4 illustrates the response of the monitor to small changes in the level of reheat applied to scrubbed flue gas. The material under investigation was AISI Type 316L stainless steel, using an uncooled probe. A reduction in reheat of approximately 8°C at Time = 2 hours on the trace has resulted in an increase in corrosion. This is shown by the decrease in the charge transfer resistance recorded by the EIM module and a corresponding rise in the level of the ZRA and ECN signals. A return to higher reheat temperature, at Time = 10 hours on the trace, caused the monitors to return to their initial levels, indicating low corrosion. The EPN monitor indicated that corrosion was minimal both before and after the excursion in reheat level. During the period of corrosion, the occurrence of individual transients in the EPN signal indicated localised corrosion. Examination of the probe surface at the end of the investigation showed it to have undergone micropitting.

The event described illustrates that a small change in operation can give rise to corrosion which would not be evident to the operator. The ability to monitor corrosion on-line and relate it to operating conditions now allows these to be changed so that plant may be run without risk of corrosion.

Off-Line Corrosion Behaviour

Corrosion in the flue duct of a small p.f. fired power generation boiler was monitored for a period of approximately 10 weeks[7]. The data obtained are summarised in Figure 5. Sections of the EPN data have been separated from the other traces to illustrate the characteristics of the technique under different corrosion regimes.

Prior to the unit being shut down at Point A in the figure, the mild steel electrochemical probe had been cooled to a constant temperature of 60°C and the EIM and ZRA monitors had shown that uniform corrosion was occurring. When the unit was shut down, the EIM signal increased to a high value while the level of the ZRA and EPN signals fell, signifying a decrease in corrosion as the deposits on the probe surface dried when the moist flue gas stream was interrupted.

While the unit remained off-line, the probe cooled and the deposits absorbed moisture from ambient air, becoming wet ("sweating") and liberating acid. In the example shown, the EPN monitor was the first technique to detect corrosion (Point B), while the EIM initially remained high and the ZRA low. The EIM and ZRA signals soon responded to the increase in corrosion as it rose to a maximum approximately three weeks into the period shown. The low level of fluctuation in the EPN trace indicates that uniform corrosion was occurring.

After the period of high corrosion, a gradual decrease was observed, until by eight weeks after shut down both the ZRA and EIM monitors recorded minimal activity. It may be concluded that corrosion had ceased because all the condensed acid on the duct surface has been consumed by the corrosion reaction or by neutralisation by alkaline components of the deposit. Electrochemical activity was still detectable by the EPN monitor at this stage, but the large transients indicated that there were only a small number of active corrosion sites, equivalent to a low rate of corrosion.

IMPROVEMENTS IN PLANT SURVEILLANCE

The trend towards higher thermal recovery has brought about a requirement for improved plant surveillance. Excessive heat losses are no longer acceptable and are being reduced at the design stage of all new equipment. This in turn requires better housekeeping, good insulation, rapid repair of any defects and improved condition monitoring. Good performance of the the new designs will be sensitive to operating conditions with care needed to acid corrosion by acidic condensates. Fortunately, means of avoiding penalties, such as loss of plant availability and/or costs for maintenance have been developed. Flue gas corrosion monitoring, as described in this paper, allows improved process control and better formulation of remedial measures. As a result, combustion equipment can now be operated with confidence under conditions which are relatively close to the threshold for the onset of dewpoint corrosion. This development coincides with an increased awareness throughout industry of the benefits associated with improved plant surveillance in general, not only corrosion monitoring, but also the use of complementary techniques such as flue gas oxygen level monitoring, improved temperature control, better burner technology and computerised data handling and display.

CONCLUSIONS

Testing in a wide variety of combustion plant has shown that it is now feasible to monitor corrosion behaviour accurately and also to respond so as to prevent excessive corrosion. The trend towards increased thermal recovery by reducing back end temperature can therefore be maintained provided that corrosion within the operating plant is carefully controlled. Improvements in monitoring techniques will thus facilitate a progressive improvement in the thermal efficiency of combustion equipment of all sizes. Suitable instrumentation is not expensive by comparison with the potential for savings. The approach has already generated considerable interest in the UK, USA and Germany and seems set to be employed on an increasing scale as awareness of its advantages grows.

ACKNOWLEDGEMENTS

The authors wish to acknowledge the assistance of colleagues in all the organisations who have contributed to the success of the work reported here. The paper is published by permission of the CEGB.

REFERENCES

1. Moore. W., 'Flue Gas Corrosion in Coal Fired Power Stations' IMechE Conference on Refurbishment and Life Extension of Steam Plant, 14-15 October 1987.

2. Stevens, R.L., and Morgan, G.C., 'Materials for High Efficiency Appliances', Int. Gas Research Conf., 1986.

3. Linnhoff B. and Turner J.A., 'Heat Recovery Networks: New Insights Yield Big Savings', Chem Eng p56 Nov 1981.

4. Holmes D.R. (Ed) 'Dewpoint Corrosion'. Ellis Horwood, Chichester, 1985.

5. Cox W.M., Forrester-Coles T.H. and Holmes D.R., Chapter 13. Reference #4.

6. Cox W.M., Phull B.S., Wrobel B.A. and Syrett B.C., Materials Performance, Vol. 25, No. 11, pp 9-17, November 1986.

7. Cox W.M., Farrell D.M., Dawson J.L., Chapter 12 Reference # 4.

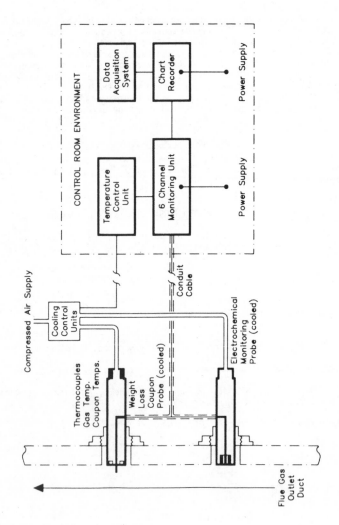

Figure 1. General layout of corrosion monitoring system, (cooled probes).

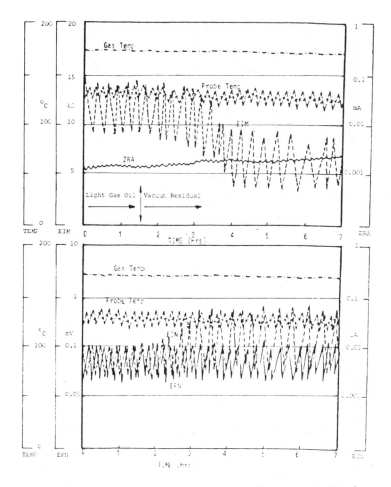

Figure 2. Corrosion in refinery plant - Change in feedstock.

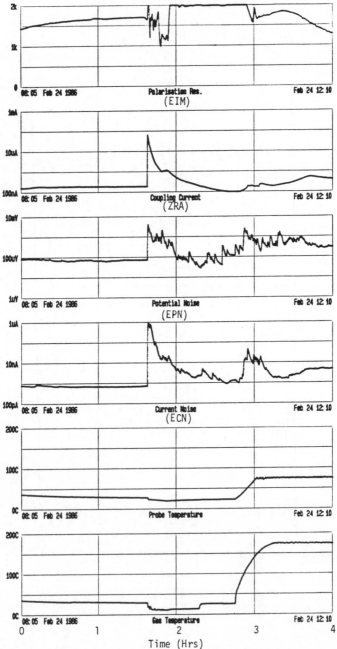

Figure 3. Corrosion on heat exchangers - Activity related to water washing.

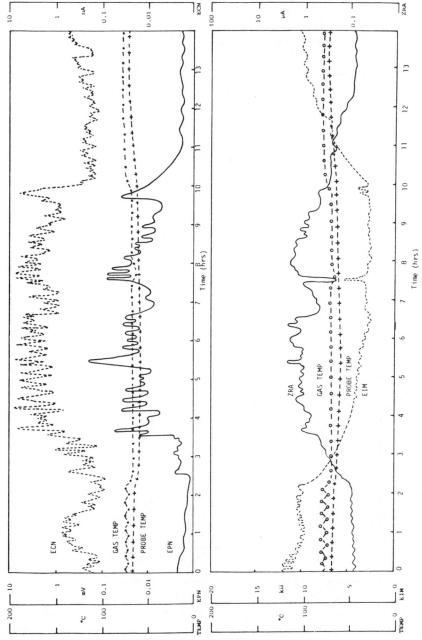

Figure 4. Corrosion monitoring in flue gas desulphurisation plant - Activity resulting from a small change in reheat.

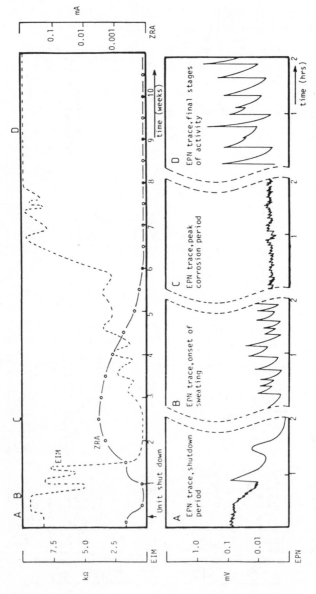

Figure 5.   Corrosion activity associated with a shut down period on a small p.f. fired boiler.
Figure courtesy of Ellis Horwood (7).

Paper 19 Continued from page 248 — Figures

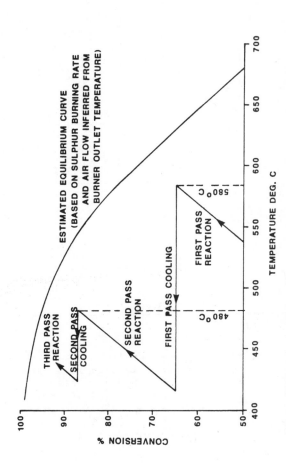

FIG.1 – ESTIMATED OPERATION OF SO₃ REACTOR, AND HENCE HEAT RECOVERY DUTIES.
(BASED ON OBSERVED TEMPERATURES AND ASSUMED 85% CONVERSION THROUGH EACH PASS)

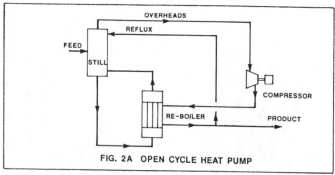

FIG. 2A OPEN CYCLE HEAT PUMP

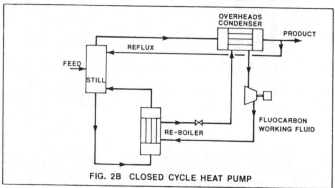

FIG. 2B CLOSED CYCLE HEAT PUMP

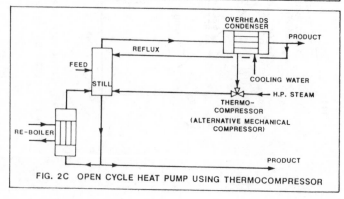

FIG. 2C OPEN CYCLE HEAT PUMP USING THERMOCOMPRESSOR

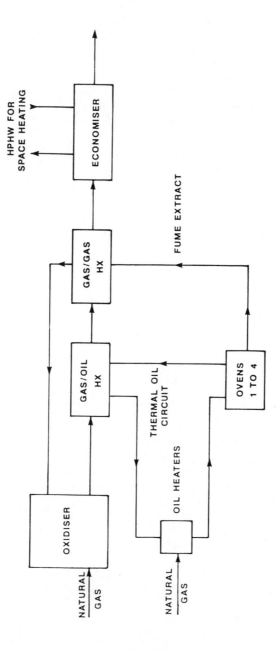

FIG 3 THERMAL OXIDISER WITH COMPREHENSIVE HEAT RECOVERY

Paper 26 Continued from page 323 – Figures

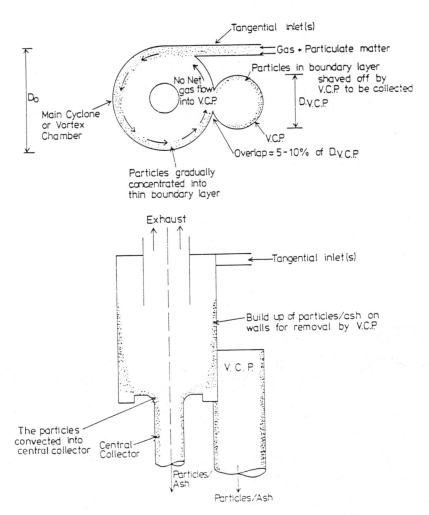

Fig. 1 Principles of Operation of Vortex
Collector Pockets (V. C. P's)

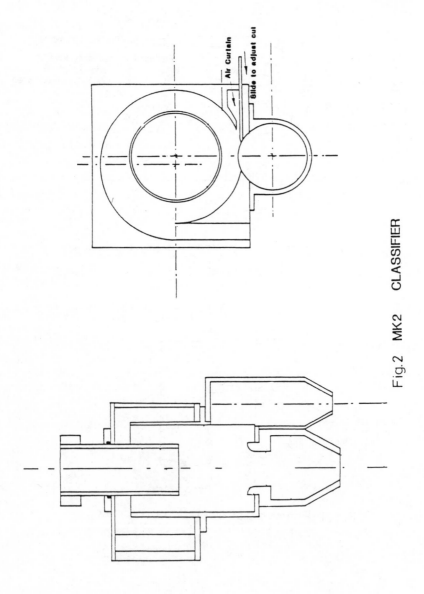

Fig. 2   MK2   CLASSIFIER

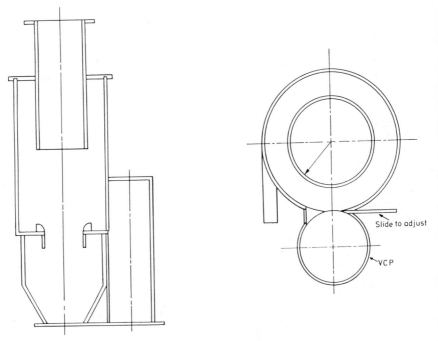

Slide to adjust

VCP

Fig. 3   Mk 3 Classifier (150 mm Exit Diameter)

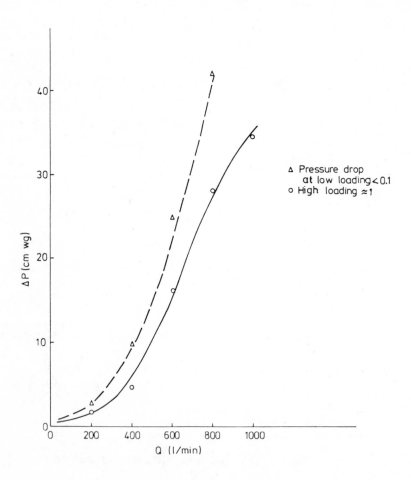

Fig. 4   Pressure Drop for Mk. 2 Classifier

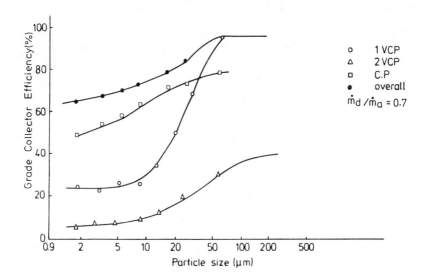

Fig. 5   Mk. 2   Classifier

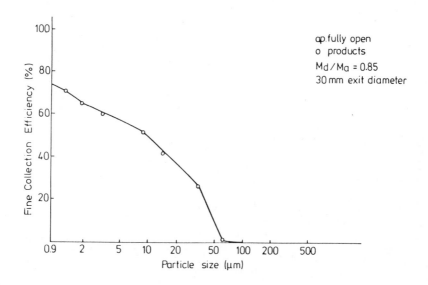

Fig. 6a Mk. 2   Classifier

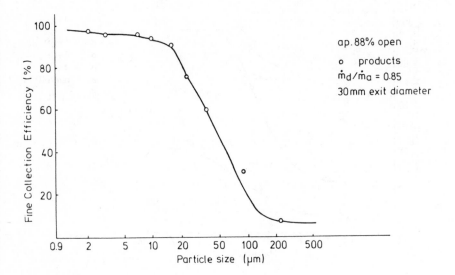

Fig. 6b Mk2   Classifier

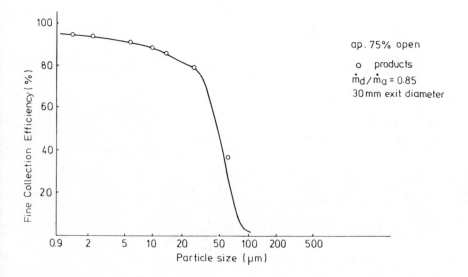

Fig.6c Mk 2   Classifier

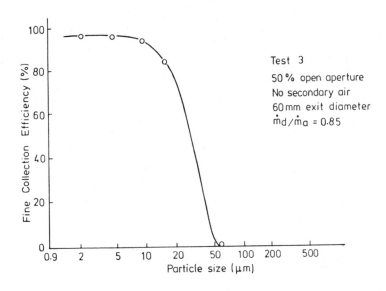

Test 3

50% open aperture
No secondary air
60mm exit diameter
$\dot{m}_d/\dot{m}_a = 0.85$

Fig 7a Mk 2 Classifier: Fine Product Efficiency Curve

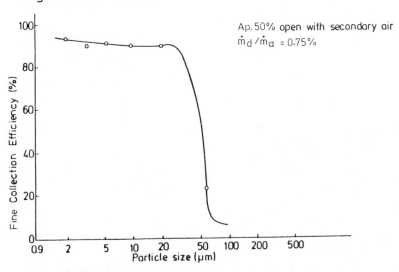

Ap. 50% open with secondary air
$\dot{m}_d/\dot{m}_a = 0.75\%$

FIG. 7b MK2 CLASSIFIER: FINE PRODUCT COLLECTION EFFICIENCY CURVE

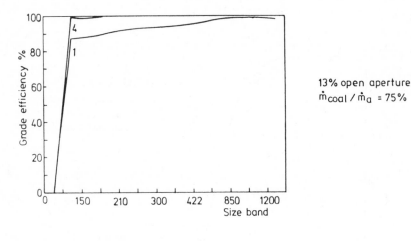

13% open aperture
$\dot{m}_{coal} / \dot{m}_a = 75\%$

1. Efficiency VCP1
4. Total efficiency

Fig. 8a   Grade Efficiency Curve for Mk 3 Classifier

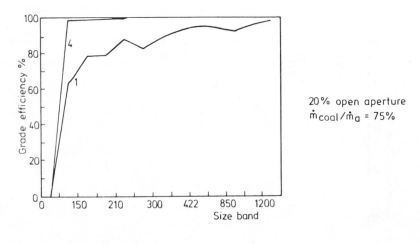

20% open aperture
$\dot{m}_{coal} / \dot{m}_a = 75\%$

1. Efficiency VCP1
4. Total efficiency

Fig. 8b   Grade Efficiency Curve for Mk 3 Classifier

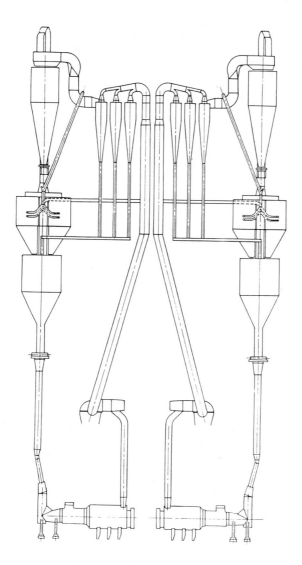

Fig. 9   Multi-cyclone Industrial plant

INDEX